Springer Undergraduate Mathematics Series

More information about this series at http://www.springer.com/series/3423

Jeremy Gray

A History of Abstract Algebra

From Algebraic Equations to Modern Algebra

 Springer

Jeremy Gray
School of Mathematics and Statistics
The Open University
Milton Keynes, UK

Mathematics Institute
University of Warwick
Coventry, UK

ISSN 1615-2085   ISSN 2197-4144 (electronic)
Springer Undergraduate Mathematics Series
ISBN 978-3-319-94772-3   ISBN 978-3-319-94773-0 (eBook)
https://doi.org/10.1007/978-3-319-94773-0

Library of Congress Control Number: 2018948208

Mathematics Subject Classification (2010): 01A55, 01A60, 01A50, 11-03, 12-03, 13-03

This Springer imprint is published by the registered company Springer Nature Switzerland AG.
The registered company address is: Gewerbestrasse 11, 6330 Cham, Switzerland

Introduction

The conclusion, if I am not mistaken, is that above all the modern development of pure mathematics takes place under the banner of number.

David Hilbert, *The Theory of Algebraic Number Fields*, p. ix.

Introduction to the History of Modern Algebra

This book covers topics in the history of modern algebra. More precisely, it looks at some topics in algebra and number theory and follows them from their modest presence in mathematics in the seventeenth and eighteenth centuries into the nineteenth century and sees how they were gradually transformed into what we call modern algebra. Accordingly, it looks at some of the great success stories in mathematics: Galois theory—the theory of when polynomial equations have algebraic solutions—and algebraic number theory. So it confronts a question many students ask themselves: how is it that university-level algebra is so very different from school-level algebra?

The term 'modern algebra' was decisively introduced by van der Waerden in his book *Moderne Algebra* (1931), and it is worth discussing what he meant by it, and what was 'modern' about it. That it still suffices as an accurate label for much of the work done in the field since is indicative of how powerful the movement was that created the subject.

The primary meaning of the term 'modern algebra' is structural algebra: the study of groups, rings, and fields. Interestingly, it does not usually include linear algebra or functional analysis, despite the strong links between these branches, and applied mathematicians may well encounter only a first course in groups and nothing else. Modern algebra is in many ways different from school algebra, a subject whose core consists of the explicit solution of equations and modest excursions into geometry. The reasons for the shift in meaning, and the ways in which structural algebra grew out of old-style classical algebra, are among the major concerns of this book. Elucidating these reasons, and tracing the implications of the transformation of algebra, will take us from the later decades of the eighteenth century to the 1920s and the milieu in which *Moderne Algebra* was written.

Classical algebra confronted many problems in the late eighteenth century. Among them was the so-called fundamental theorem of algebra, the claim that every polynomial with real (or complex) coefficients, has as many roots as its degree. Because most experience with polynomial equations was tied to attempts to solve them this problem overlapped with attempts to find explicit formulae for their solution, and once the polynomial had degree 5 or more no such formula was known. The elusive formula was required to involve nothing more than addition, subtraction, multiplication, and division applied to the coefficients of the equation and the extraction of nth roots, so it was known as solution by radicals, and Lagrange in 1770 was the first to give reasons why the general quintic equation might not be solvable by radicals.

Another source of problems in classical algebra was number theory. Fermat had tried and largely failed to interest his contemporaries in the subject, but his writings on the subject caught the attention of Euler in the eighteenth century. Euler wrote extensively on them, and what he conjectured but could not prove was often, but not always, soon proved by Lagrange. This left a range of partially answered questions for their successors to pursue. For example, Fermat had shown that odd prime numbers of the form $x^2 + y^2$ are precisely those of the form $4k + 1$, and had found similar theorems for primes of the form $x^2 + 2y^2$ and $x^2 + 3y^2$ but not for primes of the form $x^2 + 5y^2$—why not, what was going on? There was a good theory of integer solutions to the equation $x^2 - Ay^2 = 1$, where A is a square-free integer, and Lagrange had begun a theory of the general binary quadratic form, $ax^2 + bxy + cy^2$, but much remained to be done. And, famously today if less so in 1800, Fermat had a conjecture about integer solutions to $x^n + y^n = z^n, n > 2$ and had indeed shown that there were no solutions when $n = 4$, and Euler had a suggestive but flawed proof of the case $x^3 + y^3 = z^3$ that led into the theory of quadratic forms.

The nineteenth century began, in algebra, with Gauss's *Disquisitiones Arithmeticae* (1801), the book that made Gauss's name and may be said to have created modern algebraic number theory, in the sense that it inspired an unbroken stream of leading German mathematicians to take up and develop the subject.[1] The work of Gauss and later Dedekind is central to the story of the creation of modern algebra. In the 1820s Abel had wrapped up the question of the quintic, and shown it was not generally solvable by radicals. This raised a deeper question, one that Gauss had already begun to consider: given that some polynomial equations of degree 5 or more are solvable by radicals, which ones are and how can we tell? It was the great, if obscure, achievement of Galois to show how this question can be answered, and the implications of his ideas, many of them drawn out by Jordan in 1870, also uncoil through the nineteenth century and figure largely in the story.

Algebraic number theory led mathematicians to the concept of commutative rings, of which, after all, the integers are the canonical example, and Galois theory led to the concepts of groups and fields. Other developments in nineteenth century geometry—the rediscovery of projective geometry and the shocking discovery of

---

[1]Gauss also rediscovered the first known asteroid in 1801.

non-Euclidean geometry—also contributed to the success of the group concept, when Klein used it in the 1870s to unify the disparate branches of geometry.

The rise of structural mathematics was not without controversy. There was a long-running argument between Kronecker and Dedekind about the proper nature of algebraic number theory. There was less disagreement about the importance of Galois's ideas once they were properly published, 14 years after his untimely death, but it took a generation to find the right way to handle them and another for the modern consensus to emerge. By the end of the century there was a marked disagreement in the mathematical community about the relative importance of good questions and abstract theory. This is not just a chicken-and-egg problem. Once it is agreed that theory has a major place, it follows that people can work on theory alone, and the subject has to grow to allow that. By and large, equations have contexts and solving them is of value in that context, but what is the point of a theory of groups when done for its own sake? Questions about integers may be interesting, but what about an abstract theory of rings? These questions acquired solid answers, ones it is the historians' job to spell out, but they are legitimate, as are their descendants today: higher category theory, anyone?

The end of the nineteenth century and the start of the 20th see the shift from classical algebra to modern algebra, in the important sense that the structural concepts move from the research frontier to the core and become not only the way in which classical problems can be reformulated but a source of legitimate problems themselves. This was a lengthy process, and the publication of *Moderne Algebra* marks an important stage in what is still an ongoing process, one that is worth thinking about. Leo Corry (1996) has usefully distinguished between what he calls the 'body' and the 'images' of a piece of knowledge: the body of knowledge, he tells us (1996, 3) "includes theories, 'facts', open problems", the images "serve as guiding principles, or selectors", they "determine attitudes" about what is an urgent problem demanding attention, what is relevant, what is a legitimate method, what should be taught, and who has the authority to decide. It might indeed be worth separating out the social and institutional factors entirely that could be called the 'forces of knowledge': the mathematicians themselves and their institutions (universities, professional bodies, journals, and the like).

I have tried to introduce students to some of the best writing in the history of mathematics of the last 20 or 30 years, not just for the information these books and papers provide but as introductions to how history can be written, and to help students engage with other opinions and so form their own. The most relevant examples are the *The Shaping of Arithmetic after C.F. Gauss's Disquisitiones Arithmeticae* by Goldstein et al. (2007) and Corry's *Modern algebra and the rise of mathematical structures* (1996), both of which in their different ways show how much can be done when one gets away from the monotonous plod of great achievements. Kiernan's influential paper 'The development of Galois theory from Lagrange to Artin' (Kiernan 1971) is still well worth consulting. Such accounts are analytical, not merely descriptive. As Gauss famously said of his own work[2] "When the building is finished the scaffolding can no longer be seen"; whatever the merits and demerits of presenting mathematics that way little is gained by treating the history of mathematics as an attempt to rebuild each branch of mathematics as it is today, brick by brick, with names and dates attached.

In short, this is a book on the history of algebra, and as such it asks the student to think about what it is to study history. It emphasises some points that a straight mathematics course might marginalise, and it omits others that a mathematician would emphasise, as and when I judged that a history course required it. It does assume that the students can handle difficult mathematics—and happily that proved generally to be the case in the years I taught it at the University of Warwick—but it requires that they marshal arguments, based on facts, and in support of opinions, as historians must. I have tried to make it coherent, albeit selective.

It is not an attempt to write a 'complete' history of algebra in the nineteenth century. It is the result of a course of 30 lectures, and it can be taught as such. Not all the chapters correspond exactly to a lecture; some overshoot, and there is more information in the Appendices. To reduce the study of algebra in the nineteenth century and the early years of the 20th to barely 300 pages meant taking some crude decisions, and I could see no tidy way to do it. Algebra in the period, it seems, is a more heterogeneous body of knowledge than geometry was, and even real and complex analysis (once differential equation s of all kinds are reserved for another occasion). I removed most of invariant theory and the Kleinian view of geometry from discussion. The history of work on determinants, matrices, linear algebra generally, quaternions and other algebras had also to be omitted. But that only left me facing tougher decisions. I decided not to deal with Galois's second memoir, Sophie Germain and her work, higher reciprocity laws, Fermat's Last Theorem, power series methods, and the distribution of prime numbers. Several of these topics are well covered in the existing literature and would make good projects for students wishing to take the subject further. Although it has become increasingly clear in recent years that the work of Kummer, Kronecker, Hermite and others on the rich and overlapping fields of elliptic functions, modular functions and n-ary quadratic forms was a major legacy of Gauss, none of that could be described here. In fact it makes a suitable subject for research.[3] I could not, however, resist writing about the importance of Klein's famous but seldom-read *Lectures on the Icosahedron* (1884).

No undergraduate history course assumes that the students will be competent, let alone masterful, in everything that matters to the course. Students study the techniques of warfare without being able to ride a horse or fire a gun; diplomacy without being taught the skills of negotiation; nationalism without engaging in politics; labour movements without working in a field or a factory; childhood and

---

[2]See Sartorius (1856, 82).

[3]On Hermite, see Goldstein (2007) and Goldstein (2011); on Kronecker see Goldstein and Schappacher (2007, Section 4) and on the Kronecker–Klein dispute see (Petri and Schappacher 2002).

the family without becoming pregnant or rearing children. But there is a feeling that the history of a branch of mathematics should not be taught without the students acquiring something like the mastery of that branch that a straight course in it would hope to achieve. This book is an attempt not to break that connection—students should not say nonsense about quadratic reciprocity or which equations are solvable by radicals—but to open it up to other approaches.

The ethics here are those of an applied mathematics course, in which a balance is to be struck between the mathematics and the application. Doubtless this balance is struck in many different ways in different courses. Here I have tried to bring out what is important historically in the development of this or that part of algebra, and to do so it is necessary to take some things for granted, say, that this person did validly deduce this result although the proof will not be looked at. The historical purpose in examining a strictly mathematical detail has to be decided case by case.

To give one example, perhaps the gravest in the book, the treatment of quadratic reciprocity given here contains one complete proof of the general case (Gauss's third proof from 1808 in which he introduced Gauss's Lemma) and indicates how he gave an earlier proof, in the *Disquisitiones Arithmeticae*, that rested on his technique of composition of forms (although I discuss the fourth and sixth proofs in Appendix C). To explain composition of forms I indicate what is involved, and give Dirichlet's simplified account, but I merely hint at the difficulties involved in securing the deepest results without which Gauss's theory of quadratic form s loses much of its force. This gives me room to explain the history of this famously difficult subject, but deprives the students of some key proofs. My judgement was that this was enough for a history of modern algebra in which the contemporary and subsequent appreciation of these ideas is at least as important as the accompanying technicalities. I then go on to describe Dedekind's translation of the theory into the theory of modules and ideals that he created.

I also believe that sufficient understanding of some topics for the purposes of a history course is acquired by working with examples, and when these are not made explicit there is always an implication that examples help.

In 1817 Gauss wrote (see *Werke* 2, pp. 159–160) that

> It is characteristic of higher arithmetic that many of its most beautiful theorems can be discovered by induction with the greatest of ease but have proofs that lie anywhere but near at hand and are often found only after many fruitless investigations with the aid of deep analysis and lucky combinations

Certainly numerical exploration and verification are a good way to understand many of the ideas in this book. That the proofs lie deep below the surface is one of the principal impulses for the slow elaboration of modern algebra.

Sources and Their Uses

Sources are at the heart of any historical account. In the last few years a huge amount of original literature in the form of journals and books has become available over the web, and this resource will continue to grow, but inevitably much of it is in French or German. I have tried to note when translations are available (as they are for Gauss's *Disquisitiones Arithmeticae* and all of Galois's work) and I have added some of my own to the small collection that already exists. With a little effort more can be found that is in English, although that remains a problem. I have tried to indicate what was available in 2017.

Historians divide sources roughly into two kinds: primary and secondary. Primary sources are the original words upon which the historian relies—here, the papers, the unpublished notes, and the letters by mathematicians themselves. Secondary sources are subsequent commentaries and other indirect records.

It will be no surprise that secondary sources for the history of mathematics on the web are many and various. The best source for biographical information is the *Complete Dictionary of Scientific Biography*. The biographies on the St Andrews website began as digests of this work, but are now often more up-to-date and with interesting references to further reading. The third source of information is Wikipedia. This is generally accurate on mathematical topics, but a bit more hit-or-miss on historical matters. After that it's a lottery, and, of course, sources are often filled with falsehoods lazily repeated, easy to find elsewhere and hard to refute. The only option is always to give your source for any significant piece of information. For a website supply a working URL, for a book or article the author, title, and page number. Such references are to historians what purported proofs are to mathematicians: ways of making claims plausible or, when unsuccessful, exposing the error.

Solid reference works in the history of mathematics are generally quite good in describing the growth of the body of knowledge in algebra. The facts of the matter are often well known, although folk memory acts a filter: everyone knows that Gauss was a number theorist, many fewer can describe his theory of quadratic forms; everyone knows that Galois showed that the quintic is not solvable by radicals, most people guess incorrectly how he did it. The problem is that mathematicians today naturally suppose that the best mathematicians of the day respected the same imperatives that they do—have the same images of algebra, one could say—and in part this is because of the successes of the structural algebra movement. Accordingly, what a history of a body of knowledge does not do well is analyse either how the changes came about or what their implications were. The result is a curiously flat account, in which one really good idea follows another in a seemingly inevitable way, although the development of algebra has its share of controversies, of paths not taken, and dramatic shifts of interest.

Advice to Students

Throughout this book I want you, the student, to learn a number of things, each of which has two aspects. I think of these as landmarks. One aspect is mathematical: I want you to understand and remember certain results. The other aspect is historical: I want you to understand the importance of those results for the development of mathematics.

In each case you will have to do some preliminary work to get where I want you to be. You will have to absorb some mathematical terms and definitions in order to understand the result and have a good idea about its proof. You will also have to absorb some sense of the intellectual enterprise of the mathematicians and their social context to appreciate the historical dimension of their results.

I have listed some key results below, almost in the order in which they occur in the book. A good way to test your understanding as your study progresses would be to ask yourself if you have understood these results both as pieces of mathematics and as turning points in the history of mathematics. Understanding is a relative term; aim for knowing the definitions and the best examples, the key steps in the proofs of the main theorems and, equally important, a sense of the context in which the result was first proposed and what was significant about it.

All these concepts will be taught as if they were new, although some familiarity with the more elementary ones will be helpful; these include the basic definitions and theorems in group theory and ring theory. The topic of continued fractions and Pell's equation is mentioned but marginalised here; it would make an interesting project to go alongside what I have presented. The same is true of the Kleinian view of geometry, which is an important source of group-theoretic ideas and contributed significantly to the recognition of their importance. On the other hand, I assume no previous acquaintance with algebraic number theory (algebraic numbers, algebraic integers, algebraic number fields, quadratic and cyclotomic fields, ideals, the factorisation of algebraic integers into irreducibles, principal ideal domains, and the prime factorisation of ideals) nor with a modern course in Galois theory, which looks at the subject in more detail and from a more modern abstract perspective than we do here.

The treatment of the mathematics in this book is surely not the modern one, rather it is one closer to the original, but I believe that the formulations of a Dedekind or a Jordan have a lucidity and a naturalness that commends them, and that in presenting them it is possible to teach the mathematics through its history and in so doing to teach its history in a full and interesting way. In particular, it becomes possible to understand that substantial ideas in modern mathematics were introduced for good and valuable reasons.

Any living body of knowledge is different from what is revealed by an autopsy. Research mathematicians see problems, methods, and results, organised perhaps by an image of mathematics, governed also by a sense of their own abilities. They have problems they want to solve, which can be of various kinds, from just getting an answer to getting a known answer by a better method. They can be theory builders, example finders, or problem solvers—all three at different times. What a study of the history of mathematics can provide, albeit for the past, is the dynamic appreciation of mathematics where research questions, their possible solutions, their difficulty and importance are always visible.

This book, at least, has two sorts of aims. One is to describe some of the major steps on the journey from classical algebra to modern algebra. The other is to describe how some mathematicians have come to formulate, and even solve, important new problems in mathematics. These two aims are more independent than it might seem at first glance. Firstly, the advances made by the mathematicians were not aimed at creating modern algebra until the journey was more or less complete. So we might well want to think of that journey as proceeding at random, in an aimless way, and certainly not as the result of a design held for over a century or longer. Secondly, the way research tasks get formulated and even solved is largely independent of the over-arching story into which they can be made to fit. A good mathematician has a good grasp of (much of) what is known in a subject, and a way of organising this knowledge so that it generates questions. Then, with luck, he or she answers those questions, or, in failing, answers others that might, with hindsight, seem better. Again, there is a random element here. The new insight is, or is not, picked up by others and incorporated into their view of the subject, and so the whole creaky edifice advances.

We can think of a researcher (in any subject, I would think, not just mathematics) having a story they tell themselves about their part of the subject. This story has definitions, concepts, and theorems, and it culminates in things the mathematician would like to know, which may or may not be what other mathematicians would like to know. We can call this the local or personal story. The historian, or the historically-minded mathematician, fits some aspects of these local stories together to get a global story—such as the origins of modern algebra. Most likely the historian retains more of the personal stories and the mathematician more of the results, probably re-writing them in modern terms as he or she proceeds.

So understanding this book is a matter of appreciating the local stories, and allowing them to fit together in some way. This determines what your first questions have to be: What did this person know about the topic? What did he or she want to know? How do they think they could do it? If you like, think of simply getting to know several mathematicians on their own terms, without fitting them into some historical account you may have heard before.

To take these questions in order, the first one does involve you understanding the mathematics the mathematician understood. Perhaps the subject is new to you, perhaps you know it already in a more modern form. I advise resisting the temptation to tackle all the mathematics, but to give way sometimes. After all, you exercise the same judgements when finding out about many interesting things, and you have other things to do here, too. One thread to hold on to is provided by examples, rather than theorems and proofs, and another is to able to describe the significance of a result (rather than merely repeat it).

The second question is perhaps simpler. The answer will be of the form: Solve this problem; Prove this theorem; Formulate some general principle. Again, the significance of these answers is as important as the answers themselves; what chance did this or that problem have of engaging the interest of anyone else?

The third question is more about mathematical methods, but here we want to notice a few things. These mathematicians did not always succeed: not completely, and sometimes not at all. This highlights one difference between this book and some histories of mathematics: it is not about who first did what. It is about what people were trying to do, and the extent to which they succeeded. Success can take many forms, not all of them welcome: the counter-example, the refutation, the unexpected example are all advances, but they may not fit the picture people wanted to create.

In most of the book each chapter has some exercises of a mathematical kind to help you grapple with the concepts, and some questions of a more historical nature to point you in the various directions that might help you contextualise the mathematics and reveal how it was evolving. In the final third of the book the exercises fall away, because I believe that the conceptual shifts towards modern mathematics are in fact clearer and easier to see from a slight distance than in the close-up view that exercises provide.

When we look at the stories about the discovery of this or that piece of mathematics and its path to acceptance, we also want to fit them into the global story we have chosen. Here we must be careful. Many historians would object to an enterprise focussed on how some aspect of the modern world came into being, finding the question too likely to lead to tendentious answers that lose too much historical diversity. It is true that it is hard enough, and interesting enough, to say accurately how and why Lagrange, Gauss, Galois or who ever did what they did without imagining that we can explain how some part of the modern world came about. But I think that the arrival of modern algebra is itself clearly a historical event, and it is plausible to explain it in historical terms. We can do this without supposing that everything was oriented to some great event in the future.

There is a particular issue to do with how you handle information, especially when it might be wrong. To repeat a historical falsehood (say, that Jacobi was a Professor in Berlin all his working life) is a mistake (Jacobi was in fact at Königsberg for most of his career), but it doubly misleading if it is relayed as an important fact without a source.

But the more important distinction is between facts and opinions. The Jacobi error is an error of fact. If you say, however, that Klein's Erlangen programme of 1872 was very important for the growth of group theory, that's an opinion. It may be true—I think it isn't—but even if you want to argue that it is a valid opinion, it is certainly an opinion. When you find such claims in the literature, if they matter to your essay you have to cite them. You do that by writing something like "As Smith has argued (Smith 1972, p. 33) ...", or, if you disagree, "Contrary to what Smith has argued (Smith 1972, p. 33) ...." In each case (Smith 1972, p. 33) refers to an item in your references by Smith (with an initial or first name) published in 1972 (give the title and tell me how I can find it) in which this opinion is on page 33. You may also write "Smith (1972, 33) argued that ...."; in this case the subject is Smith the person, and the parenthetical reference is to an item in your bibliography. You do the same with quotations, of course. If the item is on the web, you give the URL, and a page reference if that is possible.

The fact–opinion distinction isn't easy to maintain (when does a widely shared opinion simply become a fact?) but it matters because there is not much you can do with a well-established fact. Unless you suspect that it is wrong, and research in the archives will refute it, you must accept it. But an opinion is an opinion. Here's one of mine: Dirichlet, who was a friend of Jacobi's and did teach in Berlin for many years, was hugely influential in bringing rigorous mathematics to Germany and making Berlin in particular and Prussia generally a major centre for mathematics. You can see it's an opinion: I used the words "hugely influential" and "major centre". My evidence would have to be a major shift in the quality of work done in mathematics in Prussia, most of it demonstrably linked to Dirichlet directly or through the force of his example, and a clear rise of Berlin to a position of international eminence.

This book is full of opinions, and I expect you to have yours. Even if we agree you should have arguments in support of our opinions, and these arguments might not be the same. Certainly if we disagree we should both have arguments. We can then probe each other's positions, shift our opinions, refine them or agree to disagree. But the fact–opinion distinction is central to this book, so if you are giving an opinion taken from somewhere, say where it comes from.

Can you have too many references? In principle, yes. Will you, unless you set out deliberately to do so—no!

I have provided a number of appendices. Some amplify the historical story, and could be used instead of some of the later chapters of this book or as introductions to projects. Some—those involving Gauss, Dedekind, and Jordan—also contain translations of original material. One, on permutation groups, revises some material that I did not want to assume or to teach historically, and it can be used as a reference.

Acknowledgements

I thank John Cremona and Samir Siksek for their encouragement. I thank the various referees for the comments, particularly the anonymous referee who went to the trouble to inform me of a great many misprints; every reader owes him a debt for performing that valuable service. And I thank Anne-Kathrin Birchley-Brun at Springer for her expert editorial assistance and advice.

Results Discussed in This Book

1. The Tartaglia–Cardano method for solving a cubic equation and the formula that results; a method for attacking polynomial equations of degree 4;
2. the characterisation of numbers of the form $x^2 + y^2$, $x^2 + 2y^2$, $x^2 + 3y^2$, and what is different about numbers of the form $x^2 + 5y^2$;
3. Pell's equation $x^2 - Ay^2 = 1$ for square-free integers A and its connection with the continued fraction expansion of $\sqrt{A}$;
4. Fermat's last theorem, an indication of his proof for 4th powers and of Euler's 'proof' for 3rd powers;
5. Lagrange's reduction of binary quadratic forms to canonical form, especially in cases of small positive discriminants;
6. the statement of the theorem of quadratic reciprocity and its use in simple cases;
7. a proof of the theorem of quadratic reciprocity, and an understanding of why Gauss gave several proofs;
8. cyclotomic integers and an unsuccessful connection to Fermat's Last Theorem;
9. the key features of the Gauss–Dirichlet theory of quadratic forms;
10. the quintic equation is not solvable by radicals, with an indication of Abel's proof and Galois's approach;
11. it is impossible to trisect an angle or double a cube by straight edge and circle alone;
12. Galois's theory of solvability by radicals, as explained by Galois, Jordan, Klein, and Weber;
13. Cauchy's theory of subgroups of S_n;
14. transitivity and transitive subgroups of S_n;
15. the field-theoretic approaches to Galois theory of Kronecker and Dedekind;
16. algebraic integers; ideals and prime ideals in a ring of algebraic integers;
17. ideals in rings of quadratic integers and the connection to classes of binary quadratic forms;
18. the Hilbert basis theorem;
19. the Brill–Noether $AF + BG$ theorem, the counter-example provided by the rational quartic in space;
20. primary ideals, primary decomposition in a commutative ring.

Contents

List of Figures

Chapter 1
Simple Quadratic Forms

1.1 Introduction

In this chapter we look at two topics: the question of what numbers, and specifically what primes, can be written in the form $x^2 \pm ny^2$ for small, non-square positive n?, and how to show that the equation $x^2 - Ay^2 = 1$ has solutions in integers for positive, non-square integers A. Once we have seen what mathematical conclusions we shall need, we look at how to handle the mathematics in a historical way—how to use mathematics as evidence.

1.2 Sums of Squares

When $n = 1$, the question is: What primes are sums of two squares? It is easy to see that 2 is, and working modulo 4 quickly shows that no prime of the form $4n - 1$ can be. It turns out that every prime of the form $4n + 1$ is a sum of squares, such as

$$5 = 2^2 + 1^2, \ 13 = 3^2 + 2^2, \ldots.$$

Fermat was the first to prove this. He mentioned it in a letter to Mersenne of 25 December 1640.[1] No proof of his survives, but Weil (1984, 63) plausibly suggested that the proof would have been much the same as the one Euler was later to claim.

---

[1] See *Oeuvres de Fermat* 2, 213, where he notes that the sum is unique. Fermat specifically mentioned the method of infinite descent in connection with this problem.

© Springer Nature Switzerland AG 2018

J. Gray, *A History of Abstract Algebra*, Springer Undergraduate Mathematics Series, https://doi.org/10.1007/978-3-319-94773-0_1

It is worth noting that it took Euler time to work out the details. He first wrote to his friend Goldbach about it in 1742, and told him that he had found a proof of Fermat's result in 1749, but only published his findings in 1758, and even then his proof had a gap that he plugged only in 1760 (E 228 and E 241 respectively).[2]

Euler proceeded by considering divisors. If it can be shown that every divisor of a number of the form $x^2 + y^2$ is also of that form, then the question reduces to finding all primes of this form, because we have this lemma:

Lemma 1.1. *A product of numbers of the form $x^2 + y^2$ is itself of the form $x^2 + y^2$.*

Proof.

$$(x^2 + y^2)(u^2 + v^2) = (x + iy)(x - iy)(u + iv)(u - iv) = (x + iy)(u + iv)(x - iy)(u - iv) =$$

$$(xu - yv + i(xv + yu))(xu - yv - i(xv + yu)) = (xu - yv)^2 + (xv + yu)^2.$$

Also

$$(x^2 + y^2)(u^2 + v^2) = (x + iy)(x - iy)(u + iv)(u - iv) = (x + iy)(u - iv)(x - iy)(u + iv) =$$

$$(xu + yv - i(xv - yu))(xu + yv + i(xv + yu)) = (xu + yv)^2 + (xv - yu)^2.$$

So we find

$$(x^2 + y^2)(u^2 + v^2) = (xu \pm yv)^2 + (xv \mp yu)^2.$$

Euler next proved that no prime divisor of $x^2 + y^2$ is of the form $4n - 1$ as follows. Suppose there is such a prime, say p. We can assume that x is prime to y and that both are prime to p.[3] So we have

$$x^2 \equiv -y^2 \ (\mathrm{mod}\ p).$$

But we may write $p - 1 = 2m$, where $m = 2n - 1$, so m is odd and we have

$$x^{2m} \equiv -y^{2m} \ (\mathrm{mod}\ p).$$

But by Fermat's little theorem,

$$x^{2m} \equiv y^{2m} \ (\mathrm{mod}\ p).$$

---

[2] These numbers refer to Eneström's catalogue of Euler's works; I shall generally refer to Euler's works by their E number. The catalogue, and almost of all Euler's work, is available in the invaluable Euler Archive on the web. The Euler–Goldbach correspondence has recently been published as Euler (2015), see letters 47 and 138.

[3] Here and throughout we use the now-familiar $\equiv$ symbol for congruences. It was first introduced by Gauss, as is explained below in Sect. 4.3.

From this contradiction it follows that the initial assumption must be wrong, and no prime divisor of a sum of squares is of the form $4n - 1$.

Note also that what this shows is that -1 is a square mod p, where p is an odd prime, only if p is of the form $4n + 1$. It remains to show that if p is of the form $4n + 1$ then there are integers x and y such that $x^2 + y^2 = p$. It is this detail that Euler fudged in 1758.

The best account of this material that I know is Cox's book with the very honest title: *Primes of the form $x^2 + ny^2$* (1989). There he sketches a slight reworking of Euler's argument in this form. There is what he called the descent step and then there is a reciprocity step—the names will become clear as we proceed.

Descent: Prime divisors of $x^2 + y^2$, where $\gcd(x, y) = 1$, are of the form $x^2 + y^2$.

Reciprocity: Primes of the form $4n + 1$ divide some $x^2 + y^2$ where $\gcd(x, y) = 1$.

Descent goes like this. From the identity in the above Lemma,

$$(x^2 + y^2)(u^2 + v^2) = (xu \pm yv)^2 + (xv \mp yu)^2,$$

Euler deduced that if $N = a^2 + b^2$, where a and b are relatively prime, and if $q = x^2 + y^2$ is a prime divisor of N, then N/q is also a sum of two relatively prime squares. Once that is proved, it follows that all prime divisors of an $x^2 + y^2$ are of this form, by an infinite descent argument that I'll sketch below.

To get started, notice that *if $a^2 + b^2 = (x^2 + y^2)(c^2 + d^2)$ then* by the lemma $a = (xc \pm yd)$ and $b = (xd \mp yc)$, which is a powerful clue as to how to carry out the proof. Indeed, to show that if $q = x^2 + y^2$ divides $N = a^2 + b^2$, where x and y are relatively prime, then N/q is also a sum of relatively prime squares, Euler argued as follows.[4]

Lemma 1.2. *If $q = x^2 + y^2$ divides $N = a^2 + b^2$, where a and b are relatively prime, then N/q is also a sum of relatively prime squares.*

Proof. q divides $x^2N - a^2q = (xb - ay)(xb + ay)$, and because q is prime it divides one of these factors. Without loss of generality, we can assume q divides $xb - ay$. Say $xb - ay = dq$, for some integer d.

Euler now proved that x divided $a + dy$ by showing that it divided $(a + dy)y$; this uses the fact that x and y are relatively prime. This last claim is true because $(a + dy)y = xb - dx^2$, which is obviously divisible by x. So he could set $a + dy = cx$, and deduce that

$$a = cx - dy, b = dx + cy.$$

Euler then put all this back together and deduced that

$$N = a^2 + b^2 = (x^2 + y^2)(c^2 + d^2) = q(c^2 + d^2),$$

so

$$N/q = c^2 + d^2,$$

as required (c and d are relatively prime because a and b are).

Descent arguments work by contradiction, so suppose that p is a prime dividing $N = a^2 + b^2$ and that p is not a sum of two squares. A descent argument will show that there is a prime q that divides N, is not a sum of two squares, and is smaller than p. Iterating this argument produces a contradiction.

Notice that the above Lemma shows that if pq is a sum of two squares and one of p and q is a sum of squares, then so is the other one.

Now, to arrange that any q that divides N is smaller than p we resort to a trick. If, say, $a = a_1 + kp$ then from the fact that p divides $a^2 + b^2$ we deduce that p divides $a_1^2 + b^2$. So by adding or subtracting suitable multiples of p to or from a and b we can assume that p divides $N_1 = a_1^2 + b_1^2$, where $|a| < p/2$ and $|b| < p/2$. If we throw away any common factors of a_1 and b_1 those inequalities are still true, and together they imply that $N_1 < p^2/2$. This implies that any divisor of N_1 is less than p.

Now consider any divisor q of N_1 other than p. If each one is a sum of two relatively prime squares then so is their product, and therefore so is N_1 divided by that product, which is p. Therefore at least one q is not a sum of two relatively prime squares. But it is less than p, and the descent argument applies, and the descent claim is proved: Prime divisors of $x^2 + y^2$, where $\gcd(x, y) = 1$ are of the form $x^2 + y^2$.

Reciprocity is always harder, because we have either to find x and y or to give an abstract existence proof. Cox gives this nice argument, but one not available to Euler: Fermat's little theorem implies that if $p = 4n + 1$ is prime then

$$r^{4n} - 1 \equiv 0 \bmod p$$

so

$$(r^{2n} + 1)(r^{2n} - 1) \equiv 0 \bmod p.$$

So if we can find an r such that $r^{2n} - 1 \equiv 0 \bmod p$ is *not* true then we must have

$$r^{2n} + 1 \equiv 0 \bmod p,$$

which expresses p as a divisor of a sum of squares, and we are done (by the descent step). For us, but not for Euler, it is true that only $2n$ of the $4n$ values of r mod p can satisfy $r^{2n} - 1 \equiv 0 \bmod p$, so the existence of the required r is trivial.

On the other hand, producing such an r explicitly is not trivial. Consider, for example, the case $p = 101$. We are required to find an r such that $r^{50} \equiv -1 \bmod 101$. Even in this small case it's clearly going to be time-consuming to do this; we must do something like calculate

$$r, r^2, r^4, r^8, r^{16}, r^{32}, r^{48} \equiv r^{16} \cdot r^{32} \text{ and } r^{50} \equiv r^2 \cdot r^{48} \text{ mod } 101$$

for different values of r until we get lucky—which, as it happens, we do with the natural first choice, $r = 2$. Doing a few explicit calculations of this sort should make you want to look for a theory to cut down on all the work, which in turn means looking for a property that 2 has modulo 101 that is not shared with every other integer modulo 101. The crucial property is that 2 is what is called a 'primitive root' modulo 101 (see Sect. 4.3 below) but unfortunately there is no simple way in general of finding primitive roots.

To find x and y explicitly, we write 2^{50} as $(2^2)^{25}$, and set $x/y \equiv 2^{25} \text{ mod } 101$. We are free to choose an easy value of y, such as 1 or 2, and if we choose $y = 1$ then we find $x \equiv 2^{25} \equiv 10 \text{ mod } 101$ and we deduce that 101 divides $x^2 + y^2 = 10^2 + 1^2 = 101$, which indeed it does.

Cutting a Historical Corner

In a history course we need merely to check that Euler did this, in order to be able to say that Euler *did* prove this result. If we are satisfied that there is nothing unduly novel in the proof, we need not concern ourselves with the proof ever again. That will be the case here, so, unless a historical question arises for which Euler's proof is interesting, we need not have attempted to master this mathematical step. But we can easily see that it is a piece of straight-forward (if clever) elementary algebra.

Modified Sums of Squares

Others of Fermat's discoveries were to inspire Euler to prove them and in so doing start modern number theory. The claims date from 1654, in a letter to Blaise Pascal; again, no proof of his has come down to us.[5]

> Every prime which surpasses by one a multiple of three is composed of a square and the triple of another square. Examples are 7, 13, 19, 31, 37, etc.

> Every prime which surpasses by one or three a multiple of eight is composed of a square and the double of another square. Examples are 11, 17, 19, 41, 43, etc.

Euler showed in each case that prime divisors of numbers of the form $x^2 + 2y^2$ (respectively $x^2 + 3y^2$) are again of the form $x^2 + 2y^2$ (respectively $x^2 + 3y^2$). He used the same two-step way: descent and reciprocity. It took him (on and off)

---

[5]See Fermat, *Oeuvres* 2, 310–314, esp. p. 313; quoted in Cox (1989, 8).

30 years. These remarks of Fermat establish for which primes p it is the case that -2 and -3 are squares mod p.

However, with $x^2 + 5y^2$, something unexpected happens. The identity

$$(x^2 + 5y^2)(u^2 + 5v^2) = (xu \pm 5yv)^2 + 5(xv \mp yu)^2$$

shows that the product of two numbers of the form $x^2 + 5y^2$ is again of this form. But it is not true that every divisor of a number of this form is again of that form. For example,

$$21 = 1^2 + 5 \times 2^2 = 3 \times 7,$$

but neither 3 nor 7 is of the required form. To give another example, $7 \times 23 = 161 = 6^2 + 5 \cdot 5^2$.

Fermat conjectured, but admitted that he could not prove, that primes that are congruent to 3 or 7 mod 20 have products that are of the form $x^2 + 5y^2$. The primes of the form $x^2 + 5y^2$ that are less than 300 are 29, 41, 61, 101, 109, 149, 229, 241, and 269, and Fermat did not capture them; note that they are all congruent to either 1 or 9 mod 20. Fermat could only conjecture—but not prove—this statement about the divisors of numbers of the form $x^2 + 5y^2$:

> If two primes, which end in 3 or 7 and surpass by three a multiple of four, are multiplied, then their product will be composed of a square and the quintuple of another square.[6]

As this quote reminds us, Fermat made his discoveries without the advantage of a good notation. His congruence condition, for example, can be more simply stated as the primes are congruent to 3 or 7 modulo 20.

We get an interesting glimpse of what Euler knew in 1753 from this exchange with his friend Christian Goldbach.[7] Goldbach wrote to Euler on 12 March to say that he had found that if p is a prime of the form $4dm + 1$ then p can be represented as $p = da^2 + b^2$. Euler replied on 3 April that he had noticed this theorem some time ago "and I am just as convinced of its truth as if I had a proof of it", and then gave these examples:

$$p = 4.1m + 1 \Rightarrow p = aa + bb$$

$$p = 4.2m + 1 \Rightarrow p = 2aa + bb$$

$$p = 4.3m + 1 \Rightarrow p = 3aa + bb$$

$$p = 4.5m + 1 \Rightarrow p = 5aa + bb.$$

---

[6]See Fermat's letter to Sir Kenelm Digby in 1658, in Fermat *Oeuvres*, II, 402–408 (in Latin), see p. 405, and *Oeuvres*, 3, 314–319 (in French) see p. 317. Quoted in Cox (1989, 8).

[7]See Euler (2015, letter 166) and Lemmermeyer (2007, pp. 531–532).

He then remarked that he could only prove the first of these claims, but not the rest, and to observe that the claim was only true in general when a and b are allowed to be rational numbers, giving the example

$$89 = 11 \left(\frac{5}{2} \right)^2 + \left(\frac{9}{2} \right)^2,$$

where 89 cannot be written in the form $11a^2 + b^2$ with a and b integers.[8] Later he found a proof of the second of the above claims; the others were to be established by Lagrange.

1.3 Pell's Equation

For once it is reasonable to begin with a string of negatives. The study of the equation has nothing to do with Pell, an English mathematician of the seventeenth century—the attribution started with a simple mistake by Euler. It has a lot to do with Fermat, but he was not the first to take the problem up, although he could not have known that it had already been worked on thoroughly by an Indian mathematician, Brahmagupta, who was born in AD 598. Neither Brahmagupta nor Fermat have left us their methods, and we must exercise a little charity if we credit Brahmagupta with working out all the elementary theory, but as we will see there are good reasons for believing he had a deep understanding of the problem.[9]

In any case, the problem is to find, for a given non-square integer A, integers x and y such that

$$x^2 - Ay^2 = 1.$$

The simplest cases can be done by guesswork:

- $A = 2$: we have $(x, y) = (3, 2)$;
- $A = 3$: we have $(x, y) = (7, 4)$;
- $A = 5$: we have $(x, y) = (9, 4)$.

There is a close connection between Pell's equation and the approximation of irrational square roots by rational numbers, which was developed for the first time by Lagrange in his (1769), and presented more accessibly in his *Additions* to the French translation of Euler's *Algebra* (1770).[10] Consider, for example, the question

---

[8]See Euler (2015, letter 167).

[9]A good and thorough reference for this material is Scharlau and Opolka (1984, 43–56).

[10]They are bound together in the first volume of Euler's *Opera Omnia* and in the English translation of Euler's *Algebra* published in 1840 and reprinted in 1984.

of finding rational approximations to $\sqrt{5}$. The equation

$$x^2 - 5y^2 = 1$$

implies that

$$\left(\frac{x}{y}\right)^2 = 5 + \left(\frac{1}{y}\right)^2,$$

so solutions to this equation give approximations to $\sqrt{5}$ and the approximations improve as y gets larger. To find these approximations, we use the method of continued fractions, as follows. We have

$$\sqrt{5} = 2 + \frac{1}{a_1}, \quad 0 < a_1 < 1,$$

so

$$a_1 = \frac{1}{\sqrt{5} - 2} = \frac{\sqrt{5} + 2}{1} = \sqrt{5} + 2 = 4 + \frac{1}{a_2}, \quad 0 < a_2 < 1$$

so

$$a_2 = \frac{1}{\sqrt{5} - 2},$$

and in this case the method cycles very quickly. The continued fraction expansions of $\sqrt{5}$ are

$$2, 2 + \frac{1}{4}, 2 + \frac{1}{4 + \frac{1}{4}}, \ldots,$$

which we write this way:

$$[2, 4, 4, 4, \ldots].$$

The continued fraction expansions of $\sqrt{7}$ and $\sqrt{13}$ can also be found in this way. They also cycle; indeed, it is a theorem that the expansions of $\sqrt{n}$ all cycle (when n is not a perfect square). Much less is known about the continued fraction expansions of almost any other type of irrational number.

The successive approximations to the continued fraction $[a, b_1, b_2, \ldots, b_n, \ldots]$ are

$$a, [a, b_1], [a, b_1, b_2], \ldots [a, b_1, b_2, \ldots, b_n].$$

They are called the convergents. The connection with continued fraction expansions and square roots is that the convergents alternately over- and under-estimate the square root in a fashion that converges very quickly.

Let us take the case $A = 5$. We do indeed have

$$9^2 - 5 \cdot 4^2 = 81 - 80 = 1,$$

so re-writing Pell's equation in the form $Ay^2 = x^2 - 1$ and then in the form

$$A = \left(\frac{x}{y}\right)^2 - \frac{1}{y^2},$$

and using the numbers just obtained, we have

$$5 = \left(\frac{9}{4}\right)^2 - \frac{1}{4^2},$$

which says that 9/4 is an approximation to $\sqrt{5}$. To be sure, not a good one, but we can do better. The successive convergents are

$$[2, 4, 4, 4, \ldots].$$

These approximations are successively

- $[2, 4] = 2 + \frac{1}{4} = \frac{9}{4}$
- $[2, 4, 4] = 2 + \frac{1}{4 + \frac{1}{4}} = \frac{38}{17}$
- $[2, 4, 4, 4] = 2 + \frac{1}{4 + \frac{1}{4 + \frac{1}{4}}} = \frac{161}{72}.$

The square of the last number is 5.0002 to four decimal places.

There is a useful identity:

$$(x^2 + ny^2)(u^2 + nv^2) = (xu \pm nyv)^2 + n(xv \mp yu)^2.$$

So the product of two terms of the form $x^2 + ny^2$ is again of that form. If we take $n = -5$, the useful identity says

$$(x^2 - 5y^2)(u^2 - 5v^2) = (xu - (-5)yv)^2 + (-5)(xv + yu)^2.$$

Let us take

$$(x, y) = (9, 4) = (u, v).$$

We find

$$(xu - (-5)yv) = 161, \text{ and } (xv + yu) = 72.$$

It will be helpful to write the useful identity in the form

$$(x, y) * (u, v) = (xu \pm nyv, xv \mp yu).$$

In modern terms, we have the set of integer pairs $\{(x, y) \mid x^2 + ny^2 = 1\}$ which is closed under multiplication by $*$. We don't know for sure that this set is non-empty, but if it is then the next question is: is it a group? And the question after that is: what sort of a group? (Plainly, it will be a commutative one.)

Now, it is a fact about continued fractions that they alternately over- and under-estimate the number they converge to. This is proved in most books on the subject (e.g. Khinchin 1964) and stated correctly, but not proved, in Wikipedia (the proof is not difficult). So the values $(x, y) = (38, 17)$ that we obtained are going to give us a different kind of approximation to $\sqrt{5}$, and indeed $38^2 - 5 \cdot 17^2 = -1$. Now, if it is allowed to approximate $\sqrt{5}$ that way, and solve $x^2 - 5y^2 = -1$, we could have started with smaller numbers: $(x, y) = (2, 1)$. Let us calculate

$$(2, 1) * (2, 1) = (9, 4).$$

We have worked our way back to the historical starting point. Brahmagupta's idea was that if we make a guess at values of x and y such that $x^2 - Ay^2 = 1$ for a given (non-square) integer A we can use it to generate another guess, by this $*$ method. Usually this will produce a worse pair of x and y, but we can intervene and make a better guess than the one the $*$ method provides. In this way we can guide the pairs down to a value of x and y for which, in fact, $x^2 - Ay^2 = 1$. This is what Bhāskara II did.

Brahmagupta and Bhāskara II

The Indian astronomer Brahmagupta wrote his *Brāhma-sphuṭa-siddhānta* when he was 30. It is one of the works from which the zero symbol and the decimal place value system came to Europe, along with the Hindu-Arabic numerals.[11]

Most of the book is devoted to improvements in theoretical astronomy, and some chapters are on arithmetic and on geometry. Chapter 18, however, was much more

---

[11] See Brahmagupta's method for 'Pell's' equation on the web: http://www-history.mcs.st-and.ac.uk/HistTopics/Pell.html.

algebraic. Here Brahmagupta discussed how to solve some equations of the form

$$Nx^2 + k = y^2,$$

(in modern notation) for positive integers N and small values of k (such as $\pm 1, \pm 2$ or 4); for example, when $N = 5$ and $k = 1$, the smallest solution is $x = 4, y = 9$, since

$$(5 \times 4^2) + 1 = 81 = 9^2.$$

Brahmagupta's success is a formidable achievement because trial and error speedily exhausts the hardiest of calculators without success. Particularly striking is his success with $N = 61$ and $k = 1$: here the smallest solution, which he succeeded in finding, is $x = 226,153,980$ and $y = 1,866,319,049$. It is clear that he must have had a general method, but all that have come down to us are hints that allow one to take solutions to the problem and produce others.[12]

A great advance of Bhāskara II in his *Bīja-gaṇita* (Seed computation) written in 1150, was to discover a general method for solving this problem. It is the *chakravala* or cyclic method, and starts by guessing a solution to a 'nearby' equation. It is not clear if Bhāskara II could prove that the cyclic method always works. Plofker comments (see Plofker in Katz 2007, 474)

> It is not known how, or whether, Indian mathematicians showed that this method would always provide a solution after a finite number of cycles.

The cyclic method begins by assuming that $x = a, y = b$ solves the problem $Nx^2 + k = y^2$ for a small value of k. This can always be done, because $x = 1, y = m$ is a solution of $N \cdot x^2 + (m^2 - N) = y^2$. Bhāskara then used Brahmagupta's method of combining solutions to deduce that, for any number b,

$$N(am + b)^2 + (m^2 - N)k = (bm + Na)^2.$$

This implies that $x = (am + b)/k, y = (bm + Na)/k$ is a solution of

$$Nx^2 + (m^2 - N)/k = y^2.$$

Bhāskara now chose a number m such that $am + b$ is divisible by k. This implies that $m^2 - N$ and $bm + Na$ are also divisible by k, and so

$$x = (am + b)/k, y = (bm + Na)/k$$

---

[12]See the discussion in Plofker (2009, 154–156).

are integer solutions of the equation $Nx^2 + (m^2 - N)/k = y^2$, where $(m^2 - N)/k$ is also an integer. The trick is to try to make $(m^2 - N)/k$ smaller than k and then hope to repeat this argument until a solution of the equation $Nx^2 + 1 = y^2$ is obtained.

Consider one of Bhāskara's more dramatic examples. To solve the equation

$$61x^2 + 1 = y^2$$

he needed to find a number m so that $(m + 8)/3$ is an integer and $m^2 - 61$ is as small as possible. So he took $m = 7$ and obtained $x = 5, y = 39$ as a solution of the equation $61x^2 - 4 = y^2$. Using Brahmagupta's method he then found $x = 226,153,980, y = 1,866,319,049$ as the smallest solution of $61x^2 + 1 = y^2$.

In fact, Bhāskara always stopped when $(m^2 - n)/k$ is one of the numbers $\pm 1, \pm 2$ or ± 4, for then he could apply Brahmagupta's method to find a solution to the equation $nx^2 + 1 = y^2$. It seems clear that he knew his method always worked, although he gave no proof.

Fermat too liked to tantalise his readers with very large numbers. In 1656, apparently despairing of finding French mathematicians interested in problems about numbers, he challenged the mathematicians of Europe to solve these sorts of problems, and John Wallis and Lord Brouncker in England succeeded, with an independent rediscovery of the cyclic method. For an account of their work, which gives more credit to Brouncker than hitherto, see Stedall (2000a, b).

1.4   Exercises

1. Explain why none of the numbers 2012, 2013, . . . , 2016 can be written as a sum of two squares, but 2017 can be.
2. Formulate a conjecture about which primes are of the form $x^2 - 2y^2$. Those between 3 and 200 are 7, 17, 23, 31, 41, 47, 71, 73, 79, 89, 97, 103, 113, 127, 137, 151, 167, 191, and 193.
3. Find the continued fraction expansions of $\sqrt{10}, \sqrt{13}, \sqrt{15}, \sqrt{17}, \sqrt{21}$. What do you notice if you compare the start of the continued fraction expansions of numbers of the form $\sqrt{n^2 + 1}$ and those markedly not of that form?
4. The roots of the quadratic equation $ax^2 + 2bx + c = 0$ are $\frac{-b \pm \sqrt{b^2 - ac}}{a}$. Let $a_1 = a - 2b + c, b_1 = b - c$, and $c_1 - c$ and write down the roots of the quadratic equation $a_1 x^2 + 2b_1 x + c_1 = 0$.

If you have access to a computer and can do continued fraction expansions, you will see that the continued fraction expansions of the roots of the two equations rapidly become the same. For example, if $a = 1, b = 3, c = 2$ the first pair of roots are $-3 \pm \sqrt{7}$ and the second set of roots are $\frac{-2 \pm \sqrt{7}}{3}$, but the successive integer parts go 4, 1, 4, 1, ... However, the continued fraction expansion of $3 + 2\sqrt{7}$ settles into 3, 2, 3, 10, 3, 2, 3, 10. . . . which is very different.

This suggests that the continued fraction expansion of a quadratic irrational has more to do with the quadratic equation of which it is a root than the irrational alone. Now look ahead to Lagrange's theorem on the reduction of quadratic forms to a canonical form.

Questions

1. To what extent was Fermat influenced by the mathematics of classical antiquity? Was Euler equally interested? What might your answers say about the priorities of mathematicians in the seventeenth and eighteenth centuries?
2. Why might an astronomer be interested in solving Pell's equation, and more generally in the method of continued fractions?

Chapter 2
Fermat's Last Theorem

2.1 Introduction

Here we look at Fermat's account of the equation $x^4 + y^4 = z^4$ and then at Euler's flawed but insightful account of $x^3 + y^3 = z^3$.

2.2 Fermat's Proof of the Theorem in the Case $n = 4$

In Bachet's 1621 edition of the late Greek mathematician Diophantus's *Arithmetica* we find

> Book II Prop.8
> *To divide a given square number into two squares.*
> Let it be required to divide 16 into two squares.
> And let the first square $= x^2$; then the other will be $16 - x^2$; it shall be required therefore to make $16 - x^2 =$ a square.
> I take a square of the form $(mx - 4)^2$, m being any integer and 4 the root of 16; for example, let the side be $2x - 4$, and the square itself $4x^2 + 16 - 16x$. Then $4x^2 + 16 - 16x = 16 - x^2$. Add to both sides the negative terms and take like from like. Then $5x^2 = 16x$, and $x = \frac{16}{5}$. One number will therefore be $\frac{256}{25}$, the other $\frac{144}{25}$, and their sum is $\frac{400}{25}$ or 16, and each is a square.

Fermat wrote in the margin of his copy of this edition

> On the other hand it is impossible to separate a cube into two cubes, or a biquadrate into two biquadrates, or generally *any power except a square into two powers with the same exponent*. I have discovered a truly marvellous proof of this, which however the margin is not large enough to contain.

Fermat wrote nothing else on this subject that has survived, and as no-one has ever found a proof with the means available to Fermat (as opposed to those of Sir Andrew Wiles and Richard Taylor in 1995) the most likely thing is that Fermat fell

© Springer Nature Switzerland AG 2018
J. Gray, *A History of Abstract Algebra*, Springer Undergraduate Mathematics Series,
https://doi.org/10.1007/978-3-319-94773-0_2

for a fallacious argument. If so, it seems that we will never know if he also, later, found the flaw in his "truly marvellous demonstration". The marginal note was not published until after his death, in the edition of his works made by his son Samuel in 1679. By the way, it should not, strictly, be called his last theorem, not only because it was not, until recently, a theorem but only a conjecture, but also because it is not Fermat's last unproved claim. He also raised the question of which integers can be the areas of triangles with sides that are rational numbers. Arabic writers knew of a triangle of this kind with area 5, and Leonardo of Pisa had claimed such a triangle cannot have area 1. Fermat proved this, his argument essentially settling Fermat's Last Theorem in the case $n = 4$. It is now known that the smallest possible value for the area is 5, but the general problem remains unsolved.[1]

Fermat's greatest achievement in number theory was his resolution of this case of his 'last' theorem: his proof that there are no non-zero integer solutions to the equation $x^4 + y^4 = z^4$. His interest in this problem is another illustration of how simultaneously 'modern' and 'ancient' he was. He took the problem from a note Bachet had put in Book VI of his edition of Diophantus's *Arithmetica*. Diophantus had raised problems about triangles whose areas plus or minus a number are squares, and Bachet added the remark that if A is the area of a right-angled triangle with integer sides then there is number K such that $(2A)^2 - K^4$ is a square. To see this, observe that we want to find integers K and s such that

$$\left(2pq(p^2 - q^2)\right)^2 + K^4 = s^2.$$

We have

$$\left(2pq(p^2 - q^2)\right)^2 = 4p^2q^2(p^2 - q^2)^2$$

$$= 4p^2q^2(p^4 - 2p^2q^2 + q^4) = 4p^2q^2(p^4 + 2p^2q^2 + q^4) - 16p^4q^4,$$

so we may set

$$s = 2pq(p^2 + q^2), \quad K = 2pq.$$

Fermat investigated this in the case when A is required to be a square, say m^2, and was led to look for integer solutions to $x^4 + y^4 = $ a square. He then showed that

---

[1] More precisely, a theorem of Tunnell (1983), allows one to determine all integers that are the areas of right-angled triangles with rational sides—the so-called triangular numbers—but the theorem assumes the truth of the Birch–Swinnerton-Dyer conjecture. See, among other good sources on the web, Jim Brown, Congruent numbers and elliptic curves, http://www.math.caltech.edu/~jimlb/congruentnumberslong.pdf, Keith Conrad, The Congruent Number Problem, http://www.math.uconn.edu/~kconrad/blurbs/ugradnumthy/congnumber.pdf and, for more on the early history, Norbert Schappacher, Diophantus of Alexandria: a text and its history http://www-irma.u-strasbg.fr/~schappa/NSch/Publications-files/Dioph.pdf.

there is no solution in integers to the equation $x^4 + y^4 = $ a square, which implies that there are no integer solutions to $x^4 + y^4 = $ a fourth power.

Let us look at Fermat's argument, and a helpful commentary on it by André Weil.

Fermat's Last Theorem: The Case of $x^4 + y^4 = z^4$

"Bachet: Find a triangle whose area is a given number." The area of a triangle in numbers cannot be a square. I am going to give a proof of this theorem, which I have discovered; and I did not find it without painful and laborious thinking about it, but this kind of proof will lead to marvellous progress in the science of numbers.[2]

If the area of a triangle was a square there would be two fourth powers whose difference was a square; it would follow equally that there would be two squares whose sum and difference would be squares. Consequently there would be a square number, the sum of a square and the double of a square, with the property that the sum of the two squares that make it up is likewise a square. But if a square number is the sum of a square and the double of a square its root is likewise the sum of a square and the double of a square, which I can prove without difficulty. One concludes from this that this root is the sum of the two sides of a right angle in a triangle of which one of the square components forms the base and the double of the other square the height.

The triangle will therefore be formed of two square numbers whose sums and differences are squares. But one will prove that the sum of these two squares is smaller than the first two of which one has likewise supposed that the sum and difference are squares. Therefore, if one has two squares whose sum and difference are squares one has at the same time, in integers, two squares enjoying the same property whose sum is less.

By the same reasoning, one has accordingly another sum smaller than that derived from the first, and continuing indefinitely one will always find smaller and smaller integers satisfying the same condition. But this is impossible because an integer being given there cannot be an infinity of integers which are smaller.

The margin is too narrow to receive the complete proof with all its developments.

In the same way I have discovered and proved that there cannot be a triangular number, except for one, which is also a fourth power.[3]

To follow this we are helped by a commentary by André Weil (1984, 77):

Fortunately, just for once, he had found room for this mystery in the margin of the very last proposition of Diophantus [. . .]; this is how it goes.

Take a Pythagorean triangle whose sides may be assumed mutually prime; then they can be written as $(2pq, p^2 - q^2, p^2 + q^2)$ where p, q are mutually prime, $p > q$, and $p - q$ is odd. Its area is $pq(p+q)(p-q)$, where each factor is prime to the other three; if this is a square, all the factors must be squares. Write $p = x^2, q = y^2, p+q = u^2, p-q = v^2$, where u, v must be odd and mutually prime. Then x, y, and $z = uv$ are a solution of $x^4 - y^4 = z^2$; incidentally, v^2, x^2, u^2 are then three squares in an arithmetic progression whose difference is y^2. We have $u^2 = v^2 + 2y^2$; writing this as $2y^2 = (u + v)(u - v)$, and observing that the g.c.d. of $u + v$ and $u - v$ is 2, we see that one of them must be of the form $2r^2$ and the other of the form $4s^2$, so that we can write $u = r^2 + 2s^2, \pm v = r^2 - 2s^2, y = 2rs$, and

---

[2] See Fermat, *Oeuvres* 2, 376.

[3] Fermat, *Oeuvres* 1, 340–341 and F & G 11.C8.

consequently

$$x^2 = \frac{1}{2}(u^2 + v^2) = r^4 + 4s^4.$$

Thus r^2, $2s^2$ and x are the sides of a Pythagorean triangle whose area is $(rs)^2$ and whose hypotenuse [x, JJG] is smaller than the hypotenuse $x^4 + y^4$ of the original triangle. This completes the proof "by descent".

2.3 Euler and $x^3 + y^3 = z^3$

Mathematicians seem to have enjoyed discovering that the search for proofs in number theory took them from one problem to another, and Euler at work on the case of Fermat's 'last' theorem when $n = 3$ is a good illustration of that process at work. Euler first claimed to have proved this result in a letter to Goldbach of August 1753 (see Euler 2015, letter 169), but he did not publish his work until his *Algebra* appeared in 1770. His proof is given in Chapter XV, §243, and it is a rather long, so I can only sketch it here. He modelled his study of this problem on Fermat's own published solution of the case $n = 4$, suspecting that the problems must be analogous, and Fermat had shown that there were no whole number solutions in that case by his ingenious method of infinite descent.

Euler began with a preliminary simplification of the problem. From the equation $x^3 + y^3 = z^3$ it is clear that not all the numbers x, y, and z can be even, or they would have a common factor, and we can assume without loss of generality that they do not. Nor can they all be odd, because the sum of two odd numbers is even. So either x is even or z is even, and Euler assumed z is even—the other case proceeds similarly. So he deduced that $x + y = 2p$, say, and $x - y = 2q$, say (this is not saying that p and q are primes). It follows that $x^3 + y^3 = 2p(p^2 + 3q^2)$. Here is the connection to another problem: the quadratic form $p^2 + 3q^2$.

Therefore, said Euler, $2p$ and $p^2 + 3q^2$ are both cubes.[4] The only way this fails is if p is a multiple of 3, because this is the only way that $2p$ and $p^2 + 3q^2$ are not coprime.[5]

Euler then fixed his attention on how a number of the form $p^2 + 3q^2$ can be a cube.[6] There is the identity

$$(a^2 + 3b^2)^3 = (a^3 - 9ab^2)^2 + 3(3a^2b - 3b^3)^2,$$

---

[4]Notice, with Euler, that because $2p$ is to be a cube it must be of the form $8p'^3$.

[5]I omit this case, because it proceeds similarly to the case we shall discuss. Here $z^3 = \frac{9r}{4}(3r^2 + q^2)$, so Euler could write $q = t(t^2 - 9u^2)$ and $r = 3u(t^2 - u^2)$ with u even and t odd, and obtain another infinite descent. See also Edwards *Fermat's Last Theorem*, p. 42, because it is another argument upon which no significant historical point depends.

[6]Euler, *Elements of Algebra*, p. 452.

which says that a cube of a number of the form $p^2 + 3q^2$ is itself of that form. Conversely, Euler now asserted that if a number of the form $p^2 + 3q^2$ is a cube, then it is a cube of a number of that form; that is, there are integers a and b such that $(a^2 + 3b^2)^3 = p^2 + 3q^2$.

Indeed, he said, if $p + q\sqrt{-3} = (a + b\sqrt{-3})^3$ and $p - q\sqrt{-3} = (a - b\sqrt{-3})^3$ then $p^2 + 3q^2 = (a^2 + 3b^2)^3$, which is a cube, and

$$p = a^3 - 9ab^2, \text{ and } q = 3a^2b - 3b^3.$$

Now comes the descent step. The numbers $2a$, $a - 3b$, and $a + 3b$ are relatively prime (if they were not p and q would not be) and are cubes (because their product, $2p$ is a cube, and obviously $a - 3b + a + 3b = 2a$). So the descent can begin, and the contradiction is immediate. Fermat's Last Theorem follows in the case $n = 3$—or it would if one dubious step could be made secure: the existence of the integers a and b.

On what grounds did Euler make the claim that the cube root of a number of the form $p^2 + 3q^2$ is of that form? He factorised $p^2 + 3q^2 = (p + \sqrt{-3}q)(p - \sqrt{-3}q)$ and claimed that if $p \pm \sqrt{-3}q$ is a cube it is cube of a number of the form $a \pm \sqrt{-3}b$. Now this is true, as it turns out, but not for the reasons Euler gave. Earlier in his *Algebra* (1770, Part II, Chapter XII, §182) he had considered the general question: when is $x^2 + cy^2$ a square? He factorised to get $x^2 + cy^2 = (x + \sqrt{-c}y)(x - \sqrt{-c}y)$ and treated each factor as if it were an integer[7]:

> Let, therefore, the formula $x^2 + cy^2$ be proposed, and let it be required to make it a square. As it is composed of the factors $(x + \sqrt{-c}y) \times (x - \sqrt{-c}y)$, these factors must either be squares, or squares multiplied by the same number. For, if the product of two numbers, for example, pq, must be a square, we must have $p = r^2$ and $q = s^2$, that is to say, each factor is of itself a square; or $p = mr^2$, and $q = ms^2$; and therefore these factors are squares multiplied both by the same number. For which reason, let us make $x + y\sqrt{-c} = m(p + q\sqrt{-c})$; it will follow that $x - y\sqrt{-c} = m(p - y\sqrt{-c})$; and we shall have $x^2 + cy^2 = m(p^2 + cq^2)$, which is a square.

This is interesting, partly because it is wrong, but partly because it is audacious.

Unhappily Euler's argument here is false; it is perfectly correct for ordinary whole numbers, but not—in general—for numbers of the form $x + y\sqrt{-c}$. Nevertheless his claim is true in the cases $c = 2$ and $c = 3$, although it is not for $c = 5$, because the factors of $21 = 4^2 + 5 \cdot 1$ are 3 and 7, neither of which is of the right form. This is the important difference between $x^2 + 3y^2$ and $x^2 + 5y^2$, which Euler was unable to explain.

The problem of divisors and prime factors for these numbers runs very deep. Consider $19 = 4^2 + 3 \cdot 1^2 = a^2 + 3b^2$ with $a = 4$, $b = 1$. Observe that $19 = (4 + \sqrt{-3})(4 - \sqrt{-3})$ and deduce that *if* the concept of prime makes sense for numbers of the form $m + n\sqrt{-3}$ *then* intuitively 19 is not prime among such numbers.

---

[7] Euler, *Elements of Algebra*, p. 396.

Now, in ordinary arithmetic *prime* divisors have this property: if p is prime and divides a product mn then either p divides m or p divides n. We want to extend this concept to numbers of the form $m + n\sqrt{-3}$. Certainly we can add, subtract and multiply such numbers together and show that the results are numbers of this form. However, we cannot always divide one number of this form by another and express the answer in this form, so the concept of one of these numbers being a *divisor* of another becomes interesting.

A fault is a fault. The more important point, however, is that Euler extended reasoning about integer numbers to new numbers of the form $x + y\sqrt{-c}$. At a time when imaginary quantities were still a source of controversy in mathematics, here Euler boldly proposed to discuss prime, relatively prime, square and cube numbers of this kind, treating them as if they were integers, and hence that concepts such as 'prime' would similarly apply. Moreover, the relevance of these ideas to the problem is only apparent once we have seen deeply into the problem—why might they not be red-herrings? In fact, to later mathematicians such as Lagrange, Legendre, and Gauss, the theory of 'integers' of this form was to be the way forward in number theory. It is this combination of deep intuitive perceptiveness, going for the significance of numbers of a certain form, while making elementary errors in the logic of proof that provides an illustration of Euler's amazing ability to take the 'right' risks.

A charming indication of how Euler viewed his work, his aims and his partial successes, can be gleaned from a letter he wrote to Goldbach on 4 August 1753 on his success with $x^3 + y^3 = z^3$.[8]

> There's another very lovely theorem in Fermat whose proof he says he has found. Namely, on being prompted by the problem in Diophantus, find two squares whose sum is a square, he says that it is impossible to find two cubes whose sum is a cube, and two fourth powers whose sum is a fourth power, and more generally that this formula $a^n + b^n = c^n$ is impossible when $n > 2$. Now I have found valid proofs that $a^3 + b^3 \neq c^3$ and $a^4 + b^4 \neq c^4$, where $\neq$ denotes cannot equal. But the proofs in the two cases are so different from one another that I do not see any possibility of deriving a general proof from them that $a^n + b^n = c^n$ if $n > 2$. Yet one sees quite clearly as if through a trellice that the larger n is, the more impossible the formula must be. Meanwhile I haven't yet been able to prove that the sum of two fifth powers cannot be a fifth power. To all appearances the proof just depends on a brainwave, and until one has it all one's thinking might as well be in vain. But since the equation $aa + bb = cc$ is possible, and so also is this possible, $a^3 + b^3 + c^3 = d^3$, it seems to follow that this, $a^4 + b^4 + c^4 + d^4 = e^4$, is possible, but up till now I have been able to find no case of it. But there can be five specified fourth powers whose sum is a fourth power.

It is clear from this that he was dissatisfied. Although he had found a proof, indeed one by infinite descent, it seemed to him that he would not be able to solve in this way the general case (that of $x^n + y^n = z^n$ for any $n > 2$) so we see that he had both a specific and a general aim in mind.

---

[8] See Euler (2015, pp. 532–535), and for an introduction to Elkies's discovery of fourth powers that are sums of only three fourth powers see the Wikipedia entry 'Euler's sum of powers conjecture'.

2.4 Exercises

1. Experiment with numbers of the form $a + b\sqrt{c}$ for small values of c, where a, b, c are integers, and draw up some evidence about when unique factorisation into numbers of this form holds and when it fails.

2. $19 = (4 + \sqrt{-3})(4 - \sqrt{-3})$. Define the *norm* of $m + n\sqrt{-3}$, written $N(m + n\sqrt{-3})$, as follows:

$$N(m + n\sqrt{-3}) = (m + n\sqrt{-3})(m - n\sqrt{-3}) = m^2 + 3n^2,$$

so $N(4 + \sqrt{-3}) = 19 = N(4 - \sqrt{-3})$ and $N(19) = 19^2$. Show that if a and b are numbers of the form $m + n\sqrt{-3}$ then a divides b if and only if $N(a)$ divides $N(b)$. Deduce that the only divisors of $4 - \sqrt{-3}$ are itself and 1. Are these numbers prime?

Chapter 3
Lagrange's Theory of Quadratic Forms

3.1 Introduction

It was Lagrange who sought to produce a general theory of quadratic forms, after Euler had published a number of deep and provocative studies of many examples—what would today be called 'experimental mathematics'. Here we look at one key idea in his treatment: the reduction of forms to simpler but equivalent ones. We are led to one of the great theorems in mathematics: quadratic reciprocity. It was conjectured well before it was proved for the first time, as we shall see later.

Cox's book is a very good guide to this material, as is Scharlau and Opolka *From Fermat to Minkowski*, which is a historically-based introduction to all of this material on number theory and more; Chapter III on Lagrange is very helpful.[1]

3.2 The Beginnings of a General Theory of Quadratic Forms

There comes a point where the interest of a mathematician requires a suitable generality. Some might believe that writing a number as a sum of squares is interesting (the Pythagorean theorem and all that) and find the question of what numbers are of the form $x^2 + 2y^2$ simply contrived. But even if we find that question congenial, we cannot really imagine anyone being equally interested in all the subsequent cases: $x^2 + 3y^2$, $x^2 + 5y^2$, $x^2 + 6y^2$ and so on. At some point, the question would arise: is there not, perhaps, a uniform theory about numbers of the form $x^2 + ny^2$ for all non-square integers n. We have already seen that this will be a vain hope if a very simple theory is wanted, because the theory of $x^2 + 5y^2$ seems very different from its predecessors.

---

[1] So is Andrew Granville's website: http://www.dms.umontreal.ca/~andrew/Courses/Chapter4.pdf.

J. Gray, *A History of Abstract Algebra*, Springer Undergraduate Mathematics Series, https://doi.org/10.1007/978-3-319-94773-0_3

It was Lagrange who spotted the reason for this, and so initiated the number theory of quadratic forms. He observed that there is a quadratic form that deserves to be counted alongside $x^2 + 5y^2$; it is

$$2x^2 + 2xy + 3y^2,$$

which shares with $x^2 + 5y^2$ the property that, writing them both in the form $ax^2 + bxy + cy^2$ the discriminant, $4ac - b^2$ is the same—it is 20 in each case.

Once we decide to broaden the subject and admit all binary quadratic forms (i.e. forms in x and y only), it is helpful to note a simple piece of geometry. The corresponding curves may be ellipses or hyperbolas—for example, $x^2 + 5y^2 = n$ defines an ellipse for each positive integer n—and it is clear that the question of whether there are any integers x and y satisfying, say, $x^2 + 5y^2 = 27$ is a finite one because the ellipse occupies only a finite region of the (x, y)-plane. But the same question about $x^2 - 5y^2 = 27$ will be much harder, because the hyperbola goes off to infinity in the (x, y)-plane.[2] And indeed, we have every reason to expect that there are infinitely many solutions to $x^2 - 5y^2 = 1$. We shall see that Lagrange's theory works much more easily for quadratic forms with positive discriminants than negative ones, in line with this geometric observation.

We also note that the forms $ax^2 + 2bxy + cy^2$ with $D = ac - b^2 > 0$ can be written this way:

$$ax^2 + 2bxy + cy^2 = (1/a)((ax + by)^2 + Dy^2).$$

This may not be a quadratic form with integer coefficients, but it is a sum of squares so it takes positive values exclusively if $a > 0$ and negative values exclusively if $a < 0$. We say that a form is positive if it takes only positive values (and negative if it takes only negative values), so a form with positive discriminant $ac - b^2$ is positive if and only if $a > 0$.

Now we fix some notation. We shall think of $\begin{pmatrix} x \\ y \end{pmatrix}$ as a column vector, **x**. A symmetric 2 by 2 matrix with integer entries will be written A, so a quadratic form can and will be written as

$$\mathbf{x}^T A \mathbf{x}.$$

Note, however, that this was not Lagrange's notation but a much more modern one.

---

[2]This question is solved by working modulo 5.

We have done this so we can follow Lagrange's analysis more easily. Let P be a 2 by 2 matrix, not necessarily symmetric, with integer entries and determinant ± 1, so its inverse also has integer entries. We can think of it as a change of basis matrix in $\mathbb{R}^2$ with the property that points with integer coordinates are mapped to points with integer coordinates, or even in some sense as a change of basis matrix in $\mathbb{Z}^2$. Evidently if (x, y) is a pair of integers such that $\mathbf{x}^T A \mathbf{x} = 27$, say, then $P^{-1}\mathbf{x}$ is a pair of integers such that $\mathbf{x}^T P^{-1^T} P^T A P P^{-1}\mathbf{x} = 27$. In other words, the integers represented by A and $P^T A P$ are the same: Lagrange regarded the corresponding quadratic form s as equivalent. Notice that the transformations we are using do not change the value of the determinant, $\det(P^T A P) = \det A \cdot (\det P)^2 = \det A$, so matrices with different determinants cannot be equivalent.

Here enter two irritating but unavoidable little definitional questions. Should we insist that a quadratic form be written $ax^2 + bxy + cy^2$ or $ax^2 + 2bxy + cy^2$? If we go with the former, the corresponding matrix form acquires a half-integer when b is odd; should we go with the latter we have to remember to let b be a half-integer. Lagrange began by taking the former route and used $ax^2 + bxy + cy^2$ but then switched; Gauss took the latter route and wrote $ax^2 + 2bxy + cy^2$. Likewise, is the central invariant the determinant, or indeed 4 times the determinant: $4(ac - b^2/4) = 4ac - b^2$, or its negative, $b^2 - 4ac$, with the familiar ring of school algebra? Lagrange decided that the invariant associated to $ax^2 + bxy + cy^2$ is $4ac - b^2 = D$ and called it the discriminant. The name makes sense—it discriminates—and it is *plus* 4 times the determinant. The mathematically easy case is the one with *positive* discriminant. Gauss, just to confuse matters, worked with $ax^2 + 2bxy + cy^2$ and what he called the determinant: $b^2 - ac$, which is, of course, the negative of what we would call the determinant.

Lagrange searched for a set of canonical representatives of each equivalence class.[3] He found that there were simple linear changes of variable, such as we would represent by a matrix P of determinant ± 1, which could drive down the size of the coefficient of the xy term. If we follow this using transformations of determinant 1 only, we find using

$$S = \begin{pmatrix} 1 & -1 \\ 0 & 1 \end{pmatrix} \text{ and } R = \begin{pmatrix} 0 & 1 \\ -1 & 0 \end{pmatrix}$$

that, if $A = \begin{pmatrix} a & b \\ b & c \end{pmatrix}$ then conjugating by S and R gives these results:

$$S^T A S = \begin{pmatrix} a & b - a \\ b - a & a - 2b + c \end{pmatrix} \text{ and } R^T A R = \begin{pmatrix} c & -b \\ -b & a \end{pmatrix}.$$

---

[3]His method is set out in the later sections of his *Additions*, and again in his *Recherches d'arithmétique* (1773).

At least heuristically, the switch R can be used to ensure that the xy coefficient, $2b$ is positive and $a \leq c$, while S reduces b to $b - c$ (if we assume that c is positive, as we may).

The 2 by 2 matrix with integer entries and determinant 1 form a group, denoted $SL(2, \mathbb{Z})$, which it turns out is generated by S_1 and S_2, and thus motivated, one can prove, as Lagrange did, the following theorem.

Theorem. *Lagrange's reduction theorem for positive forms. Any 2 by 2 symmetric matrix with integer entries that represents a positive form is equivalent to one of the form*

$$\begin{pmatrix} a & b \\ b & c \end{pmatrix}$$

with

$$-\frac{a}{2} < b \leq \frac{a}{2} \text{ and } a < c, \quad \text{or} \quad 0 \leq b \leq \frac{a}{2} \text{ and } a = c.$$

Such a matrix will be called the canonical form *for the corresponding quadratic form. Moreover, if the determinant $\Delta = ac - b^2$ is positive, any two matrices meeting these requirements are inequivalent.*

The proofs of both of these results are omitted, but will be found in (Scharlau and Opolka 1984, 36–38). Lagrange gave the existence part early on in his *Recherches*, but the uniqueness for forms with a positive determinant only later.[4] Note that these equivalence classes are the classes for proper equivalence, a concept introduced by Gauss later.

For forms with positive discriminant, it is conventional to look only at positive forms, and to include the negative ones by default. They are obtained by multiplying all the coefficients by -1.

Exercises

1. Check that conjugation by $SRS^{-1}RS$ (which is the same as conjugating by S first, and then by R, and so on) reduces the quadratic form $15x^2 + 50xy + 42y^2$ to $2x^2 + 2xy + 3y^2$, which is in canonical form. (Work with the matrix for the quadratic form.)
2. Check that the sequence of transformations $S^{-1}RSRS$ reduces the quadratic form $5x^2 - 20xy + 21y^2$ to $x^2 - 5y^2$, which is in canonical form.

---

[4]See Problem III, pages 723–728, for the case of positive determinant, and Problem IV, pages 728–737, with consequences explored to page 741, for the more difficult case where the determinant is negative.

When the determinant Δ of a positive form is positive the requirements in Lagrange's theorem immediately imply that

$$\frac{3a^2}{4} < \Delta$$

or

$$a < 2\sqrt{\frac{\Delta}{3}}, \text{ and } b < \sqrt{\frac{\Delta}{3}},$$

so there are only finitely many equivalence classes for each value of Δ. Indeed, when $\Delta = 1$ the only possible values for a, b, c are $1, 0, 1$ respectively, and when $\Delta = 2$ the only possible values for a, b, c are $1, 0, 2$ respectively. This means that essentially the only quadratic forms with those discriminants are the ones we have met: $x^2 + y^2$ and $x^2 + 2y^2$. So, for example, we know everything about the quadratic form $\begin{pmatrix} 89 & 55 \\ 55 & 34 \end{pmatrix}$, because its discriminant $b^2 - 4ac$ is -4, and so the number theory of $89x^2 + 110xy + 34y^2$ and $x^2 + y^2$ is essentially the same.

When $\Delta = 3$ it might seem that a new form appears: in addition to $x^2 + 3y^2$ there is $2x^2 + 2xy + 2y^2$; all the integers represented by the second of these are, of course, even. They are not the even numbers represented by $x^2 + 3y^2$. But we may reasonably feel that $2x^2 + 2xy + 2y^2$ is not a new quadratic form: its coefficients have a common multiple, and it is a disguised version of $x^2 + xy + y^2$. The theory that Lagrange, Legendre, and Gauss created only considers primitive quadratic forms, those for which the highest common factor of the coefficients is 1, so the only primitive quadratic form with $\Delta = 3$ is $x^2 + 3y^2$, and the form $x^2 + xy + y^2$ is excluded because its middle coefficient is not even.

When $\Delta = 5$ a new form does appear: in addition to $x^2 + 5y^2$ there is $2x^2 + 2xy + 3y^2$—and it is primitive. While the useful identity gives us the result that if m and n are integers represented by $x^2 + 5y^2$ then so is their product, mn, Fermat observed that there are primes such as 3, 7, and 23 whose pairwise products can also be represented by $x^2 + 5y^2$. Now we can see that these primes are represented by $2x^2 + 2xy + 3y^2$: respectively $(x, y) = (0, 1), (1, 1)$, and both $(1, -3)$ and $(2, -3)$. In fact, Euler had already conjectured that primes (other than 5) dividing $x^2 + 5y^2$ are congruent to $1, 3, 7, 9 \bmod 20$, and that

$$p = x^2 + 5y^2 \Leftrightarrow p \equiv 1, 9 \bmod 20,$$

$$2p = x^2 + 5y^2 \Leftrightarrow p \equiv 3, 7 \bmod 20.$$

Lagrange was particularly interested in the divisors of numbers represented by a given quadratic form. Indeed, his paper opens with a result that may be called a switching theorem. Note that it is a switch of attention in this sense as well, away from asking "What numbers does a given form represent?" to asking the converse: "What forms represent a given number?"

Theorem. *Lagrange's switching theorem: If an integer m divides a number represented by a quadratic form Q, then m can be properly represented by a quadratic form Q' with the same discriminant.*

Proof. The proof is no more than school algebra. We suppose (reverting to the case where the middle coefficient may be odd!) that

$$rs = ax^2 + bxy + cy^2,$$

with x and y coprime. Suppose $s = tu$, $y = tv$ and u and v are coprime. Then

$$rtu = ax^2 + bxtv + ct^2v^2.$$

Since t divides y and x and y are coprime t cannot divide x, so t divides a; say $a = et$. So

$$ru = ex^2 + bxv + ctv^2.$$

We may write $x = \alpha u + \beta v$ where α and β are integers, because u and v are coprime, so

$$ru = e(\alpha u + \beta v)^2 + b(\alpha u + \beta v)v + ct^2v^2$$

$$= e\alpha^2 u^2 + (2e\beta + b)\alpha uv + (e\beta^2 + b\beta + ct)v^2.$$

Since u and v are coprime, u must divide $e\beta^2 + b\beta + ct$. Set

$$A = eu, \; B = 2e\beta + b, \; C = \frac{e\beta^2 + b\beta + ct}{u},$$

and

$$\alpha = X, v = Y,$$

and dividing by u gives

$$r = AX^2 + BXY + CY^2.$$

A simple calculation confirms that $B^2 - 4AC = b^2 - 4ac$, and Lagrange's theorem is proved.

In fact, Lagrange went even further. To say what he did I introduce some notation: if a quadratic form (a, b, c) represents an integer m, that is, if there are integers x and y such that $ax^2 + 2bxy + cy^2 = m$, then I shall sometimes denote this by $Q \to m$. What Lagrange showed was that if $Q \to m_1m_2$ then there are quadratic form s Q_1 and Q_2 such that $Q_1 \to m_1$ and $Q_2 \to m_2$ and $D(Q_1) = D(Q_2) = D(Q)$.

In fact, if $Q = ax^2 + 2bxy + cy^2$ and $Q \to m_1$ and $Q \to m_2$, then $u^2 - Dv^2 \to m_1 m_2$. For example, if $Q = (2, 1, 3) = 2x^2 + 2xy + 3y^2$, so $D = -5$, then $Q \to 3$ and $Q \to 7$ and $u^2 - (-5)v^2 \to 21$.

This suggests that given two quadratic form s Q_1 and Q_2 of the same discriminant D such that $Q_1 \to m_1$ and $Q_2 \to m_2$ then there ought to be a quadratic form Q such that $Q \to m_1 m_2$. We shall see that Legendre was to come up with a definition of such a composite, but it had unsatisfactory properties: in particular the composite was not necessarily unique.

The next seemingly harmless result is actually crucial.

Lemma 3.1. *Let m and D be relatively prime, and m be odd, then m is properly represented by a primitive form of discriminant D only if D is a square mod m.*

The term 'properly represented' means that the relevant x and y are relatively prime.

Proof.

$$4a(ax^2 + 2bxy + cy^2) = 4(ax + by)^2 - Dy^2,$$

where $D = -4(ac - b^2)$, so if x and y are such that $ax^2 + 2bxy + cy^2 = m$ then we have

$$4am \equiv 4(ax + by)^2 - Dy^2,$$

so mod m we have

$$4(ax + by)^2 - Dy^2 \equiv 0 \bmod m,$$

which says that D is a square mod m provided that m is odd.

The lemma implies the corollary that if n is an integer and p an odd prime not dividing n then $-n$ is a square mod p only if p is represented by a primitive form of discriminant $-4n$. This is because $x^2 + ny^2$ can represent p only if $-4n$, and therefore $-n$, is a square modulo p. Therefore, if p fails this test there is no hope of a representation; however, when it passes the test there is uncertainty because the test gives necessary but not sufficient conditions whenever there is more than one inequivalent form with that discriminant—which is why $n = 5$ is much harder than $n = 1, 2, 3$.

There are significant elementary consequences of reduction theory. The fact that all binary quadratic forms with discriminant 1 are equivalent means that a number of the form $a = b^2 c$, where c is square-free, is a sum of two squares if and only if the prime factors of c are either 2 or of the form $4n + 1$. For, certainly if c is as given we know it is a sum of squares. Conversely, if a is a sum of squares (which we may assume are relatively prime) and p is a prime factor of c, then p is a factor of a sum of two squares, so by Lagrange's theorem it can be properly represented by a form with the same discriminant. But up to equivalence there is only one such form (the discriminant being 1), so p is properly represented by $x^2 + y^2$ and is therefore either 2 or of the form $4n + 1$.

A very similar argument, also due to Lagrange, shows that primes of the form $8n + 3$ can be represented by numbers of the form $x^2 + 2y^2$. He also showed that those of the form $12n + 7$ can be represented by $x^2 + 3y^2$ (because they cannot be represented by the alternative, $-x^2 + 3y^2$), and that those of the form $24n + 7$ can be represented by $x^2 + 6y^2$ (because they cannot be represented by the alternative, $2x^2 + 3y^2$).

On the other hand, there are two inequivalent forms with discriminant 5, which explains why a similar argument about $x^2 + 5y^2$ fails. What can be said, as Euler knew, is that for p an odd prime greater than 5

$$p = x^2 + 5y^2 \Leftrightarrow p \equiv 1, 9 \bmod 20$$

and

$$2p = x^2 + 5y^2 \Leftrightarrow p \equiv 3, 7 \bmod 20.$$

However, as Euler also knew, the two inequivalent forms $x^2 + 14y^2$ and $2x^2 + 7y^2$ both represent odd primes other than 7 that are congruent to one of 1, 9, 15, 23, 25, or 39 mod 56. As Cox (1989, 19) remarks, there are no simple congruence tests for these primes that will distinguish between these quadratic forms, and quite a bit of number theory from Gauss onwards is concerned with understanding why and devising alternatives.

Positive Discriminants

The same matrix analysis reduces quadratic forms with positive discriminants to canonical form and shows that for each fixed value of the discriminant there are only finitely many equivalence classes. The canonical forms are those for which

$$|a| \leq |c|, \quad 2|b| \leq a,$$

and these conditions imply that $ac < 0$ so conventionally one takes $a > 0, c < 0$. Moreover:

$$ac < b^2 \leq \frac{|\Delta|}{5},$$

from which it follows that there are only finitely many equivalence classes for each value of Δ.

The first serious problem that occurs with positive discriminant is that the reduced forms are not necessarily inequivalent. To give a trivial example, $\begin{pmatrix} 1 & 0 \\ 0 & -6 \end{pmatrix}$ and $\begin{pmatrix} -1 & 0 \\ 0 & 6 \end{pmatrix}$ are equivalent, but both are reduced. Reduction theory is easier in the negative case.

3.3 The Theorem of Quadratic Reciprocity

This theorem is one of the most important in the subject. When Gauss found his second proof of it he referred to it in his diary (27 June 1796) as the 'golden theorem'. Here I state the theorem, hint at its importance, and indicate some of the earlier work on it.

Consider the general quadratic form $ax^2 + 2bxy + cy^2$ and suppose it represents the integer N, so there are integers x and y such that

$$ax^2 + 2bxy + cy^2 = N.$$

By Lemma 3.1, for N to be represented by some quadratic form of discriminant $4(b^2 - ac)$ it is necessary that the discriminant be a square modulo N. Since 4 is a square, this condition reduces to $b^2 - ac$ being a square modulo N, so a necessary condition for this representation to hold is that the discriminant, $b^2 - ac$, is a square mod N.

This puts firmly onto the agenda the question: when is one number a square modulo another number? That is the question that the theorem of quadratic reciprocity indirectly answers. It is usually stated using what is called the *Legendre symbol*:

$$\left(\frac{a}{p} \right) = \begin{cases} 0 & p \mid a \\ 1 & p \nmid a \quad \exists x : x^2 \equiv a \bmod p \\ -1 & p \nmid a \quad \nexists x : x^2 \equiv a \bmod p \end{cases}$$

Of the interesting cases (i.e. when $p \nmid a$) we say that a is a *quadratic residue* mod p when there is an x such that $x^2 \equiv a \bmod p$, and that a is a not a *quadratic residue* mod p when there is no x such that $x^2 \equiv a \bmod p$. If we look back at the results so far we can see that we have so far been investigating the primes p such that $-1, -2, -3$ are quadratic residues mod p.

Now the theorem itself. It comes in three parts. The first part says: let p be an odd prime, then

$$\left(\frac{-1}{p} \right) = 1 \text{ if and only if } p = 4n + 1.$$

We know this from the study of primes of the form $x^2 + y^2$.

The second part says let p be an odd prime, then

$$\left(\frac{2}{p}\right) = 1 \text{ if and only if } p = 8n \pm 1.$$

We may conjecture this by looking at primes of the form $x^2 - 2y^2$.

The third and largest part concerns two odd primes, p and q, and it says that

$$\left(\frac{p}{q}\right) = \left(\frac{q}{p}\right)(-1)^{\frac{p-1}{2}\frac{q-1}{2}}.$$

More memorably, and more usefully, it says that

$$\left(\frac{p}{q}\right) = \left(\frac{q}{p}\right)$$

unless both p and q are primes of the form $4n - 1$, in which case

$$\left(\frac{p}{q}\right) = -\left(\frac{q}{p}\right).$$

Notice that this doesn't answer the question "Is q a prime mod p?", it merely relates it to the reciprocal claim: "Is p a prime mod q?"—whence the name of the theorem: *quadratic reciprocity*.

Put this together with the observation that

$$\left(\frac{ab}{p}\right) = \left(\frac{a}{p}\right)\left(\frac{b}{p}\right)$$

—in modern terms that the Legendre symbol is a homomorphism from $\mathbb{Z}$ to $\mathbb{Z}/2\mathbb{Z}$, (see the proof below—Gauss proved it in §98 of the *Disquisitiones Arithmeticae*) and we have a powerful tool. Consider for example the question: is 365 a square mod 1847? The number 1847 is prime, and we have

$$\left(\frac{365}{1847}\right) = \left(\frac{5}{1847}\right)\left(\frac{73}{1847}\right) = \left(\frac{1847}{5}\right)\left(\frac{1847}{73}\right) =$$

$$\left(\frac{2}{5}\right)\left(\frac{22}{73}\right) = -1\left(\frac{2}{73}\right)\left(\frac{11}{73}\right) = -1\left(\frac{2}{73}\right)\left(\frac{73}{11}\right) =$$

$$-1\left(\frac{2}{73}\right)\left(\frac{7}{11}\right) = -1 \times -1\left(\frac{2}{73}\right)\left(\frac{11}{7}\right) = \left(\frac{2}{73}\right).$$

So we are done once we know the value of $\left(\frac{2}{73}\right)$, which is 1 by the second part of the quadratic reciprocity theorem, and so 365 is a square mod 1847. Notice that this does not give us either of the square roots.

The proof that the Legendre symbol is a homomorphism from $\mathbb{Z}$ to $\mathbb{Z}/2\mathbb{Z}$ goes like this. It is clear that if a and b are squares mod p then so is their product, and if a is a square but b is not mod p then their product cannot be. For $a \equiv x^2$, $ab \equiv y^2 \bmod p$ implies $b \equiv \frac{y^2}{x^2} \pmod{p}$. This leaves the case where neither a nor b is a square mod p and we need to show that ab is a square. It is clear that the set $\{ax^2\}$ is the set of non-squares mod p, and so is the set $\{by^2\}$, as x and y run through the non-zero integers mod p. This means that if we pick x at random there is a y such that $ax^2 \equiv by^2 \equiv c \bmod p$, and so $ax^2 by^2 \equiv c^2 \bmod p$, and so $ab \equiv \frac{c^2}{x^2 y^2} \bmod p$. This shows that ab is a square mod p.

Given the importance of quadratic reciprocity, it is not surprising that people tried to prove it, but this proved to be hard. Famously, Legendre made a major mistake here, and the first proof is due to Gauss in 1801. Gauss went on to publish a total of six proofs (two more were left unpublished). Since than people have given many more—a figure of a hundred and fifty is sometimes mentioned—not that there are 150 different ways to prove it, but there are several and this alone is interesting: apparently this theorem is connected underneath with many other parts of mathematics. Also, fortunately, although the first proof was hard to find and is famously difficult, some of the later proofs are easy, and we will content ourselves with one of them. Meanwhile, as historians we are as interested in failures as successes, so we shall look at Legendre's account, guided here by André Weil.

First however, I note that Legendre did validly establish this important result, which he then used in his attempts on quadratic reciprocity: If a, b, c are three integers, not all of the same sign, and such that abc is square free, then the equation $ax^2 + by^2 + cz^2 = 0$ has a solution in integers (not all 0) if and only if $-bc$ is a quadratic residue mod $|a|$, $-ca$ is a quadratic residue mod $|b|$, and $-ab$ is a quadratic residue mod $|c|$. I omit the proof of this result, which is not trivial.

I turn now to Weil's commentary (Weil 1984, pp. 328–330).

What Legendre does is to distinguish eight cases according to the values of p and of q modulo 4 and to the value of (p/q); this same arrangement was observed by Gauss in his first proof of the law in question (*Disq* §136). In each of these cases, Legendre introduces an appropriate equation of the form

$$ax^2 + by^2 + cz^2 = 0$$

with $a \equiv b \equiv c \equiv 1 \pmod{4}$; this can have no non-trivial solution, since the congruence

$$ax^2 + by^2 + cz^2 = 0 \pmod{4}$$

has none. Therefore, by Legendre's criterion, $-bc$, $-ca$ and $-ab$ cannot simultaneously be quadratic residues modulo $|a|$, modulo $|b|$ and modulo $|c|$, respectively. In each case Legendre tries to choose a, b, c so that this will lead to the desired conclusion. Take firstly $p \equiv 1$, $q \equiv -1 \pmod{4}$, $(p/q) = -1$, and the equation $x^2 + py^2 - qz^2 = 0$; as $(-1/q) = -1$, we have $(-p/q) = 1$; therefore (q/p) cannot be 1; it must be -1, as required by the reciprocity law. The same reasoning applies to the case $q \equiv -q' \equiv -1$ (mod 4), $(q/q') = 1$, and to the equation $x^2 - qy^2 - q'z^2 = 0$.

Now take $q \equiv q' \equiv -1$ (mod 4), $(q/q') = -1$. In that case Legendre introduces the equation $px^2 - qy^2 - q'z^2 = 0$, where p is an auxiliary prime subject to the conditions

$$p \equiv 1 \bmod 4, \ \left(\frac{p}{q}\right) = -1, \ \left(\frac{p}{q'}\right) = -1.$$

Assuming that there is such a prime, and using what has been proved before, the conclusion $(q'/q) = 1$ follows. For this to be valid, there has to be a prime in one of the arithmetic progressions

$$\{4qq'x + m \mid x = 0, 1, 2, \ldots\}$$

where one takes for m all the integers > 0 and $< 4qq'$ which are $\equiv 1$ (mod 4) and non-residues modulo q and modulo q'. Legendre was rightly convinced that every arithmetic progression $\{ax + b\}$, with a prime to b, contains infinitely many primes. "Perhaps it would be necessary to prove this rigorously", he had written in 1785 (*Recherches*, p. 552). "We need have no doubt about it", he wrote in 1798 (*Essai*, p. 220) this time attempting to support his statement by an altogether unconvincing argument (*Essai*, pp. 12–16) which he later expanded, even more disastrously, into a whole chapter of his *Théorie des Nombres* (t. II, §XI, pp. 86–104). It was left for Dirichlet to prove the theorem of the arithmetic progression, as he did in 1837 by a wholly original method (Dir.I. 315–342) which remains as one of his major achievements; by the same method he also proved (Dir.I. 499–502) that every quadratic form $ax^2 + bxy + cy^2$ represents infinitely many primes provided a, b, c have no common divisor, as had been announced by Legendre (*Théorie des Nombres*, t. II, pp. 102–103).

In the case we have just discussed, the theorem of the arithmetic progression, taken by Legendre as a kind of axiom, gave at least some semblance of justification to his argument; but there is worse to come. Take the case of two primes p, p', both $\equiv 1$ (mod 4). Legendre seeks to treat it by means of the equation $px^2 + p'y^2 - qz^2 = 0$, where q is an auxiliary prime satisfying

$$q \equiv -1 \bmod 4, \ \left(\frac{q}{p'}\right) = 1 \ \left(\frac{p}{q}\right) = -1$$

or alternatively by means of $x^2 + py^2 - p'qz^2 = 0$, with q satisfying

$$q \equiv -1 \bmod 4, \ \left(\frac{p}{q}\right) = -1$$

(*Recherches*, pp. 519–520; *Essai*, pp. 216–217 and 220–221; *Théorie des Nombres*, t. I, pp. 233–234). Is there such a prime? Clearly its existence would follow from Dirichlet's theorem *and* the reciprocity law; but it is doubtful (as Gauss pointed out: *Disq* §297) that it could be proved otherwise. Thus Legendre's method of proof ends up unavoidably with a piece of circular reasoning out of which there seems to be no escape. This did not prevent him, in a letter of 1827 to Jacobi, from complaining, as bitterly as unjustly, that Gauss had "claimed for himself" the discovery of the reciprocity law (*Jac*.I. 398). He must have realized, however, that Gauss's criticism of his proof had its validity; in his *Théorie des Nombres* of 1830 he chose to be on the safe side by inserting (t. II, pp. 57–64) Gauss's third proof (cf. *Gau*.II. 3–8, 1808), and, for good measure (t. II, pp. 391–393) the 'cyclotomic' proof communicated to him by Jacobi at the beginning of their correspondence (*Jac*.I. 394 (1827).

3.4 Exercises

1. Find all the properly primitive reduced quadratic forms $\begin{pmatrix} a & b \\ b & c \end{pmatrix}$ with 'discriminant' $\Delta = ac - b^2 = 51$. (Recall that a form is properly primitive if the greatest common divisor of $a, 2b, c$ is 1. Hint: use the upper limit on b to find all the possible values of $ac = \Delta + b^2$, and for each value of b look for suitable values of a and c with $2a \geq |b|$.)
2. Find all 16 reduced quadratic forms with 'discriminant' 161.
3. Connect the special cases of quadratic reciprocity -1 and 2 with the quadratic forms $x^2 + y^2$ and $x^2 - 2y^2$, and explain to yourself in what way the theorem of quadratic reciprocity is an advance.

3.5 Taking Stock

These three chapters cover the bulk of what I want to say about the situation before 1800; there is also Chap. 8 on the solution of polynomial equations. Even then, there is one significant omission: the so-called *fundamental theorem of algebra*. This is the claim that every polynomial with real or complex coefficients has as many real or complex roots as its degree (counted with multiplicity) or, to state it in a more eighteenth century fashion: a polynomial (with real coefficients) has as many linear or quadratic factors as its degree permits. Put this way it belongs to a complicated story about the acceptance of complex numbers, which I have elected not to tell (for an account of early work on the fundamental theorem of algebra see Appendix A).[5]

What have these chapters established? First, that the number theory of binary quadratic forms was beginning to catch on as a subject as the eighteenth century drew to an end. Second, and this is a point of a different kind, the methods on display were those of classical algebra (changes of variable included) conducted, perhaps, with some ingenuity. The same is true of the later chapter on polynomial equations. There were hints that explanations can be found that are not simply calculations, but with little depth to them. Before we see how all this was to change in the nineteenth century, it is worth noting that even this mathematics is evidence for the historian: evidence of how mathematics was done. We shall make use of this observation repeatedly as the book progresses.

This being a book in the *history* of mathematics, now is a good time to read the article on Lagrange in the *Complete Dictionary of Scientific Biography* and to tell yourself a story about how important he was. Did he exert his influence through his work, his personality, or his power behind the scenes?

---

[5]For an account of the history of complex numbers, see (Corry 2015, Chapter 9).

An important historical transformation happened in the period from Fermat to Lagrange, and it is tempting to attribute it to Euler, although there were many influences. This is the transition from largely verbal and often geometrical exposition to predominantly symbolic (or algebraic) accounts. If you look back at how Fermat presented his insight into his 'last' theorem, and how Euler proceeded you will see the distinction clearly. This raises a difficult question about how the work of Fermat and others of his time is most adequately to be understood. What is lost if we simply replace his words with symbols, modernise his symbolic expressions, and forget his geometrical terminology?

In keeping with the didactic character of this book, let me ask the next questions explicitly.

Questions

1. As with any question of translation, can we say that Fermat and ourselves are saying the same thing, which is there beneath the way it was actually expressed, or should we admit that a web of connections is at least rearranged, and perhaps something is destroyed? For a thorough discussion of what this might involve one can consult (Goldstein 1995). We should also note that what counts as algebra, arithmetic, or geometry is fluid and determined by historical circumstances.
2. Is there a distinction to be drawn between Fermat's collection of examples and Lagrange's theory of numbers, and if so, what is it?

Chapter 4
Gauss's *Disquisitiones Arithmeticae*

4.1 Introduction

Carl Friedrich Gauss established himself as a mathematician at the age of 24 with the publication of his *Disquisitiones Arithmeticae*, which eclipsed all previous presentations of number theory and became the standard foundation of future research for a century. At its heart is a massive reworking of the theory of quadratic forms, so after an overview of the work we concentrate on how this book illuminated the many puzzling complexities still left after Lagrange's account. We then turn in the next chapter to the other major topic in Gauss's book, cyclotomy, and see how Gauss came to a special case of Galois theory and, in particular, to the discovery that the regular 17-sided polygon can be constructed by straight edge and circle alone.

Fig. 4.1 Carl Friedrich Gauss (1777–1855). Image courtesy of Dr. Axel Wittmann, copyright owner

© Springer Nature Switzerland AG 2018
J. Gray, *A History of Abstract Algebra*, Springer Undergraduate Mathematics Series,
https://doi.org/10.1007/978-3-319-94773-0_4

4.2   The *Disquisitiones Arithmeticae* and Its Importance

To introduce the *Disquisitiones Arithmeticae* I can do no better than quote from one of the best books written for many years on the history of mathematics, a full-length study of the book and its impact, edited and largely written by three of the best historians of mathematics at work today: Goldstein, Schappacher, and Schwermer's *The Shaping of Arithmetic after C.F. Gauss's Disquisitiones Arithmeticae.*[1] In their introductory essay Goldstein and Schappacher begin by saying (pp. 3–5):

> Carl Friedrich Gauss's *Disquisitiones Arithmeticae* of 1801 has more than one claim to glory: the contrast between the importance of the book and the youth of its author; the innovative concepts, notations, and results presented therein; the length and subtlety of some of its proofs; and its role in shaping number theory into a distinguished mathematical discipline.
>
> The awe that it inspired in mathematicians was displayed to the cultured public of the *Moniteur universel ou Gazette national* as early as March 21, 1807, when Louis Poinsot, who would succeed Joseph-Louis Lagrange at the Academy of Sciences 6 years later, contributed a full page article about the French translation of the *Disquisitiones Arithmeticae*:
>
>> The doctrine of numbers, in spite of [the works of previous mathematicians] has remained, so to speak, immobile, as if it were to stay for ever the touchstone of their powers and the measure of their intellectual penetration. This is why a treatise as profound and as novel as his *Arithmetical Investigations* heralds M. Gauss as one of the best mathematical minds in Europe.[2]
>
> They go on to observe that a long string of declarations left by readers of the book, from Niels Henrik Abel to Hermann Minkowski, from Augustin-Louis Cauchy to Henry Smith, bears witness to the profit they derived from it. During the nineteenth century, its fame grew to almost mythical dimensions. In 1891, Edouard Lucas referred to the *Disquisitiones Arithmeticae* as an "imperishable monument [which] unveils the vast expanse and stunning depth of the human mind", and in his Berlin lecture course on the concept of number, Leopold Kronecker called it "the Book of all Books". In the process, new ways of seeing the *Disquisitiones* came to the fore; they figure for instance in the presentation given by John Theodore Merz in his celebrated four-volume *History of European Thought in the Nineteenth Century.*[3]
>
>> Germany ... was already an important power in the Republic of exact science which then had its centre in Paris. Just at the beginning of the nineteenth century two events happened which foreboded for the highest branches of the mathematical sciences a revival of the glory which in this department Kepler and Leibniz had already given to their country. ... The first was the publication of the 'Disquisitiones Arithmeticae' in Latin in 1801. ... [Gauss] raised this part of mathematics into an independent science of which the 'Disquisitiones Arithmeticae' is the first elaborate and systematic treatise ... It was ... through Jacobi, and still more through his contemporary Lejeune-Dirichlet ... that the great work of Gauss on the theory of numbers, which for 20 years had remained sealed with seven seals, was drawn into

---

[1]Published by Springer in 2007. In the extracts that follow I have largely removed the original footnotes.

[2]*Gazette nationale ou Le Moniteur universel* 80 (1807), 312.

[3]See Merz (1896–1917), vol. I, pp. 181, 181–182 (footnote), 187–188 and 721.

current mathematical literature ... The seals were only gradually broken. Lejeune-Dirichlet did much in this way, others followed, notably Prof. Dedekind, who published the lectures of Dirichlet and added much of his own.

Gauss's book (hereafter, we shall often use the abbreviation 'the D.A.' to designate it) is now seen as having created number theory as a systematic discipline in its own right, with the book, as well as the new discipline, represented as a landmark of German culture. Moreover, a standard history of the book has been elaborated. It stresses the impenetrability of the D.A. at the time of its appearance and integrates it into a sweeping narrative, setting out a continuous unfolding of the book's content, from Johann Peter Gustav Lejeune-Dirichlet and Carl Gustav Jacob Jacobi on.

In this history modern algebraic number theory appears as the natural outgrowth of the discipline founded by the *Disquisitiones Arithmeticae*. Historical studies have accordingly focused on the emergence of this branch of number theory, in particular on the works of Dirichlet, Ernst Eduard Kummer, Richard Dedekind, Leopold Kronecker, and on the specific thread linking the D.A. to the masterpiece of algebraic number theory, David Hilbert's *Zahlbericht* of 1897. In addition, they have also explored the fate of specific theorems or methods of the D.A. which are relevant for number theorists today.

Yet a full understanding of the impact of the *Disquisitiones Arithmeticae*, at all levels, requires more than just a "thicker description" of such milestones; it requires that light be shed on other patterns of development, other readers, other mathematical uses of the book—it requires a change in our questionnaire. We need to answer specific questions, such as: What happened to the book outside Germany? What were the particularities, if any, of its reception in Germany? Which parts of it were read and reworked? And when? Which developments, in which domains, did it stimulate—or hamper? What role did it play in later attempts to found mathematics on arithmetic?

Such questions suggest narrower foci, which will be adopted in the various chapters of the present volume. In this first part, however, we take advantage of the concrete nature of our object of inquiry—a book—to draw a general map of its tracks while sticking closely to the chronology. That is to say, instead of going backwards, seeking in the *Disquisitiones Arithmeticae* hints and origins of more recent priorities, we will proceed forwards, following Gauss's text through time with the objective of surveying and periodizing afresh its manifold effects.

Fifty pages later, they conclude (pp. 57–58):

To summarize, the role of the *Disquisitiones Arithmeticae* in the constitution of number theory fifty years after its publication was two-fold. On the one hand, the D.A. provided number theory with the features of its self-organization as an academic discipline. It shaped what number theory was and ought to be: congruences and forms with integer coefficients (with their possible generalizations). This image informed advanced textbooks and helped to structure them, and it would, for an even longer time, structure classifications of mathematics. On the other hand, the D.A. had launched an active research field, with a firm grasp on number theory, algebra and analysis, supported by close and varied readings of the book. It provided the field with technical tools, and a stock of proofs to scrutinize and adapt. It also provided concrete examples of the very links between different branches of mathematics that created the field, often articulated around richly textured objects and formulae, such as the cyclotomic equation or Gauss sums. The (meta)stability of the field was not guaranteed by any unicity of purpose or concept (individual mathematicians might have their own priorities, mix differently the resources available or disregard some of them), nor by a merging into a larger domain, but by a constant circulation from one branch to another, a recycling of results and innovations. How certain branches emancipated themselves, and with which consequences for number theory and for the role of the D.A., will be the subject of the next chapter.

These are impressive claims. What was once seen as the book that created modern algebraic number theory in much the way that a stream flowing out of Lake Itasca, in Minnesota, turns into the Mississippi river (one just naturally grows into the other, getting bigger as it goes along) is now seen as structuring a landscape around it, having no overall guiding plan, and as something individual mathematicians can buy into. In fact, Goldstein, Schappacher, and Schwermer go on to argue that the standard history is correct as far as it goes, but it has marginalised a Missouri's worth of ideas that flows through complex analysis and the theory of elliptic functions. In short, and despite the quality of its predecessors, even they see it as a book that created a new field of mathematics and affected, if not indeed caused, the growth of several neighbouring disciplines within mathematics.

4.3 Modular Arithmetic

In the opening chapters of the books (sections, as Gauss called them) Gauss established a number of important results that we shall take for granted but which, taken together, form an impressive introduction to his concern for rigour. He also introduced the symbol $\equiv$ for congruences, remarking in a footnote to the second paragraph of the book, that "We have adopted this symbol because of the analogy between equality and congruence. For the same reason Legendre, in the treatise which we shall often have occasion to cite, used the same sign for equality and congruence. To avoid ambiguity, we have made a distinction." Sophie Germain was perhaps the first mathematician to appreciate this novelty, in her work on Fermat's Last Theorem, and she wrote to Gauss to say so.[4]

Here are slightly modernised statements of some of Gauss's results. Throughout, p is a prime number that does not divide the integer a.

- The congruence $ax + b \equiv c$ mod p, $a \not\equiv 0$ mod p can always be solved. This amounts to saying that the integers modulo a prime form a field; Gauss showed that all the operations of arithmetic work modulo a prime.
- Fermat's (little) theorem: $a^{p-1} \equiv 1$ mod p.
- A number a is called a *primitive root* if the least positive power of a such that $a^n \equiv 1$ mod p is $n = p - 1$. It follows that every number modulo p is a unique power of a primitive root. Gauss proved (in §55 of the *Disquisitiones Arithmeticae*) that for each prime p there is at least one primitive root. In much more modern terms this is a consequence of the more general result that the multiplicative subgroup of a finite field is cyclic. Gauss noted that no-one before him had proved this result: Lambert had stated it without proof, and Euler's attempt at a proof has two gaps. He therefore observed that "This theorem furnishes an outstanding example of the need for circumspection in number

---

[4]See Del Centina and Fiocca (2012, 693).

theory so that we do not accept fallacies as certainties" (§56). He went on to sketch a method for finding a primitive root in any given case. It will be worth keeping an eye out for later ramifications of this result.

As an example of the theorem of the primitive root, in §58 Gauss gave the fact that every number modulo 19 is a power of 2. It is easier to see that every number modulo 5 is a power of 2, and every number modulo 7 is a power of 3 (but it is not the case that every number modulo 7 is a power of 2). The theorem allows us to find which nth roots of a number modulo p are also numbers modulo p.

To find a primitive root when $p = 19$, and then to show that one always exists, we can proceed as follows. The possible orders of numbers modulo p are 1, 2, 3, 6, 9, and 18. Any number modulo p of order k satisfies the equation

$$x^k \equiv 1 \bmod 19,$$

which has at most k roots modulo 19. Therefore there is at most 1 number modulo p of order 1, at most 2 of order 2, 3 of order 3, 6 of order 6, and 9 of order 9. Elements of order 2 or 3 also satisfy the equation $x^6 \equiv 1 \bmod 19$, and elements of order 3 also satisfy the equation $x^9 \equiv 1 \bmod 19$, so it is enough to count the elements of orders 1, 6, and 9, of which there are at most 16 (and certainly less, in this case). So there must be an element of order 18, which is therefore a primitive root.

Alternatively, and more generally, we could take the product of an element of order 6 and one of order 9. It must have an order which is the least common multiple of these orders, which is 18, and so be a primitive root.

In the general case, Gauss supposed that

$$p - 1 = p_1^{\alpha_1} p_2^{\alpha_2} \ldots p_k^{\alpha_k}.$$

There are at most $(p - 1)/p_1$ numbers modulo p that are of order $(p - 1)/p_1$ and so at least one that is not; call it g_1. Let $h_1 \equiv g_1^{(p-1)/p_1^{\alpha_1}} \bmod p$, then h_1 is of order $p_1^{\alpha_1}$ because

$$h_1^{\alpha_1 - 1} \equiv g_1^{(p-1)/p_1} \not\equiv 1 \bmod p.$$

Define $h_2, h_3, \ldots, h_k$ similarly, then the order of $h_1 h_2, \ldots h_k$ is $p - 1$.

So there is nothing to worry about in the claim that primitive roots exist. This frees us to concentrate on the much more interesting question, which is what use Gauss went on to make of them—and indeed to contemplate a nice exercise in group theory well before group theory was created and which very good mathematicians before him had failed to carry out.

4.4   Gauss on Congruences of the Second Degree

The fourth section of the *Disquisitiones Arithmeticae* is a preliminary to the more important fifth section on binary quadratic form s, but it tells us a lot about what Gauss had been reading and how carefully he read. It also contains the first proof of quadratic reciprocity.

Gauss began with some remarks about composite moduli before isolating the central task of studying congruences modulo a prime. He defined residues and non-residues, and in §108 he proved the theorem that -1 is a quadratic residue of all prime numbers of the form $4n + 1$ and a non-residue of all prime numbers of the form $4n + 3$. He gave this proof, which he attributed to Euler (*Opuscula Analytica* 1, 135); earlier (§64) he had mentioned other proofs due to Euler from 1760 and 173–174.[5]

Let C be the collection of all non-zero residues mod p that are less than p; C has $\frac{1}{2}(p - 1)$ members, and this number is even if p is of the form $4n + 1$ and odd otherwise. Call two residues associated if their product is congruent to 1 modulo p. Let the number of residues that are not equal to their associates be a and those that do equal their associates be b in number, then C contains $a + 2b$ members, so a is even when p is of the form $4n + 1$ and odd when p is of the form $4n + 3$. Now the only numbers less than p that are associates of themselves are 1 and -1, so -1 is a quadratic residue if and only if a is even, i.e. if and only if p is of the form $4n + 1$.

The proof makes use of the fact, which Gauss had proved earlier, that every number less than p has an inverse modulo p that is also less than p.

Gauss then turned to the residues $+2$ and -2. He listed the primes less than 100 for which 2 is a residue: 7, 17, 23, 31, 41, 47, 71, 79, 89, 97. None, he remarked, are of the form $8n + 3$ or $8n + 5$, and wrote "Let us see therefore whether this induction can be made a certitude". To show that it can be, Gauss considered the least number t for which 2 is a residue, so there is an a such that $a^2 \equiv 2 \bmod t$. He wrote $a^2 = 2 + tu$ and showed (I omit the argument) that u will be smaller than t. But this contradicts the definition of t and so Gauss concluded, using the result above about -1, that 2 is a nonresidue and -2 a residue of all primes of the form $8n + 3$ and both 2 and -2 are nonresidues of all primes of the form $8n + 5$.

Rather more work was required to obtain the results that 2 is a residue of primes of the forms $8n + 1$ and $8n - 1$, and that -2 is a residue of primes of the forms $8n + 1$ and a nonresidue of primes of the forms $8n - 1$.

Gauss then dealt with $+3$ and -3, remarking (§120) that the theorems in this case were known to Fermat (and stated in a letter to the British mathematicians) but the first proofs were given by Euler in 1760–61, and "This is why it is still more astonishing that proof of the propositions relative to the residues $+2$ and -2 kept eluding him, since they depend on similar devices".

---

[5]The *Opuscula Analytica* is a two-volume edition of previously unpublished works by Euler that was published by the Imperial Academy of Sciences in St. Petersburg in 1783 and 1785. Most of the papers are on number theory, including topics related to quadratic reciprocity.

Then came the cases of $+5$ and -5. Inspection of cases (which Gauss called induction) led him to claim (§123) that $+5$ is a residue of all prime numbers that are residues of 5 (which are those of the form $5n + 1$, $5n + 4$, or $20n \pm 1$, or $20n \pm 9$) and a nonresidue of all odd numbers that are nonresidues of 5. "However", he said, "the verification of this induction is not so easy". I omit his demonstration.

All this was the build-up to Gauss's first proof of quadratic reciprocity, but the proof there is unilluminating, being a case-by-case examination of the different cases that arise if the theorem is approached head-on. It was described by Smith as "presented by Gauss in a form very repulsive to any but the most laborious students", although an intelligible and much shorter reworking has since been given by Ezra Brown in (1981).

4.5 Gauss's Theory of Quadratic Forms

In many ways the core of Gauss's *Disquisitiones Arithmeticae* is his theory of quadratic forms, and explicitly his composition theory for them and his classification of them into various types. His original treatment is long and can seem like on obscure exercise in classification, and to make matters worse his composition is not easy to carry out (although it can be automated, as later writers worked hard to show). So the first thing to say about it is that it was the first theory that enabled a mathematician to read off from the coefficients $a, 2b, c$ what numbers can be represented by that form. It is worth keeping an eye out, too, for the word 'character', because it is introduced here for the first time and it is the origin of the idea of characters in the representation of groups. All before the group concept had been thought of!

For Gauss a quadratic form was an expression of the form $ax^2 + 2bx + cy^2$. The quantity $D = b^2 - ac$ he called the *determinant* of the form. Because this is the negative of the determinant we would want – the one corresponding to the matrix $\begin{pmatrix} a & b \\ b & c \end{pmatrix}$ – it is best to call this the *Gaussian determinant*, and even the *discriminant*. He said that one form F_1 implies another F_2 if F_1 can be transformed into F_2 by an integer linear transformation of non-zero determinant (not necessarily ± 1). When this is the case any number represented by F_2 can be represented by F_1. In the notation introduced in the previous chapter, if the quadratic forms F_1 and F_2 are represented by symmetric matrices Q_1 and Q_2 respectively, then to say that F_1 implies F_2 is to say that there is a transformation represented by a matrix A such that $A^T Q_1 A = Q_2$.

Gauss distinguished, as his predecessors had not, between equivalence and strict equivalence of forms: in the first case the integer linear transformation has determinant ± 1, in the second, *strict*, case it has determinant $+1$. This gave him two levels of equivalence classes: equivalence and strict equivalence. Gauss called two forms *properly equivalent* if the determinant was $+1$, and *improperly equivalent* if the determinant was -1. Gauss also isolated the case when the greatest common divisor (gcd) of the coefficients a, b, c is 1 and called the corresponding form *primitive*.

He then gave a very detailed analysis of the Lagrange reduction process, showing, as an example (§175) how to find all 8 inequivalent forms with $D = -85$ (Gauss found 8 because he included the negative as well as the positive reduced forms, $-ax^2 - 2bxy - cy^2$ and $ax^2 + 2bxy + cy^2$). He also listed all the positive inequivalent forms with Gaussian determinant $b^2 - ac = D \geq -12$.

It is harder when the discriminant is positive, and Gauss introduced the concept of neighbouring forms and periods in order to handle it—this aspect of the theory will not be discussed. This was because making his theory work in concrete cases depended heavily on one being able to find representatives within each equivalence class that have the most convenient properties (see §223 for his informal criteria).

Gauss next proceeded to divide up the set of equivalence classes. Two classes represented by (a, b, c) and (a', b', c') were said to have the same *order* if and only if

$$gcd(a, b, c) = gcd(a', b', c'), \text{ and } gcd(a, 2b, c) = gcd(a', 2b', c').$$

He noted that all properly primitive classes go in one order and all improperly primitive classes go in another order (provided, of course, there are forms which are improperly but not properly equivalent). He further subdivided an order into distinct sets called the *genera* (singular *genus*). His aim here was to deal with the question raised by Lagrange's result on switching forms.

After all this (finally!, you may think) came some number theory. Gauss proved (§228) that given a properly primitive form F, not necessarily reduced, and a prime number p there are infinitely many numbers that are represented by F and not divisible by p.

The proof is simple. Let the form be $ax^2 + 2bxy + cy^2$, and p be a prime. Then p cannot divide all of a, b, and c. If it does not divide a, then whenever x is not divisible by p and y is divisible by p the quadratic form $ax^2 + 2bxy + cy^2$ is not divisible by p. A similar argument deals with the case when p does not divide c. If p divides both a and c then it does not divide $2b$ and so if x and y have values not divisible by p then $ax^2 + 2bxy + cy^2$ is not divisible by p.

Then Gauss showed (§229) that if the prime p divides D then either all of the numbers that are represented are quadratic residues mod p or none are. The proof of this result is also not difficult. Let the form F be $ax^2 + 2bxy + cy^2$, and suppose it represents m and m', say

$$m = ag^2 + 2bgh + ch^2, \quad m' = ag'^2 + 2bg'h' + ch'^2.$$

Then

$$mm' = \left(agg' + b(gh' + g'h) + chh'\right)^2 - D(gh' - g'h)^2.$$

So mm' is congruent to a square modulo D and therefore also modulo p, and so m and m' are either both residues or both non-residues modulo p, as claimed.

He called this dichotomy a *fixed relationship* for the primes p that divide D, and observed that it holds only for primes that divide D, so numbers representable by F have a fixed relationship to the prime divisors of D. With very little more work, Gauss deduced this result:

1. if $D(F) \equiv -1$ (4) then the odd numbers represented by F have a fixed relationship to 4. That is either all odd m represented by F are $\equiv 1$ (4) or all are $\equiv -1$ (4).
2. when $4|D$ the fixed relationship is: either all odd m represented by F are $\equiv 1$ (4) or all are $\equiv -1$ (4).
3. when $8|D$ the fixed relationship is that all odd m represented by F fall into one class: either all are $\equiv 1$ (8) or all are $\equiv 3$ (8), or all are $\equiv 5$ (8) or all are $\equiv 7$ (8).

For the first of these, Gauss observed in §229 that

When the determinant D of the primitive form F is $\equiv 3$ (mod. 4), all odd numbers representable by the form F will be $\equiv 1$, or $\equiv 3$ (mod. 4). For if m, m' are two numbers representable by F, the product mm' can be reduced to the form $p^2 - Dq^2$, just as we did above. When each of the numbers m, m' is odd, one of the numbers p, q is necessarily even, the other odd, and therefore one of the squares p^2, q^2 will be $\equiv 0$, the other $\equiv 1$ (mod. 4). Thus $p^2 - Dq^2$ must certainly be $\equiv 1$ (mod. 4), and both m, m' must be $\equiv 1$, or both $\equiv 3$ (mod. 4). So, e.g., no odd number other than those of the form $4n + 1$ can be represented by the form $(10, 3, 17)$.

Gauss gave more fixed relationships for $D \equiv 2$ (8) and $D \equiv 6$ (8)—I quote them but omit the proofs. The statements become clear on taking examples.

- When the determinant D of the primitive form F is $\equiv 2$ (mod 8); all odd numbers representable by the form F will be either partly $\equiv 1$ and partly $\equiv 7$, or partly $\equiv 3$ and partly $\equiv 5$ (mod 8).
- When the determinant D of the primitive form F is $\equiv 6$ (mod 8); all odd numbers representable by the form F are either only those that are $\equiv 1$ and $\equiv 3$ (mod. 8) or only those that are $\equiv 5$ and $\equiv 7$ (mod 8) .

Gauss called a fixed relationship (§230) a *character* of the form F, and the set of all fixed relationships of a form its total character.[6] He pointed out that the characters of a primitive form $ax^2 + 2bxy + cy^2$ are easy to determine from the coefficients a and c: both are obviously representable by the form but at most one is divisible by any prime dividing D because $b^2 = D + ac$, else the form would not be primitive.

By way of an example, Gauss exhibited all the positive inequivalent forms with discriminant -161. As it happens, there are 16 such classes of forms with discriminant -161, and Gauss showed that they fall into four groups of four on looking at their characters. These distinct families Gauss called the genera of the quadratic form. (Linguistic note: 'genera' comes from 'general', 'species' from

---

[6]This is the origin of the term 'character' in the theory of group representations.

'specific', so genera can be broken down into species. The terminology, including the word 'order', is taken from Linnaean biology).

Gauss produced this table (§231) as an example of his results,

Character			Representing forms of the classes			
1, 4	R7	R23	$(1, 0, 161)$	$(2, 1, 81)$	$(9, 1, 18)$	$(9, -1, 18)$
1, 4	N7	N23	$(5, 2, 33)$	$(5, -2, 33)$	$(10, 3, 17)$	$(10, -3, 17)$
3, 4	R7	N23	$(7, 0, 23)$	$(11, 2, 15)$	$(11, -2, 15)$	$(14, 7, 15)$
3, 4	N7	R23	$(3, 1, 54)$	$(3, -1, 54)$	$(6, 1, 27)$	$(6, -1, 27)$

Here, each row contains all the inequivalent reduced forms of the same *genus*.

For example, given the form $(10, 3, 17) = 10x^2 + 6xy + 17y^2$, observe that it is primitive. Form the discriminant, which in this case is $3^2 - 10 \times 17 = -161$. This tells us that representable odd numbers will either all be congruent to 1 mod 4 or all be congruent to 3 mod 4. Factorise the discriminant : $-161 = -7 \times 23$. This tells us that the fixed relationship with 7 and 23 is crucial.

The way to understand this table, and the method of fixed relations in general, is to view it as a series of questions asked, in this case, of a form with discriminant congruent to -1 mod 4 and any odd number m:

- Is m congruent to 1 mod 4?
- Is m congruent to a square mod 7?
- Is m congruent to a square mod 23?

Only if the answers to these questions are all 'Yes', or if the answers are one 'Yes' and two 'No's, can the odd number m be representable by a form of the corresponding genus.

The deeper implications of this table, and of its proof, rest heavily on Gauss's theory of the composition of forms, as we shall see later on. First, Gauss claimed that exactly half the possible genera occur—those for which there is an even number of 'No's, and second that all the genera occur for which there is an even number of 'No's. We shall examine Gauss's reasons for this later on, but for now notice that this means that an odd number m is representable by a form of discriminant -161 only if an even number of 'No's are returned to the above questions.

Notice, by the way, that there are examples of inequivalent forms in the table of the form (a, b, c) and $(a, -b, c)$. Such pairs of forms represent the same numbers, because

$$(a, b, c)(x, y) = ax^2 + 2bxy + cy^2 = (a, -b, c)(-x, y).$$

The inequivalent classes in a genus have the same complete character as Gauss called it, which means that they represent the same numbers modulo D. For example, Gauss simply stated that the inequivalent reduced forms $(2, 1, 81)$ and $(9, 1, 18)$ are in the same genus. In fact, they both represent 85, for example, and give out squares modulo 7 and modulo 23, as does the form $(9, -1, 18)$ but not

$(1, 0, 161)$. But this is not the case for forms in distinct genera. This means that the question "Is the number n representable by a form of discriminant D?" is partially answered either by discovering that the answer is no, or by exhibiting the genus of forms for which it is possible. But in the second (affirmative) case it remains to see which of the forms in that genus do the trick. It leaves open for further research the question of how the genera can be told apart—this would, alas, take us too far afield.

4.6 Exercises

1. Find which of the numbers 5, 9, 13, 17, 25, 29, 33 are representable by a form of discriminant -161, and when the number is representable determine the form that represents it.
2. Show that the numbers 101, 103, and 109 are not representable by any form of discriminant -161 but that the number 107 is representable, and determine the form that represents 107.

   These exercises make clear that if a number is representable by a form of a given genus it may be represented only by one form of that genus.
3. We shall now see that there are odd numbers that meet Gauss's necessary requirements to be representable by a quadratic form but are not representable. This shows that his criteria are not sufficient.

   (a) There are two equivalence classes of forms with discriminant -13: $F_1 = (1, 0, 13)$ and $F_2 = (2, 1, 7)$. By Gauss's analysis, we have two tests to apply to odd numbers m that might be representable by one of these forms: is $m \equiv 1 \bmod 4$? and is m congruent to a square modulo 13?
   (b) Check that the form $F_1 = (1, 0, 13)$ represents odd numbers that answer both questions with a 'Yes' and that the form $F_2 = (2, 1, 7)$ represents numbers that answer both questions with a 'No'.
   (c) Gauss's criteria say that odd numbers representable by F_2 must be congruent to 3 mod 4 and not squares mod 13, because 7 is plainly representable. Among the numbers that fulfil these criteria are 7, 11, and 15 but while 11 is representable, 15 cannot be written in the form $2x^2 + 2xy + 7y^2$.

Chapter 5
Cyclotomy

5.1 Introduction

In the seventh section of his book, Gauss transformed what had been a branch of geometry and trigonometry into a branch of arithmetic with profound implications for algebra that were to influence Abel and Galois. His discoveries concerning the pth roots of unity for primes p in this part of the *Disquisitiones Arithmeticae* were the first to excite other mathematicians. We have already seen that Poinsot praised Gauss's work highly in 1807 and alerted French readers to the forthcoming French translation of Gauss's book. In the re-edition of his treatment of the solution of equations (Lagrange 1808, 274), Lagrange did the same, calling Gauss's work "excellent". He singled out the treatment of cyclotomy, which he called "original" and "ingenious", and the use in that theory of the concept of a primitive root, which he called an "ingenious and happy idea" (1808, 275).

Gauss's study has its origins in some more modest observations of Lagrange, which he had published in the course of his Mémoire (1770) on the solution of equations. There he had noted that the study of the equation

$$x^n - 1 = 0,$$

where n is composite, say $n = pq$, reduces to the study of an equation of degree p and one of degree q. For, on setting $x^q = y$ one obtains the equation $y^p = 0$. Let this equation be solved, and α be one of its roots, then the original equation reduces to the study of $x^q = \alpha$ and on setting $x = \sqrt[q]{\alpha}$ to the equation $t^q = 1$. So it is enough to consider equations of the form $x^n = 1$ where n is a prime.

Lagrange passed over the case $n = 2$ and wrote $n = 2p + 1$. This gave him the cyclotomic equation, after a division by $x - 1$, in the form

$$x^{2p} + x^{2p-1} + \cdots + x^2 + x + 1 = 0. \tag{5.1}$$

© Springer Nature Switzerland AG 2018
J. Gray, *A History of Abstract Algebra*, Springer Undergraduate Mathematics Series,
https://doi.org/10.1007/978-3-319-94773-0_5

He divided through by x^p and introduced the variable $y = x + x^{-1}$. This enabled him in principle to write Eq. (5.1) in powers of y. He deduced that when n is of the form $2^\lambda 3^\mu 5^\nu$ the cyclotomic equation can be solved by radicals. If the solution of cubic equations is allowed, he added, then the cyclotomic equation can be solved by radicals when n is of the form $2^\lambda 3^\mu 5^\nu 7^\omega$. But one is halted at $n = 11$, he said, by the appearance of a quintic equation.

This remark highlights an amusing ambiguity in the subject of solving equations by radicals. Even the formula for the solution of the cubic equation exhibits cube roots of arbitrary quantities, which may, for example, be complex. The search for a formula involving radicals that would solve equations of higher degree, say n, must surely be willing to accept pure nth roots. But in that case, why not accept the ten roots other than 1 of the equation $x^{11} - 1 = 0$? The answer would seem to be that these roots are to be expressed by a formula involving the coefficients of the equation, and there was no such formula. On this supposition, cube roots can be taken because the formula for the solution of the equation $x^3 + px + q = 0$ in the special case where $p = 0$ reduces to $\sqrt[3]{-q}$. That said, the distinction is obscure and disappears in the nineteenth century.

Lagrange then noted a way to solve equations of the form $x^n - 1 = 0, n > 5$, had been opened up by de Moivre. It follows from his work, said Lagrange (p. 248), that if $x + x^{-1} = 2 \cos \varphi$ then $x^n + x^{-n} = 2 \cos n\varphi$. This leads to the equations

$$x^2 - 2x \cos \varphi + 1 = 0, \text{ and } x^{2n} - 2x^n \cos \varphi + 1 = 0.$$

These have the solutions[1]

$$x = \cos \varphi \pm i \sin \varphi, \text{ and } x^n = \cos n\varphi \pm i \sin n\varphi,$$

where the same sign must be taken in each case, and clearly any solution of the first equation is a solution of the second.

Lagrange then observed that all the solutions are distinct and, except for $x = 1$, complex. Moreover, when n is prime they are all powers of one of them. It is a little more complicated to see that if n is composite, all the roots will be powers of some α^m, where α is a root and m is prime to n. He concluded his account with the explicit solutions of $x^n - 1 = 0, n = 1, 2, \ldots, 6$.

In the seventh and final section of the (published) *Disquisitiones Arithmeticae* Gauss dealt at length with roots of unity, concentrating on the case where they are the roots other than 1 of the equation $x^p - 1 = 0$ where p is a prime, which is to say that they are the roots of the cyclotomic polynomial

$$c_p(x) = x^{p-1} + x^{p-2} + \cdots + x + 1 = 0.$$

(We shall drop the suffix p when there is no risk of ambiguity.)

---

[1]Lagrange wrote $\sqrt{-1}$ where we have i.

He proved that the cyclotomic equation is irreducible, and that it can be understood through the successive adjunction of the roots of equations of certain degrees that divide $p - 1$, as we shall see.[2]

Dedekind later commented that:

> Here enters for the first time the concept of irreducibility (art. 341) which has become crucial for the whole orientation of later algebra; although Gauss makes only a limited use of it (art. 346), I have no doubt that this fundamental principle has also led him to the discovery of particular points and that he has preferred the synthetic presentation only for the sake of brevity; certainly this is what the important words (art. 365) let us conclude: "and we can prove with complete rigour that these higher equations cannot by any means be avoided or reduced to lower ones [etc.]." The truth of the assertion contained in them is easy to prove at the present stage of algebra, namely since the development of Gauss's thoughts by Abel and Galois. In fact, from the seed laid by Gauss, a science is born, which one ... could perhaps describe as the science of the algebraic affinities between numbers, or, if one wants to use an expression chosen by me, as the science of the affinities between fields.[3]

The cases $p = 17$ and $p = 19$ were the ones he used for illustrative purposes, and they make a good introduction. However, the calculations involved can be rather daunting, so I shall start with a less intimidating example, but one with its own irritations. We shall see that Gauss introduced these ideas precisely to guide the calculations and explain some regularities he had detected in the algebra—and it is the algebra and the number theory of these objects that interested him, not their trigonometry or geometry. As he put it, he aimed to make it "abundantly clear that there is an intimate connection between this subject [cyclotomy] and higher Arithmetic".

5.2 The Case $p = 7$

In this case the cyclotomic equation is $c(x) = x^6 + x^5 + \cdots + 1 = 0$. A typical root is x^n for some $n \in \{1, 2, 3, 4, 5, 6\}$. Gauss had earlier called a number r mod p with the property that every number mod p is a power of r a *primitive root* mod p so he now looked for a primitive root mod 7. Such an element is 3: written multiplicatively, the cyclic group $G_7^* = \{1, 2, 3, 4, 5, 6\}$ is generated by 3, and modulo 7 the powers of 3 are:

$$\{3^1 \equiv 3, \ 3^2 \equiv 2, \ 3^3 \equiv 6, \ 3^4 \equiv 4, \ 3^5 \equiv 5, \ 3^6 \equiv 1\}.$$

---

[2]Dedekind was the first to establish the irreducibility of equations of the form $x^{\varphi(m)} - 1 = 0$, where m is not prime and $\varphi(m)$ is the number of numbers relatively prime to m; see Dedekind (1857).

[3]See Dedekind (1873, pp. 408–409) quoted in (*Shaping*, 115). Article 365 will occupy us later, when we look at the reception of Wantzel's work.

The *exponent* of a is defined to be the value of n such that $3^n \equiv a \bmod 7$. In this case the exponents and the values are too often the same for this to be a good teaching example, but nonetheless they display the corresponding elements of the group, now written additively, as G_6^+. (Additively, because $3^j \times 3^k = 3^{j+k}$.)

The element n in the (multiplicative version of the) group is now written as 3^j, with $n = 3^j$. So in this group we calculate 6×4 by noting that $6 = 3^3$ and $4 = 3^4$ so

$$6 \times 4 \equiv 3^3 \times 3^4 \equiv 3^{3+4} \equiv 3^7 \equiv 3 \bmod 7,$$

and indeed $6 \times 4 \equiv 3 \bmod 7$. Here we have used Fermat's little theorem, that $n^6 \equiv 1 \bmod 7$. The point of passing from n numbers to j numbers is that the corresponding exponents add (just like logarithms) so pattern spotting becomes easier mod 7. When we write G in additive notation we shall call the elements j-numbers, and denote it G_6^+; in multiplicative notation we call the elements n-numbers. Note that the identity element in G_6^+ is 6, so from now on I shall replace it by 0 when appropriate.

In modern language, we have a group isomorphism $G = G_7^* \to G_6^+$ given by $n \mapsto j$ where $n = 3^j$.

We can display the isomorphism between these groups this way:

$n =$	1	2	3	4	5	6
$j =$	6	2	1	4	5	3

The subgroup of G_6^+ of order 2 generated by 3 is, of course, $\{3, 0\}$, and its cosets are $\{4, 1\}$ and $\{5, 2\}$. The corresponding roots are $\{x, x^6\}$, $\{x^3, x^4\}$ and $\{x^2, x^5\}$. We pass from one to the next by replacing x by x^3.

Let $y = x + x^6$, which we shall write as $y_1 = x + x^{-1}$ because $x^7 = 1$. Then it is a simple matter of algebra to check that

$$y_1^2 = x^2 + 2 + x^{-2}$$

and

$$y_1^3 = x^3 + 3(x + x^{-1}) + x^{-3} = x^3 + 3y + x^{-3},$$

and writing the given cyclotomic equation as

$$x^3 + x^2 + x + 1 + x^{-1} + x^{-2} + x^{-3} = 0$$

we find that we have

$$y_1^3 + y_1^2 - 2y_1 - 1 = 0.$$

We can check that the same equation is satisfied by $y_2 = x^2 + x^5$ and by $y_3 = x^3 + x^4$.

Or we can argue that

$$y_1 + y_2 + y_3 = -1, \; y_1 y_2 + y_2 y_3 + y_3 y_1 = -2, \; \text{and} \; y_1 y_2 y_3 = 1, \tag{5.2}$$

and come to the same conclusion.

Now, the defining equations for y_1, y_2, y_3 are quadratic equations for what we can call x_1, x_2 and x_3 respectively, and y_2 and y_3 are rational functions of y_1. So adjoining one of them to the rational field adjoins all three, and so the original cyclotomic equation, which is irreducible over the rationals, factors when any y is adjoined. Indeed, the original equation of degree 6 becomes:

$$(x^2 - x y_1 + 1)(x^2 - x y_2 + 1)(x^2 - x y_3 + 1) = 0,$$

as we can check this by expanding it and using Eq. (5.2) above.

Notice that the map $x \mapsto x^k, k = 1, \ldots, 6$, induces a permutation of the roots that sends x^a to $(x^a)^k$, so any permutation of the roots is determined by its effect on x. So the group of permutations of the roots is G_7^*. The group acts transitively on the roots.

5.3 The Case $p = 19$

We take $p = 19$, as in Gauss, *Disquisitiones Arithmeticae*, §353. The 18 roots of

$$c(x) = x^{18} + \cdots + x + 1 = 0,$$

can be permuted, but not arbitrarily. Once the image of one root is given, the images of the others are fixed because they are all known rationally once one is given. Now, all the roots are known explicitly, they are of the form

$$\cos \frac{2\pi k}{19} + i \sin \frac{2\pi k}{19} = e^{2\pi i k/19}, \; 1 \le k \le 18,$$

so Gauss denoted them by their n value (but note that I have introduced the bars above the numbers to keep a potential ambiguity at bay):

$$\bar{1}, \bar{2}, \ldots, \bar{n}, \ldots \bar{18},$$

where $\bar{k}$ stands for $e^{2\pi i k/19}$.

Gauss found that 2 is a primitive root mod 19. The successive powers of 2 mod 19 are

$$2^1 \equiv 2, \ 2^2 \equiv 4, \ 2^3 \equiv 8, \ 2^4 \equiv 16, \ 2^5 \equiv 13, \ldots, 2^{18} \equiv 1 \text{ mod } 19.$$

He now used the additive form of the group to hunt for subgroups.

Note that knowing the exponent of a enables us (as it did Gauss) to compute the order of a, which is the least power, n, of a such that $a^n \equiv 1$ mod 19 (note that by Lagrange's theorem the order of a must divide 18, so it is either 1, 2, 3, 6, 9, or 18). For, if $a = 2^j$ then $a^n = 2^{jn}$, which will be equivalent to 1 mod 19 if and only if jn is a multiple of 18. As an example, $4 \equiv 2^2$, so we expect 4 to have order 9, and indeed $4^9 \equiv 2^{2 \cdot 9} \equiv 2^{18} \equiv 1$, while $4^{2n} \not\equiv 1$ for any smaller power of 4. In the same spirit we easily find that $11 \equiv 2^{12}$, so the order of 11 is 3, because $18|12 \times 3$ and 3 is the least number n such that $12n$ is a multiple of 18. We find, however, that $13 = 2^5$, so the order of 13 is also 18.

The comparison with logarithms and exponents, $2^j \cdot 2^k = 2^{j+k}$, makes calculations much easier. It was for this reason that Gauss introduced the exponents and kept track of the roots not by the value of $\bar{k}$ but by the corresponding exponent n such that $2^n \equiv k$ mod 19.

Gauss (§353) focussed on the fact that $8 = 2^3$ has order 6, and that the six powers of 8 are $\{8, 64 \equiv 7, 18, 11, 12, 1\}$, and that these elements form a group (the group generated by 8 mod 19).

The factorisation $19 - 1 = 2 \cdot 3 \cdot 3$ suggested to Gauss that it would be worthwhile looking at, for example, the 3rd power of the primitive root $2^3 = 8$ and its 6 powers. He observed that one recovers this set whatever element of it we begin with, denoted it (6, 1), and called it a *period*. The 6 in the notation was chosen because the period has 6 elements, the 1 because it has the element 1 in it. The other periods turn out to be

$$(6, 2) = \{2, 3, 5, 14, 16, 17\} = (6, 3), \text{ and } (6, 4) = \{4, 6, 9, 10, 13, 15\}.$$

Of these (6, 1) is a subgroup of the permutations of the roots, the others its cosets. Its cosets are obtained by increasing the j numbers by 1 and 2 respectively. Gauss called them (6, 2) and (6, 4) because each has 6 elements and the first coset contains 2 (it also contains 3) and the second coset contains 4. This corresponds to replacing n by $2n$ and $4n$ respectively, so the roots corresponding to (6, 2) are $(x^2)^8 = x^{16}, (x^2)^7 = x^{14}$, etc: $\{x^2, x^3, x^5, x^{14}, x^{16}, x^{17}\}$. Finally, the roots corresponding to the coset (6, 4) are $\{x^4, x^6, x^9, x^{10}, x^{13}, x^{15}\}$.

Gauss proved that the periods are well defined (independent of any choices made), that they are either disjoint or agree completely, and that their union is the whole set of roots. Periods with the same number of roots he called similar (in the case we are looking at, this number is 6) and he showed how to multiply them together. This amounts to showing that the cosets themselves form a group.

Because Gauss was interested in factorising the equation $x^{18} + \cdots + x + 1 = 0$ by adjoining roots of equations of lower degree, he looked at certain sums of the periods. The sum of the roots in the subgroup $(6, 1)$ is $P = x^8 + x^7 + x^{18} + x^{11} + x^{12} + x$. He called the sum of the roots corresponding to $(6, 2)$ P', and the sum of the roots corresponding to $(6, 4)$ P''. Straight-forward algebra shows that

$$P^2 = 2P + P' + 2P'' + 6,$$

and a little more algebra shows that

$$P^3 + P^2 - 6P + 7 = 0.$$

It is also true that P' and P'' satisfy this cubic equation. This remarkable result arises because the combination of the roots has been selected according to a rational plan that reflects the underlying structure of the roots.

By adjoining a root of this equation, say P (and therefore the others) he then could proceed to the equation for the periods within $(6, 1)$, which are $(2, 1)$, $(2, 7)$, and $(2, 8)$.

So, still following Gauss, suppose we adjoin to the rational numbers the strange-looking irrational numbers P, P', P''. We now repeat the trick, working with the group $(6, 1)$. It has an element of order 2 that we pick up as before by passing to the additive form. The group is $\{3, 6, 9, 12, 15, 0\}$, and we pick the subgroup $\{9, 0\}$. The corresponding roots are x^{18} and x^1, so Gauss called this subgroup $(2, 1)$ because it has 2 elements and one of them is associated to x^1. Let's call the sum of these roots Q. The cosets of this subgroup in the group $(6, 1)$ are $\{3, 12\}$ and $\{6, 15\}$ in additive notation, the corresponding sets of roots are $\{x^8, x^{11}\}$ with sum Q' and $\{x^7, x^{12}\}$ with sum Q''. In Gauss's notation, this makes the cosets $(2, 8)$ and $(2, 7)$. The sum of these is P, of course. We also find that $QQ' + Q'Q'' + Q''Q = -P' - 1$ and that $QQ'Q'' = P' + 2$. So each of Q, Q', Q'' satisfy the cubic equation

$$x^3 - Px^2 - (P' + 1)x - (P' + 2) = 0,$$

all of whose coefficients have been obtained already.

It remains to observe that the roots corresponding to the subgroup $(2, 1) = \{9, 0\}$ satisfy the equation

$$x^2 - Qx + 1 = 0,$$

which is the quadratic equation whose roots were 1 and 18, so they are known once Q is known. In this way all the 18 non-trivial 19th roots of unity were found by solving two cubic and one quadratic equation.

The Regular 17-gon

A similar process applies with any prime number p. When $p = 17$ the result is that all the 17th roots of unity are found by solving a sequence of four quadratic equations. Solving a quadratic equation is a matter of extracting a square root, and so a task that can be carried out, in classical terms, by a straight edge and circle construction. So the regular 17-gon became the first regular n-gon known to be so constructible since Greek times, and indeed this construction works every time $p = 2^n + 1$ is prime.

In this case, the cyclotomic polynomial is first factored into two polynomials of degree 8 by adjoining $\sqrt{17}$. The roots of the equation $x^2 + x - 4 = 0$ are $\frac{1}{2}(-1 \pm \sqrt{17})$. Gauss called $-\frac{1}{2}(-1 + \sqrt{17})$ (8, 1), and the other root (8, 3), and looked first at the equation

$$x^2 - (8, 1)x - 1 = 0.$$

He called its roots (4, 1) and (4, 9), and chose for (4, 1) the root

$$\frac{1}{2}((8, 1) + \sqrt{(8, 1)^2 + 4}) = \frac{1}{2}(12 + \sqrt{12 + 3(8, 1) + 4(8, 3)}),$$

whose numerical value he determined to be 2.0494811777. Then he repeated the analysis with (8, 3) in place of (8, 1) and obtained (4, 3) and (4, 10).

Next, Gauss considered the equation $x^2 - (4, 1)x + (4, 3) = 0$, with roots (2, 1) and (2, 13), and gave their numerical values as 1.8649444588 and 0.1845367189. Now, the roots of the cyclotomic equation corresponding to (2, 1) are ζ and $\zeta^{16} = \zeta^{-1}$, because they are the roots of $x^2 - (2, 1)x + 1 = 0$, and they work out to be $0.9324722294 \pm 0.3612416662i$. The other powers of ζ he deduced from the other equations in the tree of factorisations.

5.4   Exercises

1. Verify that 3 is a primitive root modulo 17, and that 2 is a primitive root modulo 19.
2. Find the powers of every number modulo 17; find the powers of every number modulo 19.
3. Explain why requiring $p = 2^n + 1$ to be prime forces $n = 2^m$.

   Hint: To factor $x^a + 1$, write $a = 2^k b$, where b is odd and set $x^{2^k} = y$, so $x^a + 1 = y^b + 1$ and show that $y + 1$ is a factor of $y^b + 1$. So the only expression of the form $x^a + 1$ that could be prime must be one where $a = 2^k$.

Chapter 6
Two of Gauss's Proofs of Quadratic Reciprocity

6.1 Introduction

In this chapter I discuss two of Gauss's proofs of quadratic reciprocity: one (his second) that uses composition of forms, and the other (his sixth) that uses cyclotomy. The sixth was the last proof of this theorem he published, although he went on to leave two more unpublished.

Famously, the theorem has acquired a great number of proofs—André Weil estimated as many as 150, a number more or less confirmed by the list in Lemmermeyer (2000, Appendix B)—and although not all of them are very different some come from deep and suggestive connections in the subject. Thus the Russian mathematician Yuri Manin has written (in Lemmermeyer 2000, 22):

> When I was very young I was extremely interested in the fact that Gauss found seven or eight proofs of the quadratic reciprocity law. What bothered me was why he needed seven or eight proofs. Every time I gained some more understanding of number theory I better understood Gauss's mind. Of course he was not looking for more convincing arguments— one proof is sufficiently convincing. The point is, that proving is the way we are discovering new territories, new features of the mathematical landscape.

I regard this chapter as one of the harder ones in the book (some help is offered in the next chapter) and it is unusual for a book to present its most demanding material early, so some explanation is in order. Gauss's work was difficult in many ways: it is technical and depends on complicated expressions; it makes novel insights and connections; it introduces new concepts. For many readers it is likely that it goes outside their student education (which raises the question of what is it doing in a history of modern algebra, which is a well-defined subject) and that also makes it hard. All of these problems are reasons for including this material. If Gauss's rewriting of number theory is to be appreciated as the huge change in mathematics that it was then we have to understand its impact, and that involves considering just how difficult it was. How firmly one grasps the picture is a question of reading what follows in various ways; I have tried to offer an outline, some detail, and a

© Springer Nature Switzerland AG 2018

J. Gray, *A History of Abstract Algebra*, Springer Undergraduate Mathematics Series, https://doi.org/10.1007/978-3-319-94773-0_6

simpler, more modern, account. But what Gauss did was not to take one more step in the discovery of mathematics, it was to open up a new field. Inevitably, his work was difficult—only rich visions count—and this chapter offers one opportunity to consider what mathematics really is.

6.2   Composition and Quadratic Reciprocity

Gauss's second proof relies on his theory of the composition of forms. Although Gauss was far from clear on the point, his sharp distinction between proper and improper equivalence was key to showing that composition of forms is—to use a wild anachronism—almost a group operation. The qualification 'almost' is important, because it hints at the heart of the problem. As is discussed in more detail in Appendix B, Gauss investigated when the composite of two forms (note that they need not have the same discriminant) can be defined, and discovered both that it cannot always be defined, and that when it can the answer is not unique (it can always be varied by a transformation of the X, Y variables of the composite). More precisely (§§235, 236), he showed that for two forms to have a composite it is necessary and sufficient that the ratios of the discriminants of the forms are squares. He then showed that when composition is defined it is associative (the proof is particularly long and tiring). Only then did he show that composition is well-defined on proper equivalence classes: composition at the level of classes *is*, we would say, a group operation. Here it turned out to be essential that proper equivalence is used: the result is false if improper equivalence is admitted, a fact that had halted Legendre's investigations.

Recall that Gauss said that two strict equivalence classes of forms (with the same 'determinant') have the same order if and only if

$$gcd(a, b, c) = gcd(a', b', c'), \text{ and } gcd(a, 2b, c) = gcd(a', 2b', c').$$

The genus of a class is its complete set of characters, and these are determined by the determinant and its prime factors.

Gauss introduced no symbols for composition, class, order, or genus, but it will help if we have some, so I shall write $C(f)$ for the class of a form f, $O(f)$ for the order of a form f and $G(f)$ for its genus, and denote the composition of the forms f and f' by juxtaposition ff'.

Because we are presently taking the definition of the composition of two forms on trust, and because there is no simple expression for the coefficients of ff' in terms of the coefficients of f and f', we can simply state the results concerning the composition of forms that Gauss obtained. In *Disquisitiones Arithmeticae* §245, Gauss proved that if

$$O(f) = O(g) \text{ and } O(f') = O(g') \text{ then } O(ff') = O(gg'),$$

so composition respects order.

Then in §§247 and 248 Gauss proved that $G(f)$ and $G(f')$ determine $G(ff')$, and concluded in §249 that if

$$C(f) = C(g), \ C(f') = C(g'), \ O(f) = O(g), \ O(f') = O(g')$$

and

$$G(f) = G(g), \ G(f') = G(g')$$

then

$$C(ff') = C(gg').$$

A little later, §252, he deduced that for a given value of the 'determinant' and for a given order, every genus contains the same number of classes. He also noted that different orders may contain different numbers of classes.

He then turned his attention to what he had called ambiguous classes. These arise when a form F can imply a form F' both properly and improperly. When this happens, he noted that there is then a form G so that F implies G and G implies F' and G is improperly equivalent to itself. A form (a, b, c) is improperly equivalent to itself if a divides $2b$, $b \neq 0$, and these forms Gauss called *ambiguous*. After a lengthy analysis he concluded (in §260) that "if the number of properly primitive classes of determinant D is r, and the number of properly primitive ambiguous classes of this determinant D is n, then the number of all properly primitive classes of the same determinant that can be produced by the duplication of a similar class will be r/n."[1]

He then proved this theorem (§261): Half of all the assignable characters for a positive nonsquare determinant can correspond to no properly primitive genus and, if the determinant is negative, to no properly primitive positive genus. He observed that

> it does not yet follow from this that half of all the assignable characters actually correspond to properly primitive (positive) genera, but later we shall establish the truth of this profound proposition concerning the most deeply hidden properties of numbers

He then (§262) deduced the theorem of quadratic reciprocity. I omit most of Gauss's proof of the theorem for -1, 2, and -2, and quote only his proof for general odd p (a prime) and q. But since it uses one part of that, I quote it first. Recall that Gauss wrote $a\,Rp$ for what we would write with the Legendre symbol as $\left(\dfrac{a}{p}\right) = 1$, and $a\,Np$ for $\left(\dfrac{a}{p}\right) = -1$. He also used Rp to stand for testing whether a number is a residue modulo p.

---

[1] Recall that a form (a, b, c) is properly primitive if the greatest common divisor of a, $2b$, c is 1.

II. -1 is a residue of any prime number p of the form $4n + 1$. For the character of the form $(-1, 0, p)$, as of all properly primitive forms of determinant p will be Rp and therefore $-1\,Rp$.

. . .

VIII. Any prime number p of the form $4n + 1$ is a nonresidue of any odd number q that is a nonresidue of p. For clearly if p were a residue of q there would be a properly primitive form of determinant p with the character Np.

IX. Similarly if an odd number q is a nonresidue of a prime number p of the form $4n + 3$, $-p$ will be a nonresidue of q; otherwise there would be a properly primitive positive form of determinant $-p$ with character Np.

X. Any prime number p of the form $4n + 1$ is a residue of any prime number q which is a residue of p. If q is also of the form $4n + 1$ this follows immediately from VIII; but if q is of the form $4n + 3$, $-q$ will also be a residue of p (by II) and so $p\,Rq$ by IX.

XI. If a prime number q is a residue of another prime number p of the form $4n + 3$, $-p$ will be a residue of q. For if q is of the form $4n + 1$ it follows from VIII that $p\,Rq$ and so (by II) $-p\,Rq$; this method does not apply when q is of the form $4n + 3$, but it can easily be resolved by considering the determinant $+pq$. For, since of the four characters assignable for this determinant Rp, Rq; Rp, Nq, Np, Rq; Np, Nq, two of them cannot correspond to any genus and since the characters of the forms $(1, 0, -pq)$, $(-1, 0, pq)$ are the first and fourth respectively, the second and third are the characters which correspond to no properly primitive form of determinant pq. And since by hypothesis the character of the form $(q, 0, -p)$ with respect to the number p is Rp, its character with respect to the number q must be Rq and therefore $-p\,Rq$. Q.E.D.

If in propositions VIII and IX we suppose q is a prime number, these propositions joined with X and XI will give us the fundamental theorem of the preceding section.

It may help to re-write Gauss's claims in a modern notation, in which they say

1. VIII:

$$p \equiv 1 \ (4) \text{ and } \left(\frac{q}{p}\right) = -1 \Rightarrow \left(\frac{p}{q}\right) = -1;$$

2. IX:

$$p \equiv -1 \ (4) \text{ and } \left(\frac{q}{p}\right) = -1 \Rightarrow \left(\frac{-p}{q}\right) = -1;$$

3. X:

$$p \equiv 1 \ (4) \text{ and } \left(\frac{q}{p}\right) = 1 \Rightarrow \left(\frac{p}{q}\right) = 1;$$

4. XI:

$$p \equiv -1 \ (4) \text{ and } \left(\frac{q}{p}\right) = 1 \Rightarrow \left(\frac{-p}{q}\right) = -1.$$

To prove them, we argue first that given a quadratic form of discriminant $-D$, then if $-D \equiv \square \ (m)$ there are x and y such that $x^2 + Dy^2 \equiv 0 \ (m)$. We define

$Q_1(x, y) = x^2 + Dy^2$, which has discriminant $-D$, and observe that $Q_1(x, y) \rightarrow km$, so there is another quadratic form of discriminant $-D$ such that $Q_2 \rightarrow m$.

Now, to prove VIII we consider a quadratic form of discriminant $-p$. We look at the fixed relationships, and because $p \not\equiv -1$ (4) we can only ask of the odd numbers represented by the quadratic form if they are congruent to squares mod p. Suppose $\left(\dfrac{p}{q}\right) = 1$. Then there is a quadratic form of discriminant $-p$ that represents q, but this gives a contradiction with $q \not\equiv \square \, (p)$.

The other arguments proceed similarly, as you should check. Then, knowing the theorem for $\left(\dfrac{-1}{p}\right)$, you can deduce the theorem of quadratic reciprocity.

Precisely Half the Assignable Characters Occur

The proof of this "deeply hidden property numbers", as Gauss called it, is the most difficult result in the *Disquisitiones Arithmeticae*. It involved him in a "digression" into the theory of ternary forms, and when the proof was finished he remarked (*DA* §267) that

> these theorems are among the most beautiful in the theory of binary forms because, despite their extreme simplicity, they are so profound that a rigorous demonstration requires the help of many other investigations.

It is not possible to survey Gauss's proof here. Even Dedekind declined to do so, writing that

> It is impossible for us to communicate the proof, which Gauss has based on the theory of ternary quadratic forms

before going on to give Dirichlet's later proof (see Dedekind (1871, 407)). The theorem remains central to many presentations of algebraic number theory, and I believe it is the origin of the idea that the best mathematics has depth (see Gray (2015))—an idea first adopted by German mathematicians after Gauss.

6.3   Smith's Commentary on Gauss's Sixth Proof

In his *Report on the theory of numbers* the English mathematician Henry Smith gave a comprehensive account of number theory as it stood in 1859. In §§20–21 he surveyed Gauss's fourth and sixth proofs of quadratic reciprocity.[2] To follow Smith's account of the sixth proof, in §21, we need to know that in his account of the fourth proof Smith introduced the function

$$1 + r^k + r^{4k} + r^{9k} + \cdots + r^{(n-1)^2 k} = \psi(k, n),$$

---

[2]For a translation of Gauss's sixth proof, see Appendix C.

where $r = \cos\frac{2\pi}{n} + i\sin\frac{2\pi}{n}$ and n is an odd number. The function $\psi(k,n)$ plays an important role in Gauss's fourth proof. Note that $\psi(1,n) = i^{(n-1)^2/4}\sqrt{n}$. Smith now wrote as follows.

21. Gauss's Sixth Demonstration. — This demonstration depends on an investigation of certain properties of the algebraical function

$$\xi_k = \sum_{s=0}^{s=p-2} (-1)^s x^k \gamma^s.$$

in which p is a prime number, γ a primitive root of p, k any number prime to p, and x an absolutely indeterminate symbol. These properties are as follows :–

1. $\xi_k^2 - (-1)^{\frac{1}{2}(p-1)}p$ is divisible by $\frac{1-x^p}{1-x}$,
2. $\xi_k - \left(\frac{k}{p}\right)\xi_1$ is divisible by $1 - x^p$,
3. If $k = q$ be a prime number, $\xi_1^q - \xi_q$ is divisible by q.

From (1) we may infer that $\xi_1^{q-1} - (-1)^{\frac{1}{4}(p-1)(q-1)}p^{\frac{1}{2}(q-1)}$ is divisible by $\frac{1-x^p}{1-x}$; and, by combining this inference with (1) and (2), we may conclude that

$$\xi_1(\xi_1^q - \xi_q) - (-1)^{\frac{1}{2}(p-1)}p\left((-1)^{\frac{1}{4}(p-1)(q-1)}p^{\frac{1}{2}(q-1)} - \left(\frac{q}{p}\right)\right)$$

is also divisible by $\frac{1-x^p}{1-x}$; that is to say,

$$(-1)^{\frac{1}{2}(p-1)}p\left((-1)^{\frac{1}{4}(p-1)(q-1)}p^{\frac{1}{2}(q-1)} - \left(\frac{q}{p}\right)\right)$$

is the remainder left in the division of the function $\xi_1(\xi_1^q - \xi_q)$ by $\frac{1-x^p}{1-x}$. But every term in that function is divisible by q ; the remainder is therefore itself divisible by q. We thus obtain the congruence

$$(-1)^{\frac{1}{4}(p-1)(q-1)}p^{\frac{1}{2}(q-1)} \equiv \left(\frac{q}{p}\right) \bmod q,$$

which involves the equation

$$\left(\frac{p}{q}\right)\left(\frac{q}{p}\right) = (-1)^{\frac{1}{4}(p-1)(q-1)}.$$

Gauss has given a purely algebraical proof of the theorems (1), (2), and (3), on which this demonstration depends. The third is a simple consequence of the arithmetical property of the multinomial coefficient, already referred to in Art. 10 of this Report; to establish the first two, it is sufficient to observe that $\xi_k^2 - (-1)^{\frac{1}{2}(p-1)}p$ and $\xi_k - \left(\frac{k}{p}\right)\xi_1$ vanish, the first, if x be any imaginary root, the second, if x be any root whatever, of the equation $x^p - 1 = 0$. If, for example, in the function ξ_k we put $x = r = \cos\frac{2\pi}{p} + i\sin\frac{2\pi}{p}$, we obtain the function $\psi(k,p)$, which satisfies, as we have seen, the two equations $\psi(k,p)^2 = (-1)^{\frac{1}{2}(p-1)}p$, and $\psi(k,p) = \left(\frac{k}{p}\right)\psi(1,p)$. It is, indeed, simplest to suppose $x = r$ throughout the whole demonstration, which is thus seen to depend wholly on the properties of the same trigonometrical function ψ, which presents itself in the fourth demonstration; only it will be observed that here no necessity arises for the consideration of composite values of n in the function $\psi(k, n)$; nor for the determination of the ambiguous sign in the formula (A). In this specialized form, Gauss's sixth proof has been given by Jacobi (in the 3rd edit. of Legendre's *Théorie des Nombres*, vol. ii. p. 391), Eisenstein (*Crelle's [Journal]*, vol. xxviii. p. 41), and Cauchy (*Bulletin de Férussac*, Sept. 1829, and more fully *Mém. de l'Institut*, vol. xviii. p. 451, note iv. of the Mémoire), quite independently of one another, but apparently without its being at the time perceived by any of those eminent geometers that they were closely following Gauss's method. (See Cauchy's Postscript at the end of the notes to his Mémoire; also a memoir by M. Lebesgue in *Liouville's [Journal]*, vol. xii. p. 457; and a foot-note by Jacobi, *Crelle's [Journal]*, vol. xxx. p. 172, with Eisenstein's reply to it, *Crelle's [Journal]*, vol. xxxv. p. 273.)[3]

Some comments on Smith's commentary are in order. It can be written more simply using the congruence notation and writing out a few intermediate steps, as follows. I shall also write $c_p(x)$ for $\frac{1-x^p}{1-x}$, and α for $(-1)^{\frac{1}{4}(p-1)(q-1)} p^{\frac{1}{2}(q-1)}$.

1. $\xi_k^2 \equiv (-1)^{\frac{1}{2}(p-1)} p \bmod c_p(x)$,
2. $\xi_k - \left(\frac{k}{p}\right) \xi_1$ is divisible by $1 - x^p$, which is divisible by $\frac{1-x^p}{1-x}$, so

$$\xi_k \equiv \left(\frac{k}{p}\right) \xi_1 \bmod c_p(x),$$

3. If $k = q$ be a prime number, $\xi_1^q \equiv \xi_q \bmod q$.

From (1) by raising both sides to the power $\frac{1}{2}(q-1)$, we may infer that $\xi_1^{q-1} \equiv \alpha \bmod c_p(x)$. In (2) we set $k = q$ and obtain $\xi_q \equiv \left(\frac{q}{p}\right) \xi_1 \bmod c_p(x)$. We write the long expression that we want to prove as

$$\xi_1(\xi_1^q - \xi_q) \equiv (-1)^{\frac{1}{2}(p-1)} p \left(\alpha - \left(\frac{q}{p}\right)\right) \bmod c_p(x)$$

and expand it as

$$\xi_1^{q+1} - \xi_1\xi_q \equiv (-1)^{\frac{1}{2}(p-1)} p\alpha + (-1)^{\frac{1}{2}(p-1)} p \left(\frac{q}{p}\right).$$

Now

$$\xi_1^{q+1} \equiv \xi_1^2\xi_1^{q-1} \equiv \xi_1^2\alpha \equiv (-1)^{\frac{1}{2}(p-1)} p\alpha \bmod c_p(x),$$

---

[3]This profusion of footnotes tells us that there was a priority dispute between these men, all of whom were in any case following Gauss.

and

$$-\xi_1\xi_q \equiv -\left(\frac{q}{p}\right)\xi_1^2 \equiv -\left(\frac{q}{p}\right)(-1)^{\frac{1}{2}(p-1)}p \mod c_p(x),$$

so the long expression is established. So indeed

$$\xi_1(\xi_1^q - \xi_q) \equiv (-1)^{\frac{1}{2}(p-1)}p\left(\alpha - \left(\frac{q}{p}\right)\right) \mod c_p(x).$$

But by (3) every term on the right-hand side is congruent to zero modulo q; the left-hand side is therefore congruent to zero modulo q. We thus obtain the congruence

$$\alpha \equiv \left(\frac{q}{p}\right) \mod q.$$

And, of course, $p^{\frac{1}{2}(q-1)} \equiv \left(\frac{p}{q}\right) \mod q$, so we do deduce the equation

$$\left(\frac{p}{q}\right)\left(\frac{q}{p}\right) = (-1)^{\frac{1}{4}(p-1)(q-1)}.$$

6.4 Exercises

[These exercises follow (Scharlau and Opolka 1984, 65–70).] Define what is today called a Gauss sum

$$g_j = \left(\frac{1}{p}\right)\zeta^j + \left(\frac{2}{p}\right)\zeta^{2j} + \cdots + \left(\frac{p-1}{p}\right)\zeta^{p-1},$$

where $\zeta = e^{2\pi i/p}$ is a primitive pth root of unity.

1. Show that $(g_1)^2 = \left(\frac{-1}{p}\right)p$.
2. Show that $(x + y)^q \equiv x^q + y^q \mod q$.
3. Deduce that

$$(g_1)^q \equiv g_q \mod q,$$

   where q is an odd prime.
4. Deduce that

$$\left(\frac{-1}{p}\right)^{\frac{1}{2}(q+1)}p^{\frac{1}{2}(q+1)} \equiv \left(\frac{q}{p}\right)\left(\frac{-1}{p}\right) \mod q,$$

and

$$\left(\frac{-1}{p}\right)^{\frac{1}{2}(q-1)} p^{\frac{1}{2}(q-1)} \equiv \left(\frac{q}{p}\right) \bmod q.$$

5. Use Euler's criterion, which says that for every a relatively prime to q

$$a^{\frac{1}{2}(q-1)} \equiv \left(\frac{a}{q}\right) \bmod q,$$

to deduce that

$$\text{if } p \equiv 1 \bmod 4 \text{ then } \left(\frac{p}{q}\right) = \left(\frac{q}{p}\right),$$

and

$$\text{if } p \equiv 3 \bmod 4 \text{ then } \left(\frac{-1}{q}\right)\left(\frac{p}{q}\right) = \left(\frac{q}{p}\right).$$

6. Deduce the law of quadratic reciprocity.

Chapter 7
Dirichlet's *Lectures* on Quadratic Forms

7.1 Introduction

Gauss's first proof of quadratic reciprocity was given in Section IV of the *Disquisitiones Arithmeticae*. Gauss went on to publish five more, leaving a further two unpublished; he also knew that these different proofs hinted at important connections to as-yet undiscovered parts of mathematics. But insofar as some of these proofs were intended to explore or illustrate these connections they were of varying levels difficulty and not all suitable for beginners. The simplest is the third proof, which was adopted by Peter Gustav Lejeune Dirichlet in lectures that he gave in the 1850s and which formed part of his book *Vorlesungen über Zahlentheorie* (*Lectures on Number Theory*) that did so much to bring number theory to a wide audience of mathematicians.[1] We look at this proof here, and then turn to look at Dirichlet's *Lectures* more broadly.

7.2 Gauss's Third Proof of Quadratic Reciprocity

First, Gauss proved a result nowadays known as Gauss's lemma. Fix a prime number p and consider the set $P = \{1, 2, \ldots p - 1\}$. Let

$$A = \{1, 2, \ldots \tfrac{1}{2}(p - 1)\} \text{ and } B = \{\tfrac{1}{2}(p + 1), \tfrac{1}{2}(p + 3), \ldots, p - 1\},$$

so $P = A \cup B$. Let $k \in P$, and consider $kA = \{k, 2k, \ldots k(p - 1)\}$ taken mod p. Some elements of kA belong to A, some, say μ in number, to B. Gauss first showed

---

[1] For the third proof see Gauss (1808), reprinted in Gauss, *Werke* 2, 1–8, and for an English translation see Smith's *Source Book*, which is accessible via Google books.

© Springer Nature Switzerland AG 2018

J. Gray, *A History of Abstract Algebra*, Springer Undergraduate Mathematics Series, https://doi.org/10.1007/978-3-319-94773-0_7

that $k^{(p-1)/2} \equiv \pm 1 \bmod p$ according as μ is even or odd, and so k is or is not a quadratic residue of p according as μ is even or odd.

The theorem then followed after some fairly straight-forward arguments. On the left I look at a simple particular case, on the right at the general case.

We assume throughout that

$$p = 2p' + 1, \quad q = 2q' + 1.$$

First Part of the Proof

$p = 11, P = \{1, \ldots, 10\}$

$p' = 5, q = 7$

$A = \{1, \ldots, 5\}, \quad B = \{6, \ldots, 10\}$

$7A = \{7, 3, 10, 6, 2\}$

$\quad = \{2, 3\} \cup \{6, 7, 10\}$

$A_7 = \{2, 3\}, \quad B_7 = \{6, 7, 10\}$

$card\{6, 7, 10\} = 3$

$\bar{B}_7 = \{11 - 6, 11 - 7, 11 - 10\} =$

$\prod a_j \bar{b}_k = 2 \cdot 3 \cdot (11 - 6) \cdot (11 - 7)(11 - 10) = 5! \Rightarrow$

$\prod a_j \bar{b}_k = 2 \cdot 3 \cdot (11 - 6) \cdot (11 - 7)(11 - 10)$

$\equiv (-1)^3 2 \cdot 3 \cdot 6 \cdot 7 \cdot 10 \pmod{11}$

But $\prod 2 \cdot 3 \cdot 6 \cdot 7 \cdot 10 \equiv 5! 7^5 \pmod{11}$

So $7^5 \equiv (-1)^3 \equiv \left(\frac{7}{11}\right) \bmod 11$

$P = \{1, \ldots, p - 1\}$

q

$A = \{1, \ldots, p'\}, \quad B = \{p' + 1, \ldots, p - 1\}$

$qA = \{q, 2q, \ldots, p'q\}$

$A_q \cup B_q$, where $A_q = qA \cap \{1, \ldots p'\}$

and $B_q = qA \cap \{p' + 1, \ldots p - 1\}$

$card B_q = \mu$

Consider $\bar{B}_q = \{p - b_j | b_j \in B_q\}$

$\prod a_j \bar{b}_k = p'! \Rightarrow$

$\prod a_j \bar{b}_k \equiv \prod a_j b_k (-1)^\mu \pmod{p}$

But $\prod a_j b_k \equiv p'! q^{p'} \pmod{p}$

So $q^{p'} \equiv \left(\frac{q}{p}\right) \pmod{p}$

Check: the squares mod 11 are $\{1, 4, 9, 5, 3\}$, so 7 is not a square mod 11.

Second Part of the Proof

Write $[x]$ for the integer part of $\frac{x}{11}$

We have $7k = p[7k] + r_k, 1 \le k \le 5$

The remainders lie between 0 and p

Partition them into the a_j, those below $\frac{11}{2}$

and the b_k, those above;

Let the sum of the former be A and of the latter B

Let $M = \sum_k [7k]$, then

$A = 5, B = 23, M = 7$

and $11M + A + B = 105 = \frac{120}{8} \cdot 7$

Write $[x]$ for the integer part of $\frac{x}{p}$

We have $qk = p[qk] + r_k, 1 \le k \le p'$

respectively those below and above $\frac{p}{2}$

Let $M = \sum_k [kq]$, then

$pM + A + B = \frac{p^2 - 1}{8} q$

But

$$\{a_1, \ldots, a_\lambda, p - b_1, \ldots, p - b_\mu\} = \{1, 2, \ldots, p'\},$$

so their sum is

$$\frac{p^2 - 1}{8} = \mu p + A - B,$$

and so

$$\frac{p^2 - 1}{8}(q - 1) = (M - \mu)p + 2B,$$

and we only need to find if μ is even or odd. We have

$$\mu \equiv M + \frac{p^2 - 1}{8}(q - 1) \bmod 2.$$

Now assume that q is odd, positive and less than p, and so that $\mu \equiv M \bmod 2$.

Look at the sequence $\lfloor 7k \rfloor$, $1 \le k \le 5$
It goes 0, 1, 1, 2, 3

Look at the sequence $\lfloor kq \rfloor$, $1 \le k \le p'$
It goes $0, \ldots, q'$
because $\frac{p'q}{p} = \frac{q-1}{2} + \frac{p-q}{2p}$.

Look at where each sequence steps up. How many steps of q are just less than an integer t and how many steps of q just exceed t? When the step occurs we have

$$\frac{sq}{p} < t < \frac{(s+1)q}{p},$$

so

$$s < \frac{tp}{q} < s + 1,$$

so the step occurs after a run of $s = \left\lfloor \frac{tp}{q} \right\rfloor$ terms that have the value $t - 1$. The number of terms taking the top value, q', is $p' - \left\lfloor \frac{q'p}{q} \right\rfloor$. We compute M by multiplying the number of terms taking a given value by that value, and we obtain

$$-\left\lfloor \frac{p}{q} \right\rfloor - \left\lfloor \frac{2p}{q} \right\rfloor - \cdots - \left\lfloor \frac{q'p}{q} \right\rfloor + q'\frac{p-1}{2} = p'q'.$$

Set

$$N = \left\lfloor \frac{p}{q} \right\rfloor + \left\lfloor \frac{2p}{q} \right\rfloor + \ldots + \left\lfloor \frac{q'p}{q} \right\rfloor$$

and we obtain

$$M + N = \frac{p-1}{2} \cdot \frac{q-1}{2},$$

which, said Dirichlet, is obviously valid for any two odd positive primes because it is symmetric in p and q and one must be less than the other. To finish off, note that if p is a positive odd prime and q is odd and not divisible by p then

$$\left(\frac{q}{p}\right) = (-1)^M.$$

If also q is a positive prime, then

$$\left(\frac{p}{q}\right) = (-1)^N.$$

So

$$\left(\frac{p}{q}\right)\left(\frac{q}{p}\right) = (-1)^{M+N} = (-1)^{p'q'},$$

which is *quadratic reciprocity*.

Weil, in his little book *Number Theory for Beginners* (1979, 55), proved quadratic reciprocity in more or less this way. His version of the Gauss lemma was this: Let $p = 2n+1$ and call a Gaussian set modulo p a set of coset representatives $u_1, \ldots y_n$ for the subgroup $H = \{\pm 1\}$ in $G = \mathbb{Z}/p\mathbb{Z}$, so any number is congruent to a unique $\pm u_j$. In G fix a non=zero element a and define e_j by $au_j = e_j u_k$. Then a is or is not a quadratic residue mod p according as the product $e_1 \ldots e_n = 1$ or -1.

The quadratic character of 2 mod p follows from taking $a = 2$ in the Gauss lemma. The quadratic reciprocity theorem follows in the same way, with $a = q$.

7.3   Dirichlet's Theory of Quadratic Forms

Peter Gustav Lejeune-Dirichlet was not just the most important German mathematician in the 1830s, 40s, and 50s, he was one of the most important mathematicians of the time. He not only made Gauss's number theory more accessible, he significantly deepened it by bringing in analytic methods that have their origin in some observations of Euler's. His results, and his insights into them, have created a subject that is more active than ever today, almost two centuries later. He also brought rigorous epsilon–delta analysis to Germany and by being much more careful than Cauchy helped place it securely at the heart of analysis, and towards the end of his life he was a powerful influence on Bernhard Riemann.

His finest achievement was to give an analytic formula for the class number of binary quadratic forms, which unfortunately we cannot describe here. It opened up the field of Euler products and the zeta function, and it belongs intimately with his proof that every arithmetic progression $\{a + kb \mid (a, b) = 1, k = 1, 2, \ldots\}$ contains infinitely many primes.

Dirichlet was born in 1805 in Duren, which is now part of Germany but was then a French territory under the rule of Napoleon, and to study mathematics in 1822 he took himself to Paris, which was then by far the most important place for the subject. There he became a tutor to the children of General Foy, and this introduced him to the elite circles of Parisian society as well as members of the Académie des Sciences. In this way he came to know Joseph Fourier, who was the perpetual secretary of the Académie, and his (1829) is the first major paper on the question of whether a Fourier series converges to the function it represents. But he had already identified number theory as his deepest love, and he was able to show that Fermat's Last Theorem for $n = 5$ is false: there are no non-zero integer solutions to $x^5 + y^5 = z^5$. Or rather, he was able to get close to establishing this conclusion, and Legendre completed it, sad to say in a manner that seemed to give Legendre more credit and to diminish Dirichlet's contribution.

In 1823 General Foy died, but by then Dirichlet had met and impressed Alexander von Humboldt, who was already an influential figure in Prussian society, and when Dirichlet returned to Germany in 1827 Humboldt was helpful in obtaining a professorship for him at the newly-founded University of Berlin, which was conferred in 1831. On the way home, Dirichlet visited Gauss in Göttingen. There seem to be no record of what they discussed, but Dirichlet wrote to his mother that Gauss had received him kindly and that he had formed a better impression of Gauss than he had expected.[2] In 1855, Dirichlet moved to Göttingen as Gauss's successor.

In his introduction to his first edition of Dirichlet's *Lectures*, Dedekind wrote that Dirichlet had originally given these lectures in Göttingen between 1855/56, and that he had heard the most important of them himself when a student between 1855 and 1858. Like any lecturer, Dirichlet had to decide what, within his chosen topic, he was going to include, where he was going to leave his listeners and readers, and how he was going to get them there. His preferred mode as a writer was to set up everything that one was going to need first, and then to use this information as the need arose. So he supplied a chapter on divisors of numbers: greatest common divisors, prime and composite numbers, the Euler ϕ function—the number $\phi(m)$ is the number of positive integers less than and relatively prime to m, so $\phi(5) = 4$—and related topics. Then came a chapter on modular arithmetic. Then came a chapter on quadratic reciprocity, with full proofs.

Chapter 4 was on binary quadratic forms, which, following Gauss, he wrote

$$ax^2 + 2bxy + cy^2,$$

and abbreviated to (a, b, c). To study, for example, $4x^2 + 3xy + y^2$ he studied its double, $8x^2 + 6xy + 2y^2$ and derived the properties of the given form from those of its double.

---

[2]See Kummer's obituary of Dirichlet, an extract of which is in English in (Scharlau and Opolka 1984, 144–147).

He noted that

$$ax^2 + 2bxy + cy^2 = \frac{1}{a}\left((ax+by)^2 - (b^2-ac)y^2\right),$$

so $b^2 - ac$ cannot be a perfect square else the form would be a product of linear factors. He called $b^2 - ac$ the *determinant* of the form, and denoted it D—be careful, it is the *negative* of the determinant of the matrix $\begin{pmatrix} a & b \\ b & c \end{pmatrix}$, so $D = -(ac - b^2)$. Henceforth D will not be a square.

He wrote out explicitly what happens to the form $ax^2 + 2bxy + cy^2$ under the linear transformation

$$x = \alpha x' + \beta y'$$

$$y = \gamma x' + \delta y',$$

and noted that each number representable by the second form can be represented by the first form. Moreover, if the transformed form is $a'x^2 + 2b'xy + c'y^2$ then the relationship between the determinants is

$$D' = b'^2 - a'c' = (\alpha\delta - \beta\gamma)^2 D.$$

Because the multiplying factor is a square, D and D' have the same sign. If $(\alpha\delta - \beta\gamma)$ is positive the transformation is said to be proper, otherwise improper. Transformations can be followed by others, and the product of an even number of improper transformations is proper.

Two equivalent forms represent the same numbers. A sufficient condition is that one can be transformed into the other by a transformation of the same determinant, for both D/D' and D'/D must be integer squares. So D and D' must be ± 1. If $D = 1$ the equivalence is proper, otherwise improper.

Exercises

1. Show that if (a, b, c) and (a', b', c') have the same determinant, and $a = a'$, and $b \equiv b' \bmod a$ then the two forms are properly equivalent, and that a suitable transformation is $\begin{pmatrix} 1 & \beta \\ 0 & 1 \end{pmatrix}$.
2. Show that (a, b, c) and $(c, -b, a)$ are properly equivalent.

Forms can be improperly equivalent to themselves. This happens when there are two forms both properly and improperly equivalent to each other; each is then improperly equivalent to itself. The improper transformation establishing this equivalence is then necessarily of the form $\begin{pmatrix} \alpha & \beta \\ \gamma & -\alpha \end{pmatrix}$. When moreover $\gamma = 0$ it follows that $\alpha = \pm 1$ and $\alpha\beta = 2b$, which Dirichlet, quoting Gauss, called a *forma anceps*, usually if poorly translated as an ambiguous or two-sided form. Dirichlet proved that a form improperly equivalent to itself is equivalent to a two-sided form.

Gauss had shown the importance of concentrating on proper equivalence, and Dirichlet followed him. In the study of (proper) equivalence classes of forms he singled out two central questions:

1. How to decide when two forms of the same determinant are equivalent;
2. How to find all the transformations that send one of two equivalent forms to the other (discussed in the *DA*, §§178, 179, for forms of negative determinant).

The first was discussed in the *DA*, §173, for forms of negative determinant, when the answer is found by reducing them to canonical form. Gauss also took note of whether the equivalence, if it exists, is proper, improper, or both.

Recognising that these questions might seem too dry, he motivated them by showing how the theory of the representation of numbers by quadratic forms reduced to them (which Gauss had discussed in the *DA*, §180, for forms of negative determinant). Consider the equation

$$ax^2 + 2bxy + cy^2 = m.$$

Dirichlet's method is a simplification of Gauss's. He noticed that it is enough to look for numbers x and y which are relatively prime (these are called proper representations). Therefore there are integers η and ξ such that $x\eta - y\xi = 1$. The transformation $\begin{pmatrix} x & \xi \\ y & \eta \end{pmatrix}$ sends (a, b, c) to the equivalent form (m, n, l)—check the first coefficient! Because $n^2 - ml = D$ and D is not a square, we have $m \neq 0$ and so we have the important result that D is a quadratic residue of m. Furthermore, n is a root of the congruence $z^2 \equiv D \bmod m$. It is, of course, possible to find other integers η' and ξ' such that $x\eta' - y\xi' = 1$, but Dirichlet showed that the resulting n' satisfies the congruence $n' \equiv n \bmod m$. Dirichlet therefore concluded that the representation problem reduces to his two central questions, because we may follow this procedure:

1. is D is a quadratic residue of m? If not, m is not representable. If it is a residue
2. find all the roots, n, of $z^2 \equiv D \bmod m$, and with each n look at (m, n, l). If m is representable then (a, b, c) and (m, n, l) are equivalent, so one must be able to answer the question: are they equivalent? If they are not, m is not representable. If they are equivalent
3. find all the transformations sending one to the other (the second of the central problems). Each such transformation gives an x and a y such that $ax^2 + 2bxy + cy^2 = m$.

Dirichlet noted that it is not easy to find all transformations mapping one form to another, but it is enough to find one, for if t and t' are two such transformations then $t't^{-1}$ maps the first form to itself.[3] Now he let σ be the greatest common divisor of $a, 2b, c$, so all representable numbers are necessarily divisible by σ. He called σ the divisor of the form.

Exercises

1. $2b/\sigma$ is even if and only if σ is the greatest common divisor of $a, 2b, c$.
2. $2b/\sigma$ is odd if and only if $\sigma/2$ is the greatest common divisor of $a, 2b, c$.

In the first case, the simplest form with determinant D is $(\sigma, 0, \frac{-D}{\sigma})$. In the second case, the simplest form with determinant D is $(\sigma, \frac{1}{2}\sigma, \frac{\sigma^2-4D}{4\sigma})$. The simplest form $(1, 0, -D)$ is called the principal form with determinant D.

Dirichlet denoted the greatest common divisor of the three numbers a, b, c (careful, not $2b$!) by τ. In the first case $\tau = \sigma$; in the second case $\tau = \sigma/2$. If $\tau = 1$ the form is said to be primitive, and if $\sigma = 1$ properly primitive or of the first kind. If $\tau = 1$ and $\sigma = 2$ it is improperly primitive. If $\tau > 1$ the form is said to be derived from the corresponding primitive form. Since equivalent forms have the same divisor, which may therefore be said to be the divisor of a class of forms, and following Gauss Dirichlet called the collection of all forms with the same determinant and divisor an *order*.

At this point I shall skip over a topic in Dirichlet's lectures, which comes straight from Gauss's *Disquisitiones Arithmeticae*. In §62 Dirichlet discussed the way of finding a transformation of a form (with determinant D and divisor σ) to itself reduces to solving the equation $t^2 - Du^2 = \sigma^2$. He noted that this is not too hard when D is negative because the equation has only finitely many solutions, but much harder when D is positive because the equation now has infinitely many solutions, and so he decided to do the easier case first. I skip the argument in §62 and now resume his account.

He began his account of quadratic forms with negative determinants with the reduction theory of Lagrange, as refined by Gauss to a theory of proper equivalence. A form (a, b, c) is said to be *reduced* when

$$c \geq a \geq 2|b|.$$

Every form is equivalent to a reduced form, so the question of interest is: when are two reduced forms equivalent? Two pages of elementary algebra established that this only happens when either the forms are $(a, a/2, c)$ and $(a, -a/2, c)$ and the

---

[3]The bull at a gate method for finding a matrix P of determinant 1 such that $P^T A P = B$ or $P^T A = B P^{-1}$, given symmetric matrices A and B, does not lead to a usable system of equations.

transformation is $\begin{pmatrix} 1 & -1 \\ 0 & 1 \end{pmatrix}$, or (a, b, a) and $(a, -b, a)$ and the transformation is $\begin{pmatrix} 0 & -1 \\ 1 & 0 \end{pmatrix}$. The proof made essential use of the fact that the determinant of the form is negative.

It follows at once that the number of equivalence classes is finite for each negative determinant, and Dirichlet drew the classical consequences:

1. every positive prime of the form $4n + 1$ is expressible as a sum of two squares in essentially a unique way;
2. every positive prime of the form $8n + 1$ or $8h + 3$ is expressible as a sum of a square and the double of a square in essentially a unique way;
3. every positive prime of the form $3n + 1$ is expressible as a sum of a square and the triple of a square in essentially a unique way.

Then came the first of the more troubling cases:

1. every prime of the form $20n + 1$ or $20n + 9$ is representable in four ways by the form $(1, 0, 5)$;
2. every prime of the form $20n + 3$ or $20n + 7$ is representable in four ways by the form $(2, 1, 3)$.

Dirichlet now turned to the harder case of forms with positive determinants and gave a complete solution there too. But, as he noted, the methods are very different and I shall not follow him there (it leads through Pell's equation and continued fractions). But it is important to note that this problem was solved as well. In particular, although the number of reduced forms is finite, reduced forms can be equivalent in complicated ways, and the divide presaged by $t^2 - Du^2 = \sigma^2$ returns to the story.

Chapter 5 of the lectures was devoted to a fresh account of one of Dirichlet's greatest discoveries: the existence of an analytical formula for the class number (the number of equivalence classes of forms of a given determinant).[4] It would take us too far into difficult mathematics to cover this—see the account in Scharlau and Opolka (1984, pp. 109–143) which the authors modestly call a sketch—but we should note several historically important facts:

1. there is a formula;
2. it was found after a lot of difficult, delicate analysis;
3. its justification is close to the one Dirichlet also gave for the theorem that there are infinitely many primes in arithmetic progressions $\{an + b : \gcd(a, b) = 1\}$;
4. the formula does not answer the question: How many determinants are there with a given class number?

---

[4]First published in Dirichlet (1839/1840).

7.4   Taking Stock

In the last four chapters we have seen the origins of modern number theory. Gauss's *Disquisitiones Arithmeticae* gave a coherent structure to what had been rather more scattered results, with an acknowledged starting point, a theory, and some very difficult, deep ideas by the end. It helped establish number theory, certainly in Germany, as a highly prestigious branch of mathematics. It is worth reflecting on this: how is it that some topics in mathematics become important?

We have already seen that many ideas which Gauss presented were already in the literature. But note that on some occasions he found that their supposed 'proofs' were flawed, in the case of quadratic reciprocity badly so. His successful work helped establish that there *is* a theory. By giving it an elementary starting point Gauss gave number theory something that the advanced work of Lagrange had lacked, and by taking the subject well beyond where his illustrious predecessor had left it Gauss gave the subject of quadratic forms a new depth and a greater insight. In fair part, his success was due to his theory of the composition of forms.

This brings us to a key point: what is so good about composition of forms, especially when it is so difficult to carry out? We can observe, as historians, that even the best of Gauss's contemporaries and successors skirted round this point. But somehow it underpins the theory of characters (in a way that Lagrange's ideas had not) and character theory is a major step in answering the basic question: which quadratic forms $ax^2 + bxy + cy^2$ of a given discriminant represent a given number N? Gauss did not give a complete answer, but the key to the importance of the *Disquisitiones Arithmeticae* is there. Had it answered every question, including the basic one, it would have killed the subject, but its depth and obscurity acted as a lure. It said: First master this book, then extend it. So two good questions to answer would be to determine what Gauss's character theory accomplished, and how the composition theory of forms enters into, or enables, character theory.

It has not been possible in this course to make a comparison of Legendre's book on number theory and the one by Gauss, and to my knowledge no-one has ever properly done so. But we have seen one grievous flaw in Legendre's work (quadratic reciprocity) and one success that perhaps Gauss should have been more positive about: a near discovery of composition of forms. Legendre's book sold well in French for decades, and not just because his French was easier to read than Gauss's Latin: it contained real insights. One should not forget it in any account of the rise of number theory.

But there was more to the *Disquisitiones Arithmeticae* than quadratic forms. The theory of cyclotomy was almost wholly new, and it made great sense of many earlier results. We can profitably compare Gauss's account of what radicals enter the resolution of a cyclotomic equation with Lagrange's investigation of what general polynomial equations are solvable by radicals on the one hand, and, in due course, with Galois theory. In each case the aim of the mathematician was to explain how some calculations work.

The historian has a major clue (and a mathematical friend) in the work of Dirichlet. It is clear that if a mathematician of his calibre felt it right to write an elementary book explaining Gauss's work then he thought it was both difficult and valuable. His *Lectures* do dispel the mysteries of the *Disquisitiones Arithmeticae* (but look for what he did not write about too) and if we look at his account we can begin to get a clearer sense not only of the subject Gauss re-created but of what was important about it.

Appendix: Cubic and Quartic Reciprocity

Cubic and quartic reciprocity, as their names suggest, are the generalisations to 3rd and 4th powers respectively, of quadratic reciprocity. On the face of it, there is no reason to believe that there will be such a generalisation. For example, the theory of quadratic forms does not generalise to cubic forms, so it was a matter of discovery and numerical experiment to see if the theory generalises. In fact, it does, and it was to prove very interesting and challenging, but we cannot pursue it here.

However, we should note that the way Gauss found to generalise quadratic reciprocity to 4th powers in 1831 made essential use of the introduction of complex numbers, specifically integers of the form $m + in$, $m, n \in \mathbb{Z}$. The law, he discovered, is most simply expressed and proved in terms of primes in this ring (nowadays called the ring of Gaussian integers), and in this ring certain ordinary primes cease to be prime. For example $5 = (2 + 1)(2 - i)$, and generally primes of the form $4n + 1$, being sums of two squares, say $x^2 + y^2$, factorise as $(x + iy)(x - iy)$.

Chapter 8
Is the Quintic Unsolvable?

8.1 Introduction

We now turn to what today is regarded as a different branch of *algebra*, the solution of polynomial equations, although, as we have seen, Gauss's work on the 'higher arithmetic' was not automatically regarded as being part of algebra (and some of its deepest connections were made with analysis). We shall find that polynomial algebra also evolved in the direction of deepening conceptual insight, so here too we witness one of the origins of the transformation from school algebra to modern algebra.

Explicit solutions to quadratic, cubic and quartic equations were well known by the late eighteenth century, but there progress had stopped; there was no formula known for solving equations of the fifth degree. In 1770, Lagrange attempted to understand how the methods used to solve equations of low degree could either be generalised to equations of the 5th degree or shown not to help.

8.2 Solution of Equations of Low Degree

The Solution of the Cubic Equation

The solution of the cubic equations by Italian mathematicians at the start of the sixteenth century has been described in many histories of mathematics.[1] Here we have space only for the underlying mathematics. Consider the similarity in form

---

[1] See (Barrow-Green, Gray, and Wilson, 2018) for a recent account. In the early years, and perhaps until Gauss, it was assumed that the equations considered have real coefficients.

© Springer Nature Switzerland AG 2018
J. Gray, *A History of Abstract Algebra*, Springer Undergraduate Mathematics Series,
https://doi.org/10.1007/978-3-319-94773-0_8

between these two equations:

$$x^3 + px + q = 0$$

$$(u + v)^3 - 3uv(u + v) - (u^3 + v^3) = 0.$$

The first of these is the cubic equation whose solution is sought; the second is an algebraic identity, true for all values of u and v. So, if it is possible to arrange matters so that p is $-3uv$, and q is $-(u^3 + v^3)$, then the formal similarity between the equations will ensure that x is $u + v$. So it remains only to discover those values of u and v that yield $3uv = -p$ and $u^3 + v^3 = -q$. This can be done by noticing that u^3 and v^3 are the roots of the quadratic equation $t^2 + qt - p^3/27 = 0$ and so

$$u = \sqrt[3]{(-q/2) + \sqrt{(q/2)^2 + (p/3)^3}},$$

$$v = \sqrt[3]{(-q/2) - \sqrt{(q/2)^2 + (p/3)^3}}.$$

The solution to the original cubic is then $x = u + v$.

Or rather, it would be if we knew which cube root to take. Let us take u and v each to be one of the above values, and agree that ω is a primitive cube root of unity ($\omega \neq 1, \omega^3 = 1$).

Then we have these candidates for roots to consider:

$$\omega^k u + \omega^m v, \quad k, m = 0, 1, 2.$$

A little algebra shows that we must have $k + m \equiv 0 \pmod 3$, so the solutions of the cubic are

$$u + v, \quad \omega u + \omega^2 v, \quad \omega^2 u + \omega v.$$

In terms of the value of the formula, there are several different types of cubic. The occurrence of complex numbers is counter-intuitive. Consider, as Cardano did in Chapter XIII of his *Ars Magna* (1543), the book in which the theory of cubic equations was written down for the first time, this cubic:

$$x^3 - 8x + 3 = 0.$$

One root is $x = -3$, and the other two are $x = \frac{1}{2}(3 \pm \sqrt{5})$, so they are all real. But the formula seems to involve square roots of complex numbers, because in this case $p = -7$ and $q = 6$ so

$$\left(\frac{q}{2}\right)^2 + \left(\frac{p}{3}\right)^3 = 9 + \left(\frac{-7}{3}\right)^3 = -\frac{100}{27}$$

is negative. But for cubic equations with only one real root but two complex roots the formula gives the real root without involving complex numbers. This strange behaviour might be an annoying feature of the formula, and one might hope that another formula could be found that did not have this defect, but that was not to be: the solution to cubics with three real roots leads through complex numbers. This is the phenomenon of the *irreducible case*, and was not to be explained until the nineteenth century (see Sect. 9.4 below).

What is more, even if the roots are integers the formula may disguise this magnificently. Consider this famous example given by Cardano (1543, Ch. XI)

$$x^3 + 6x - 20 = 0.$$

The only real root of this equation is $x = 2$, but the formula gives the unpromising

$$x = \sqrt[3]{10 + \sqrt{108}} + \sqrt[3]{10 - \sqrt{108}}.$$

There is no systematic way of reducing this expression to 2, but noting that $\sqrt{108} = 6\sqrt{3}$ one can make the guess that the cube roots might be of the form $a + b\sqrt{3}$, and indeed $(1 + \sqrt{3})^3 = 10 + 6\sqrt{3}$.

You can check (in the next Exercise) that if the roots of the cubic equation

$$x^3 + px + q = 0$$

are x_1, x_2, and x_3, then (after choosing a sign for the square root)

$$\sqrt{\left(\frac{q}{2}\right)^2 + \left(\frac{p}{3}\right)^3} = \frac{(x_1 - x_2)(x_2 - x_3)(x_3 - x_1)}{6\sqrt{-3}}.$$

So the radical is a polynomial in the roots, and one that takes precisely two values as the roots are permuted. Moreover, the six possible values of

$$\sqrt[3]{-\frac{q}{2} - \sqrt{\left(\frac{q}{2}\right)^2 + \left(\frac{p}{3}\right)^3}}$$

are the six possible values of

$$\frac{x_1 + \omega x_2 + \omega^2 x_3}{3}.$$

Here too a radical has been expressed as a polynomial in the roots.

Exercises

1. Check the details of the above claim by using the fact that $x_1 + x_2 + x_3 = 0$ to express $p = x_1x_2 + x_2x_3 + x_3x_1$, $q = -x_1x_2x_3$, and $r = (x_1 - x_2)(x_2 - x_3)(x_3 - x_1)$ in terms of x_1 and x_2. Now show that

$$\left(\frac{q}{2}\right)^2 + \left(\frac{p}{3}\right)^3 = \frac{-1}{108}r^2.$$

2. Show that by a linear change of variable any cubic equation $ax^3 + bx^2 + cx + d = 0$, $a \neq 0$, can be written in the form

$$x^3 + px + q = 0.$$

3. Make the substitution $x = y + \frac{\alpha}{y}$ and show that the resulting equation can be written as a quadratic equation in y^3 for a suitable value of α, which you should state. Solve this equation for y^3 and hence find y and therefore x. Note that there are always three distinct cube roots of a non-zero number and two distinct square roots, so you have six expressions for the roots of the original cubic equation. Can you sort out which expressions are roots and which are not?

The Solution of the Quartic Equation

Cardano acquired a servant or pupil in the late 1530s called Lodovico Ferrari, who seems to have been so violent at times that Cardano was frightened of him, but who turned out to be a brilliant mathematician who always acknowledged a lifelong debt to Cardano.[2] It was Ferrari who discovered how to solve the quartic (4th power) equation; his solution was published in Chapter **XXXIX** of the *Ars Magna*. The equation can always be put in the form

$$x^4 + px^2 + qx + r = 0.$$

Ferrari had the idea of putting x^4 on one side and all the others on the other side and then trying to complete the square for the terms in x^2. So he wrote

$$x^4 = -px^2 - qx - r, \qquad (8.1)$$

---

[2] See Cardano, 'Vita Lodovici Ferarii Bonoinesis', in *Opera Omnia* 9, 568–569, and Morley, H. *Life of Cardan* 1, 266.

and added $2zx^2 + z^2$ to each side. This gave him a square for the terms on the left hand side. So he wrote

$$x^4 + 2zx^2 + z^2 = (x^2 + z)^2 = (2z - p)x^2 - qx + (z^2 - r),$$

and looked for the value of z that makes the right-hand side a perfect square. This is found, by the $b^2 - 4ac$ test, to be a root of

$$q^2 = 4(2z - p)(z^2 - r), \qquad (8.2)$$

which expands into

$$8z^3 - 4pz^2 - 8rz + 8pr - q^2 = 0.$$

Alternatively, we write

$$(x^2 + z)^2 = (\alpha x - \beta)^2 \Leftrightarrow$$

$$\alpha^2 = 2z - p, \ 2\alpha\beta = q, \ \beta^2 = z^2 - r,$$

so

$$q^2 = 4(2z - p)(z^2 - r).$$

Also

$$\alpha = \sqrt{2z - p}, \ \beta = \sqrt{z^2 - r},$$

so Eq. (8.1) remains irreducible on adjoining z, but not on adjoining $\sqrt{2z - p}$ or $\sqrt{z^2 - r}$ (notice that if we adjoin one of these square roots we adjoin the other, by Eq. (8.2)).

The Eq. (8.2) is a cubic in z and so, by Cardano's earlier work, solvable! Taking the (any?) value of z that is a solution of this equation, one solves the corresponding equation for x^2 and takes square roots to get the required values of x.

There's an interesting little count to do here. Solving a cubic equation involves taking a square root and a cube root (and $2 \times 3 = 6 = 3!$) and now we know that solving a quartic equation involves taking a square root and a cube root and then two square roots (and $2 \times 2 \times 2 \times 3 = 24 = 4!$).

Polynomial Equations and Partially Symmetric Expressions in Their Roots

The first key insight into the process of solving an equation by radicals was to observe that given a polynomial equation of degree n with n distinct roots, there might be expressions in the roots that are somewhat symmetric. That is to say that while the group of symmetries of the n roots is the full permutation group S_n, there might be expressions in the roots that are invariant under some but not all these permutations. The sum of the roots and the product of the roots do not help, because they are invariant under all permutations of the roots and are, of course, the coefficients respectively and up to sign of the x^{n-1} term and the constant term of the given equation. However, there might be expressions in the roots that take more than 1 value under the action of S_n but, say, k less than $n!$ values. These expressions should satisfy polynomial equations of degree k, and if they do and if $k < n$ then these equations should contribute to the attempt to solve the given equation by radicals. It is the purpose of the next few paragraphs to spell out the details of this insight.

Consider the equation

$$x^n - a_1 x^{n-1} + \cdots + (-1)^{n-1} a_{n-1} x + (-1)^n a_n = 0. \tag{8.3}$$

Let its roots be $u_1, u_2, \ldots u_n$, which we shall assume are distinct. Then, as is well known,

$$\sum u_j = a_1, \quad \sum_{j<k} u_j u_k = a_2, \ldots, u_1 u_2, \ldots u_n = a_n.$$

The crucial fact is that any algebraic expression in the roots which is fully symmetric (that is, invariant under every permutation of the roots) is a rational expression in the coefficients $a_1, a_2, \ldots, a_n$ of Eq. (8.3). This implies that any partially symmetric expression in the roots satisfies a polynomial equation of some degree whose coefficients are rational expressions in the coefficients $a_1, a_2, \ldots, a_n$ of the original Eq. (8.3). More precisely, suppose α is an algebraic expression in the roots of the Eq. (8.3) that takes only k values as the roots are permuted, say $\alpha_1 = \alpha, \alpha_2, \ldots, \alpha_k$. Let

$$\beta_1 = \sum \alpha_j, \beta_2 = \sum_{j<k} \alpha_j \alpha_k, \ldots, \beta_k = \alpha_1 \alpha_2 \ldots \alpha_k.$$

Then the αs satisfy the polynomial equation

$$x^k - \beta_1 x^{k-1} + \cdots + (-1)^{k-1} \beta_{k-1} x + (-1)^k \beta_k = 0, \tag{8.4}$$

and each coefficient β_j is fully symmetric and so rational in the coefficients of the original Eq. (8.3).

It follows that partially symmetric expressions in the roots taking k values, polynomial equations of degree k obtained from the original equation, and subgroups of the full permutation group S_n with an orbit of order k and therefore of index $n!/k$ in S_n are intimately related.

8.3 Lagrange (1770)

Lagrange was the first to suspect that the quintic might not be solvable by radicals. His analysis hinged on the construction of intermediate expressions which he called resolvents; we shall need to consider their definition and their worth.

It may help to start with an example that is almost too elementary, the quadratic equation

$$aX^2 + bX + c = 0,$$

whose roots are, of course,

$$\frac{-b \pm \sqrt{b^2 - 4ac}}{2a}.$$

If we denote the roots by x and y then the only quantity in the solution formula that is not a rational expression in the coefficients is the square root, which is this expression in terms of the roots of the equation:

$$\sqrt{b^2 - 4ac} = a(x - y).$$

This is not symmetric in the roots, but it is rational in the roots.

The merit in thinking this way does not become clear until we turn to cubic equations. We denote the roots by x, y, and z, and the solution formula involves a square root, as we have just seen:

$$\sqrt{\left(\frac{q}{2}\right)^2 + \left(\frac{p}{3}\right)^3} = \frac{(x - y)(y - z)(z - x)}{6\sqrt{-3}}$$

and, as the expression on the right shows, it too is rational in the roots of the given cubic equation. Indeed, that expression takes only two values as the roots are permuted, so, by the remark about partially symmetric expressions, it must satisfy a quadratic equation, which is why it can be expressed as a square root involving the coefficients of the equation. Lagrange's insight was that this is crucial to the existence of solutions of the cubic and suggestive of an approach that might work for equations of higher degree (or show that no such approach exists).

To understand Lagrange's approach, set $s = x + \alpha y + \alpha^2 z$ and $t = x + \alpha z + \alpha^2 y$, where α is a primitive cube root of 1: $\alpha^3 = 1$. The expressions s and t are what are called *Lagrange resolvents* (Lagrange called them *réduite* or reduced) and each of them take 6 values as x, y, z are permuted.[3] However,

$$s^3 = x^3 + y^3 + z^3 + 6xyz + 3(x^2y + y^2z + z^2x)\alpha + 3(xy^2 + yz^2 + zx^2)\alpha^2,$$

---

[3] We assume for the present that these quantities do not vanish.

and there is a similar expression for t^3 (write it down). So $s^3 + t^3 = A$ and $s^3 t^3 = B$ are both invariant under all 6 permutations of x, y, z, and so they are rational expressions in a, b, and c, the coefficients of the cubic equation. This means that s^3 and t^3 are the roots of a quadratic equation

$$X^2 - AX + B = 0$$

with coefficients that are rational in a, b, c. This equation is called a *resolvent equation*, and it means that s^3 and t^3 can be found from the coefficients of the original equation by nothing more than the taking of a square root.

To obtain the quadratic equation more systematically we proceed as follows. Denote the elements of S_3 by $g_0 = e, g_1, \ldots, g_5$, and write $g_j s$ for the effect of the permutation g_j on s; for example, $(23)s = t$. Write down the equation of degree 6

$$\prod_{j=0}^{5}(X - g_j s) = 0.$$

This is a symmetrical expression in the permutations of x, y, z in s and so therefore are the coefficients in the equation. Now it turns out that this equation of degree 6 is in fact this quadratic equation in X^3:

$$(X^3 - s^3)(X^3 - t^3) = 0.$$

So (and here we get closer to Lagrange) we can start with the cubic equation, obtain a quadratic equation whose coefficients are rational functions of the coefficients of the cubic equation, pick a solution of that quadratic equation (say s^3), and take its cube root: the resulting expression will be of the form $s = x + y\alpha + z\alpha^2$, a rational expression in the roots of the given cubic equation. Do the same with t, and we have the three equations

$$x + y + z = -a$$

$$x + \alpha y + \alpha^2 z = s$$

$$x + \alpha^2 y + \alpha z = t$$

from which the roots x, y, z can be found (one has to check that the correct choice of cube roots has been made).

So the value of a Lagrange resolvent is that on the one hand knowing it enables all the roots of the cubic equation to be found, and on the other hand it can be found from a polynomial of degree two whose coefficients are rational functions of the coefficients of the cubic equation we want to solve. This suggests that there might be a programme to solve a polynomial equation of arbitrary degree n: find a resolvent (an expression in the roots and roots of unity) that satisfies a resolvent equation of degree less than n whose coefficients are rational functions of the coefficients of the equation we want to solve. If we solve this equation, we find the roots of the equation we want to solve. So the question becomes: can we find a Lagrange resolvent? All that we know initially is that it must have some sort of symmetric property—but that's rather vague.

Lagrange on the Solution of the Quartic Equation

We can now sketch Lagrange's very lengthy analysis of the quartic equation[4]

$$x^4 + mx^3 + nx^2 + px + q = 0 \tag{8.5}$$

with roots a, b, c, d. He began with an account of Ferrari's method. To deal with the correct choices of signs, he argued that the radical expression for the roots satisfied an equation of degree 12 that, as he showed after a long calculation ending on p. 261, was of the form

$$(x^4 + mx^3 + nx^2 + px + q)^3 = 0.$$

Lagrange then focussed on the fact that Ferrari's method depends on the introduction of an auxiliary cubic equation. He let y be a root of that cubic equation, and introduced a variable z so that he could put the quartic equation (8.5) in the form

$$(x^2 + \frac{mn}{2} + y)^2 - z^2 \left(x + \frac{my - p}{2z^2}\right)^2 = 0.$$

This is a difference of two squares, and so he deduced that two of the roots of the quartic satisfy one of the next two equations, and the other two roots satisfy the other equation, whatever root, y, of the cubic equation is chosen:

$$x^2 + \frac{mx}{2} + y + z\left(x + \frac{my - p}{2z^2}\right) = 0 \tag{8.6}$$

$$x^2 + \frac{mx}{2} + y - z\left(x + \frac{my - p}{2z^2}\right) = 0. \tag{8.7}$$

Without any loss of generality, the first one gave him expressions for $a + b$ and ab, and the second one for $c + d$ and cd. From these he deduced that

$$z = \frac{c + d - a - b}{2}, \quad y = \frac{ab + cd}{2}.$$

---

[4]See pp. 254–305 of the *Oeuvres* edition.

He noted that the expression for y had precisely 3 values:

$$\frac{ab+cd}{2}, \quad \frac{ac+bd}{2}, \quad \frac{ad+bc}{2},$$

and he now argued very much as follows.

To eliminate unnecessary fractions, Lagrange worked with $2y = ab + cd$ rather than y. Symmetry considerations dictate that the three values of $2y$, which are

$$u = ab + cd, \quad v = ac + bd, \quad w = ad + bc,$$

are the roots of a cubic equation with rational coefficients:

$$u^3 - Au^2 + Bu - C = 0,$$

in which A, B, and C are expressions in a, b, c, d.

It is easy to see that $A = u + v + w = n$, and a little more work to discover that

$$B = uv + vw + wu = mp - 4q,$$

and

$$C = uvw = (m^2 - 4n)q + p^2.$$

This implies that u, v, and w are the roots of the cubic equation

$$z^3 - nz^2 + (mq - 4q)z + (m^2 - 4n)q + p^2 = 0. \tag{8.8}$$

In the special case where $m = 0$ u, v, and w are the roots of the equation

$$z^3 - nz^2 - 4qz - 4nq + p^2 = 0, \tag{8.9}$$

and so $y = u/2$, $v/2$, and $w/2$ satisfy the equation

$$8z^3 - 4nz^2 - 8qz - 4nq + p^2 = 0.$$

This is the equation just found above, and so we deduce that the roots of Eq. (8.8) are

$$\alpha = \frac{1}{2}(tu + vw), \quad \beta = \frac{1}{2}(tv + uw), \quad \gamma = \frac{1}{2}(tw + uv).$$

But it reduces when we adjoin λ and μ, which are as follows:

$$\lambda = \pm\sqrt{2y - p}, \quad \mu = \pm\sqrt{y^2 - r}.$$

Now we get

$$\left((x^2 + \alpha) + (\lambda x + \mu)\right)\left((x^2 + \alpha) - (\lambda x + \mu)\right) = 0.$$

The roots of each factor are simply found by taking the appropriate square roots.

Lagrange indeed showed explicitly that if $u = ab + cd$ is given then, because $abcd = q$, the quantities ab and cd are the roots of

$$t^2 - ut + q = 0.$$

He pursued this line of argument to the end, and showed that the roots of the quartic are expressible in terms of

$$t_1 = (a + b - c - d)^2, \quad t_1 = (a + c - b - d)^2, \quad t_1 = (a + d - b - c)^2,$$

as

$$\frac{\sqrt{t_1} + \sqrt{t_2} + \sqrt{t_3}}{4},$$

$$\frac{\sqrt{t_1} - \sqrt{t_2} - \sqrt{t_3}}{4},$$

$$\frac{-\sqrt{t_1} + \sqrt{t_2} - \sqrt{t_3}}{4},$$

$$\frac{-\sqrt{t_1} - \sqrt{t_2} + \sqrt{t_3}}{4}.$$

Lagrange pointed out that it is not necessary to take care of the signs so long as they are chosen in the fashion indicated. Other choices of sign give the negatives of the roots.

Lagrange looked at all the known methods for solving the quartic equation, and showed that they all depended on the existence of partially symmetric functions of the roots. He then summarised his findings in these words (Lagrange 1770/71, 304):

> We have not only brought together these different methods and shown their connection and interdependence, but we have also – and this is the main point – given the reason, a priori, why some of them lead to resolvents of degree three and others to resolvents of degree six that can be reduced to degree three. This is due, in general, to the fact that the roots of these resolvents are functions of the quantities t, u, v, w, which, like the function $tu + vw$, take on just three values under all the permutations of these four quantities; or which, like the function $t + u - v - w$, take on six values that are pairwise equal but have opposite signs; or which take on six values that can be separated into three pairs such that the sum or product of the values in each pair is not changed by the permutations of the quantities t, u, v, w .... The general solution of quartic equations depends uniquely on the existence of such functions.

Lagrange then looked for a generalisation of this method to the case of quintic equations. He was aware that there was not much of a pattern to draw on. The quartic equation particularly was only solved by some happy accidents, because the equation of degree 6 that arose could in fact be solved. To solve a quintic equation, in Lagrange's opinion, one would write down the equation satisfied by an expression of the form

$$x + y\alpha + z\alpha^2 + w\alpha^3 + v\alpha^4,$$

where α is a primitive 5th root of unity. This equation would break up into equations of degree 24—but what then? On the one hand, this equation of degree 24 is special, but why should it be a disguised equation of degree some factor of 24? One might hope that, because the cyclic group of order 5 sits inside the symmetry group of the pentagon (call it P_5, a group of order 10) one could bring that equation down to order 12, using

$$x + y\alpha + z\alpha^2 + w\alpha^3 + v\alpha^4, \text{ and its conjugate } x + y\alpha^4 + z\alpha^3 + w\alpha^2 + v\alpha.$$

Perhaps this could be connected to the alternating group A_5 in S_5, which surely involves nothing more than square roots of the discriminant $\prod_{i<j}(x_i - x_j)$, and indeed P_5 is even a subgroup of A_5. But it is of index 6 in A_5 and there is no subgroup in between. Even this analysis surpasses what Lagrange could have done, but his is the credit for suspecting that in any case the quintic polynomial equation cannot be solved by radicals.

He expressed his somewhat downbeat conclusions this way (Lagrange 1770/71, 403):

> Here, if I am not mistaken, are the true principles for the solution of equations and for the best analysis of its conduct. As one can see, everything reduces to a kind of combinatorial calculus by which one finds a priori the results to be expected. It may seem reasonable to apply these principles to equations of the fifth and higher degrees, of which the solution is at present unknown, but such application requires too large a number of examinations and combinations whose success is very much in doubt. Thus, for the time being, we must give up this work. We hope to be able to return to it at another time. At this point we shall be satisfied with having set forth the foundations of a theory which appears to us to be new and general.

In fact, Lagrange never returned to the topic and his next long paper on the topic of solving equations was on the method of finding numerical approximations to their roots, which some historians have suggested is further evidence that he despaired of solving the quintic algebraically. In fact, such an approach is consistent with the belief that even if the equations can be solved by radicals the solution expressions would be unintelligible, as the example of the cubic equation has already suggested.

8.4 Exercises

Lagrange did not think of groups of permutations in any sense—that idea came later, as we shall see—but in terms of expressions taking not many different values as its variables are permuted. The next exercises are designed to help you explore this piece of mathematics, to which we shall return. After Lagrange, mathematicians interested in the problem of solving the quintic equation became interested in finding functions (algebraic expressions) in 2, 3, or 4 symbols taking not many values. For example, the expression $u + 2v$ takes two values when u and v are permuted, $u + 2v$ and $v + 2u$, but the expressions $u + v$ and uv take only one value each: $u + v = v + u$, $uv = vu$.

1. Find expressions in symbols u, v and w that take respectively 1, 2, 3, 6 values when u, v and w are permuted. [Solutions: uvw, or any other symmetric expression, such as $u + v + w$; $(u - v)(v - w)(w - u)$; for example, $(u - v)^2$; $u + 2v + 3w$.]
2. Find expressions in symbols u, v, w and t that take respectively 1, 2, 3, 4, 6, 8, 12 values when u, v, w and t are permuted.
3. What has this got to do with subgroups of S_2, S_3, S_4 respectively? For example, what permutations of u, v, w do not alter the value of $(u - v)(v - w)(w - u)$.

It may help to point out what only becomes clear after the work of Galois, when we, but not Lagrange, look for answers in the properties of the group S_4—that we do so is a measure of the depth and success of Galois's insight. The alternating group A_4 is generated by three-cycles, the pairwise products of the transpositions, such as $(123) = (13)(12)$. There are many subgroups of order 4, we single out this one:

$$H_4 = \{e, (12)(34), (13)(24), (14)(32)\},$$

which later became known as the Klein group. The group we obtained that permutes the roots t, u, v, w but leaves each of u, v, and w fixed as well as the discriminant of the quartic is the Klein group. So adjoining the roots of the cubic equation found by Ferrari corresponds to passing to the Klein group in the chain of subgroups of S_4. This group still acts transitively on the roots, so even with these quantities adjoined the equation remains irreducible.

It is interesting to compare this with Camille Jordan's account, contained in the first treatment of the subject that fully appreciated Galois's ideas, and which was published exactly a century later, in 1870.[5] (We shall look at Jordan's work in more detail later, see Chap. 13.)

Consider the quartic equation

$$x^4 + px^3 + qx^2 + rx + s = 0.$$

---

[5] Jordan, *Traité des substitutions et des équations algébriques* 1870, §387.

Let x_1, x_2, x_3, x_4 be its roots; H the group of order 8 obtained from the substitutions $(x_1, x_2), (x_3, x_4), (x_1, x_3)(x_2, x_4)$. A function φ_1 of the roots of the given equation and invariant under the substitutions of H depends on an equation of the third degree (§366). This solved, the group of the given equation reduces to a group K obtained from those of its substitutions which are common to the group H and its transforms by the various substitutions of G. One sees easily that these common substitutions are the four obtained from $(x_1, x_2)(x_3, x_4), (x_1, x_3)(x_2, x_4)$. But these substitutions commute with the subgroup H' of order 2, obtained from the powers of $(x_1, x_2)(x_3, x_4)$. Therefore a function of the roots that is invariant under the substitutions of H' only actually depends on an equation of the second degree. This solved, the group of the given equation reduces to H'. This equation then decomposes into two equations of the second degree, having respectively for roots x_1, x_2 and x_3, x_4 (§357). It is enough, moreover, to solve one of these equations to reduce the group of the given equation to the identity.

Among the various functions that one can take for φ_1, the most convenient is the function $(x_1 + x_2 - x_3 - x_4)^2$ adopted by Lagrange. One easily finds the coefficients of the equation of degree 3 on which it depends:

$$Y^3 - (3p^2 - 8q)Y^2 + (3p^4 - 16p^2q + 16q^2 + 16pr - 64s)Y - (p^3 - 4pq + 8r)^2 = 0.$$

Let v_1, v_2, v_3 be its roots, one has

$$x_1 + x_2 - x_3 - x_4 = \sqrt{v_1}; \ x_1 - x_2 + x_3 - x_4 = \sqrt{v_2};$$

$$x_1 - x_2 - x_3 + x_4 = \sqrt{v_3}; \ x_1 + x_2 + x_3 + x_4 = -p;$$

whence

$$x_1 = \frac{1}{4}(-p + \sqrt{v_1} + \sqrt{v_2} + \sqrt{v_3}), \ldots.$$

One also has

$$\sqrt{v_1}\sqrt{v_2}\sqrt{v_3} = (x_1 + x_2 - x_3 - x_4)(x_1 - x_2 + x_3 - x_4)(x_1 - x_2 - x_3 + x_4)$$

$$= -p^3 + 4pq - 8r,$$

whence

$$x_1 = \frac{1}{4}\left(-p + \sqrt{v_1} + \sqrt{v_2} + \frac{-p^3 + 4pq - 8r}{\sqrt{v_1}\sqrt{v_2}}\right).$$

The other roots are obtained by changing the signs of the two independent radicals $\sqrt{v_1}$, $\sqrt{v_2}$.

8.5   Revision on the Solution of Equations by Radicals

These exercises review the solution process for the cubic and quartic equation step by step. The term 'known' number refers to elements of the field we are working in, and so it may change from step to step; the term 'new' number is a number that is not in the field we are working in and must be adjoined. Note that the correct answer to the repeated question "Are these numbers 'known' or 'new'?" is likely to be not a clear-cut 'Yes' or 'No' but a qualified "Almost certainly 'Yes' but under some conditions 'No'." When you think this is the case, you are not asked to examine the special cases.

1. The Cubic Equation

1. Write down the equation of the typical cubic equation in the form $x^3 + px + q = 0$.
2. (a) Set $x = y + \frac{\alpha}{y}$ and find the value of α that makes the resulting equation one in powers of y^3.
   (b) Write the equation as a quadratic in y^3: are the coefficients 'known' numbers or 'new' ones?
   (c) Solve the equation for y^3 and write the solutions explicitly in terms of a and b: are the solutions 'known' numbers or 'new' ones?
3. (a) For each value of y^3 obtained above, write down all the corresponding cube roots: are these cube roots 'known' numbers or 'new' ones?
   (b) Use the equation $x = y + \frac{\alpha}{y}$ to find the solutions of the original cubic equation: are these solutions 'known' numbers or 'new' ones?
4. List the fields you have been working over at each stage, describe each extension as you come to it in terms of the corresponding auxiliary equation, and note if you adjoined one or all the roots at each stage.

2. The Quartic Equation

1. Write down the equation of the typical quartic equation in the form

$$x^4 + px^2 + qx + r = 0.$$

2. (a) Write the equation as $x^4 = -px^2 - qx - r$, add $2zx^2 + z^2$ to each side, and rewrite the equation in the form $x^4 + 2zx^2 + z^2 = R$ for an explicit expression R that you should write down.
   (b) Write down the condition that R is a perfect square, and arrange it as a cubic equation for z that you should write down: are the coefficients 'known' numbers or 'new' ones?
   (c) Let the solutions of the equation for z be ζ_1, ζ_2, and ζ_3: are the solutions 'known' numbers or 'new' ones?
3. (a) For each ζ_j obtained above, write down the equation for x: are the coefficients 'known' numbers or 'new' ones (with respect to the original field)?
   (b) Write R in the form $(ux + v)^2$ and so determine expressions for u and v in terms of ζ_j: are u and v 'known' numbers or 'new' ones?
4. Take square roots and obtain two quadratic equations for x that you should write down explicitly: are the coefficients 'known' numbers or 'new' ones (with respect to the field you are now working over)?
5. For each ζ_j obtained above, write down the corresponding values of x: are the solutions 'known' numbers or 'new' ones?

6. List the fields you have been working over at each stage, describe each extension
   as you come to it in terms of the corresponding auxiliary equation, and note if
   you adjoined one or all the roots at each stage.

The following irrelevant fact may satisfy some idle curiosity: the cubic equation
$ax^3 + bx^2 + cx + d = 0$ reduces to this:

$$x^3 + \left(\frac{3ac - b^2}{3a^2}\right) x + \frac{2b^3 - 9abc + 27a^2 d}{27a^3} = 0.$$

3. A Remark on the Symmetric Polynomials

A polynomial in the n variables $x_1, x_2, \ldots, x_n$ is said to be symmetric if its value
is unaltered by any permutation of the variables. Thus when $n = 3$ the polynomial
$x_1^2 + x_2^2 + x_3^2$ is symmetric but the polynomial $x_1^2 + x_2^2 - x_3^2$ is not—it is altered in
value by the permutation that exchanges x_2 and x_3, for example.

The basic symmetric polynomials are $\sigma_1 = x_1 + x_2 + \cdots x_n$, $\sigma_2 = x_1 x_2 +$
$x_1 x_3 + \cdots + x_{n-1} x_n, \ldots \sigma_n = x_1 x_2 \ldots x_n$. It is a fundamental fact about symmetric
polynomials that any symmetric polynomial can be written as a sum of the basic
symmetric polynomials, and indeed in a unique way (up to order).

It is well known that the coefficients of the polynomial in the equation $x^n +$
$a_1 x^{n-1} + \cdots + a_n = 0$ are given, up to a sign, by the basic symmetric polynomials
in the roots $x_1, x_2, \ldots, x_n$ of the equation: $a_j = (-1)^j \sigma_j$. Newton stated, without
proof, that the powers of the roots are symmetric in the roots, and this was proved
by Maclaurin. The more general claim—the 'fundamental fact'—was stated and
proved by Waring, and then by Lagrange (1808, Note VI). Here I give the very
elegant proof due to Gauss (1816).

Gauss imposed an ordering on the variables: x_j comes before x_k if j comes
before k. So x_3 comes after x_2 and before x_5, for example. Let us call this an
alphabetical ordering. Gauss extended it to an ordering on monomials of the same
degree, essentially by writing out each monomial in full (writing, for example,
$x_1^4 x_2^2 x_3$ as $x_1 x_1 x_1 x_1 x_2 x_2 x_3$) and applying alphabetic order to them. In this way the
polynomial can be written as a sum of terms of each degree, each sum arranged in
alphabetical order.

It will be enough to deal with polynomials all of whose monomials are of
the same degree, because Gauss's process works with each degree separately. So
suppose that the monomials in a symmetric polynomial p in $x_1, x_2, \ldots, x_n$ of degree
n are arranged in alphabetical order. If it contains the monomial $x_2^4 x_5^2 x_1$ then it must
contain all monomials of the form $x_j^4 x_k^2 x_l$, because it is symmetric, and so it contains
$x_1^4 x_2^2 x_3$, which is the one that comes first in the ordering among terms of this form.
Suppose moreover that this is the term that comes first in the alphabetical ordering
on the polynomials. Now consider the polynomial

$$p - \sigma_1^2 \sigma_2 \sigma_3.$$

It is symmetric, and it contains no term of degree 7 of the form $x_j^4 x_k^2 x_l$ or any term that comes earlier in the alphabetical ordering. The polynomial p is finite, so a finite number of operations of this form reduce it to a sum of the basic symmetric polynomials, as required.

To see that the decomposition into a sum of the basic symmetric polynomials is unique, suppose that it was not, and that the polynomial p is obtained in distinct ways as these polynomials in the basic symmetric polynomials: $F(\sigma_1, \sigma_2, \ldots, \sigma_n)$ and $G(\sigma_1, \sigma_2, \ldots, \sigma_n)$. Then $F - G$ represents the zero polynomial. Apply alphabetical ordering to the variables $\sigma_1, \sigma_2, \ldots, \sigma_n$ and look at the first monomial of top degree. It cannot be equal to any other monomial, because, when written out in $x_1, x_2, \ldots, x_n$ it contains some terms not found in any other monomial. Therefore $F = G$ and the representation is unique.

Chapter 9
The Unsolvability of the Quintic

9.1 Introduction

The other momentous advance in algebra around 1800 was the discovery that the quintic equation cannot be solved by an algebraic formula. How can a negative of this kind be proved? What is it to analyse how a problem can be solved? The two mathematicians who did this are Ruffini and Abel, with a little help from Cauchy. Before we look at their work—and Abel's in particular—let us consider what it is to solve an equation algebraically.

We have a polynomial equation

$$a(x) = x^n + a_1 x^{n-1} + \cdots + a_{n-1} x + a_n = 0, \tag{9.1}$$

with rational coefficients, which we shall assume is irreducible and has no repeated roots.

We suppose we have found an "auxiliary" polynomial, $a_1(x) = 0$, whose roots we know; say they are $\alpha_1, \alpha_2, \ldots, \alpha_k$ and we adjoin them to the rational numbers, so the field extension of $\mathbb{Q}$ we are working with, call it K_1, consists of rational expressions in $\alpha_1, \alpha_2, \ldots, \alpha_k$. These are quotients of one polynomial in the αs divided by another, so they are expressions that are algebraic in the rational numbers and the coefficients of the auxiliary polynomial.

We now repeat this procedure with a second auxiliary polynomial whose coefficients are in K_1 and obtain a new field, K_2, and carry on doing this until we obtain a field that contains the roots of the polynomial equation we want to solve, $a(x) = 0$.

For this to count as solving the original equation, we require that the auxiliary polynomials be easier to solve that the original equation (else we can 'solve' the equation in one step by letting $a(x) = 0$ be its own auxiliary polynomial!). For it to count as a procedure that solves the original equation by radicals, we stipulate

© Springer Nature Switzerland AG 2018
J. Gray, *A History of Abstract Algebra*, Springer Undergraduate Mathematics Series,
https://doi.org/10.1007/978-3-319-94773-0_9

that every auxiliary equation be of the form $t^n - p = 0$ for some n, where p is an element of the field we are extending. We say that the new number $p^{1/n}$ is adjoined to the field. (The exponent n can vary from step to step, but we can always assume that it is prime because an extension obtained from $t^{mn} - p = 0$ can be regarded as an extension obtained from $t^m - p = 0$ followed by an extension obtained from $t^n - p^{1/m} = 0$.)

We also insist that the roots of the original polynomial $a(x) = 0$ are obtained after finitely many steps, and also that the field we reach in this way is the smallest field containing the roots and not some larger one. The solution methods for cubic and quartic equations conform to this picture if we allow that in the case of the irreducible cubic, the procedure leads via the adjunction of a complex cube root of unity to a real subfield.

9.2   Ruffini's Contributions

It is one thing to suspect that something cannot be done in mathematics, even to have good reasons to be suspicious, but another to prove that it cannot be done. The first to attempt a proof that the general quintic equation cannot be solved by radicals was the Italian mathematician Paolo Ruffini. His proof was not accepted by most mathematicians at the time, and historians and mathematicians have disagreed ever since about how small or large are the gaps in his arguments.[1] He published his ideas in a two-volume book in 1799 and a number of later papers in which he rebutted the objections of Malfatti, who could not accept Ruffini's negative conclusion, and adopted the suggestions of his friend Abbati, who accepted the proof of unsolvability and made some simplifications.[2]

For reasons of brevity I forgo a detailed discussion, but draw attention to one notable feature of Ruffini's work: the explicit attention to families of permutations. Ruffini established that there was no Lagrange resolvent for the general quintic equation that itself satisfied an equation of degree less than 5. Nor could such a resolvent be found that satisfied a pure quintic (an equation of the form $x^5 - \alpha = 0$) if the resolvent were to be rational in the roots of the original equation. He succeeded by focussing on the sets of permutations that can arise, which sets he called *permutazione*, and we can think of as permutation groups. He divided them into various types, such as transitive and intransitive. If a function of n roots is left invariant by p permutations then Ruffini called p the degree of equality—it is the order of the permutation group that leaves the function invariant, and he noted without proof that p divides $n!$. In particular, he described all the subgroups of S_5, and on the basis of this concluded correctly that there are no subgroups of S_5 of orders 15, 30, or 40. Therefore, there are no functions of 5 variables having only 8,

---

[1] See (Kiernan 1971) and for a careful and sympathetic modernised account (Ayoub 1980).
[2] The most thorough account of Ruffini's work remains (Burkhardt 1892).

4, or 3 values. Abbati's contribution was to reach the same conclusions with a more systematic method. Ruffini also proved that if a subgroup of the permutations on five objects has an order divisible by 5 then it has an element of order 5.

Ruffini's success was in transferring the question of solvability by radicals from the question about adjoining successive roots (which was a question about the existence of resolvents) to a question about the corresponding families of permutations that fix the newly adjoined elements, and then largely dealing with the new question by an exhaustive analysis of these families.

That said, there were obscurities in Ruffini's work. He never made it clear if what are called accessory irrationals must be adjoined, such as complex roots of unity— these are elements that lie outside the smallest field containing the roots. Recall that this seemed to be necessary when solving a cubic with all its roots real (and indeed it is). Therefore Ruffini was unclear about whether successive adjunctions of radicals must end exactly in the smallest field containing the roots of a given quintic, or merely contain it—in fact containment is enough, and strict equality is not required. Sylow commented that Ruffini assumed with insufficient proof that the radicals in any solution must be rational in the roots of the given equation.[3] More damagingly, his book was long and frequently obscure, and it met with a poor reception outside Italy, with the exception of Cauchy, who strongly endorsed it.[4] I will explain at the end of this chapter why the obscurities in Ruffini's work matter mathematically, but the most likely explanation for its poor reception was its novelty and its great length, rather than any failing it might have to modern eyes.

Ruffini's approach was soon to be superseded by that of Niels Henrik Abel, who accomplished in a few pages what Ruffini had taken two volumes to do. Abel wrote in posthumously published memoir that

> The first and, if I am not mistaken, the only one who, before me, has sought to prove the impossibility of the algebraic solution of general equations is the mathematician Ruffini. But his memoir is so complicated that it is very difficult to determine the validity of his argument. It seems to me that his argument is not completely satisfying.[5]

However, it is not clear if Abel knew of Ruffini's work at all when he published his own accounts in 1824 and 1826.

9.3 Abel's Work

When Abel began his work on the solvability of the quintic, he initially thought he had found a general formula, but on being challenged to use it to solve some sample equations he realised that he had failed, and he became convinced instead that the

---

[3] See Abel, *Oeuvres complètes* 2, 293.

[4] Cauchy and Ruffini were also strongly linked by their enthusiasm for right-wing Catholic doctrine.

[5] See Abel, *Oeuvres complètes* 2, 218.

task cannot be done. His proof of this claim—that the quintic equation cannot be solved by radicals—was one of the papers he took with him on his trip to Germany and had published as a separate pamphlet at his own expense (it is (Abel 1824)).

Fig. 9.1 Niels Henrik Abel (1802–1829). Photo courtesy of the Archives of the Mathematisches Forschungsinstitut Oberwolfach

When that account proved unconvincing, he wrote a second, (Abel 1826). In it, Abel first defined some terms. He said that a function of some quantities is rational if it is obtained by addition, subtraction, multiplication and division on these quantities. If root extraction is also involved, he said that the function is algebraic.

Abel then considered what such a formula would look like if it existed. To set out the theory in a general form, Abel began with an equation of degree k:

$$c_0 + c_1 y + c_2 y^2 + \cdots + c_{k-1} y^{k-1} + y^k = 0. \tag{9.2}$$

He then supposed that the last step in implementing such a formula would be the taking of nth roots, i.e. solving the equation

$$z^n - p = 0,$$

where p is a known quantity, and n is a prime number. Then the roots of equation (9.2) will have the form

$$y = q_0 + p^{1/n} + q_2 p^{2/n} + \cdots + q_{n-1} p^{(n-1)/n},$$

where the qs are numbers we have obtained already, probably as a consequence of earlier adjunctions. We can even write, as Abel did,

$$y = p_0 + R^{1/m} + p_2 R^{2/m} + \cdots + p_{m-1} R^{(m-1)/m},$$

by replacing R with R/p_1^m, but nothing much is gained by doing so.

The qs have also been obtained by adjoining nth roots, perhaps with other values of n, so they are also of this form. If we steadily unpack each of the qs and p we eventually reach expressions that only involve the coefficients of the equation, which are, of course, symmetric functions of the roots. But first Abel showed that at each stage what is adjoined cannot also satisfy an equation of degree $n - 1$ (for the appropriate n).

Abel put his expression for a root, y, into Eq. (9.2) and argued (in Sect. 9.2 of his paper) that if we expand it out and collect like terms the equation reduces to something of this form:

$$r_0 + r_1 p^{1/n} + r_2 p^{2/n} + \cdots + r_{n-1} p^{(n-1)/n} = 0,$$

where the rs are rational functions in p and the qs and the coefficients of Eq. (9.2). Now, for this to be true, he said, we must have

$$r_0 = 0, \ r_1 = 0, \ \ldots, \ r_{n-1} = 0.$$

This is true because we are assuming (and it can be proved) that $p^{1/n}$ satisfies an irreducible equation of degree n, namely $z^n - p = 0$. If it also satisfies an equation of degree $n - 1$ then all the coefficients of that equation must be zero. The argument is fiddly, and I omit it, but the basic idea is to derive a contradiction from the existence of common roots. Care must be taken because the equations cannot be assumed in advance to be irreducible.

Abel now introduced a primitive nth root of unity, α, and noted that the values of y which are obtained when $p^{1/n}$ is replaced by $\alpha^j p^{1/n}$, $j = 0, 1, \ldots n - 1$, are also roots of the equation. This is because it satisfies the equation

$$r_0 + r_1 \alpha p^{1/n} + r_2 \alpha^2 p^{2/n} + \cdots + r_{n-1} \alpha^{n-1} p^{(n-1)/n} = 0,$$

which is zero because the r_j are all zero.

Let us denote the value of y corresponding to $p^{1/n}$ by y_1, the value of y corresponding to $\alpha p^{1/n}$ by y_2, and generally the value of y corresponding to $\alpha^j p^{1/n}$ by y_{j+1}. We have just seen that $y_1, y_2, \ldots, y_n$ are n roots of Eq. (9.2).

So Abel now had a system of n equations for the n values of nth roots:

$$y_1 = q_0 + p^{1/n} + q_2 p^{2/n} + \cdots + q_{n-1} p^{(n-1)/n},$$

$$y_2 = q_0 + \alpha p^{1/n} + \alpha^2 q_2 p^{2/n} + \cdots + \alpha^{n-1} q_{n-1} p^{(n-1)/n},$$

$$\cdots,$$

$$y_n = q_0 + \alpha^{n-1} p^{1/n} + q_2 p^{2/n} + \cdots + \alpha q_{n-1} p^{(n-1)/n}.$$

This system can be solved, yielding the $q_j p^{j/n}$s as linear functions of the ys and powers of α. So the qs that enter the formula for the solution are rational functions of the roots of the equation. In fact

$$q_j = n^{j-1} \frac{y_1 + \alpha^{-j} y_2 + \cdots + \alpha^{-(n-1)j} y_n}{(y_1 + \alpha^{-1} y_2 + \cdots + \alpha^{-(n-1)} y_n)^j}.$$

An exactly similar argument, which Abel barely bothered to write down, then establishes that at every stage what is adjoined is a rational function of the roots of the original equation. He concluded that

> [Abel's theorem on the form of a solution] If an equation is solvable by radicals one can always put one root in such a form that all the algebraic functions of which it is composed can be expressed as rational functions of the roots of the given equation

To understand this remark, see that it is true when the given equation is a cubic. A root of a cubic equation is a sum of two cube roots (each multiplied by a cube root of unity) each of which contains a term rational in the coefficients of the cubic equation to which is added or subtracted a square root of a term that is itself rational in the roots of the equation. What is at stake here is the first strictly algebraic expression with which the formula begins (the one deepest inside the nested roots). It is a radical in the coefficients, but a rational function in the roots (typically $\prod_{j<k}(x_j - x_k)$) and some roots of unity.

In Sect. 9.4, Abel turned to the quintic equation, and concluded his proof as follows: Take the first of these intervening expressions—again, that means the first symmetric expression in the coefficients that is subjected to root extraction. There must be one, else multi-valued functions of the roots could not arise. Call it $v = R^{1/m}$ (if need be by relabelling), where R is a rational function of the coefficients, v is a rational function of the roots, and m, which need not be 5, is a prime number. Let the roots be considered as independent variables. As they are permuted R does not change and $R^{1/m}$ takes m different values. So m must divide 5!, and by a recent theorem of Cauchy, which Abel had discussed in Sect. 9.3 of his paper, this forces $m = 2$ or $m = 5$ because m is prime. Abel now showed that neither alternative is coherent. You can consult the details in a subsection below, but it is enough to know for our purposes that Abel succeeded in this matter.

Abel added an 'Analysis of the preceding memoir' to the paper when it was republished in a French journal (as (Abel 1826)). Usually these things were quite short: Abel added almost 7 pages to his 21-page memoir. He now observed explicitly that an algebraic function will be of the form $p' = f(x_1, x_2, \ldots, p_1^{1/m_1}, p_2^{1/m_2}, \ldots)$, where f is a rational function, the p_j are rational functions of $x_1, x_2, \ldots$, and the n_j are prime numbers. Such a function Abel called algebraic of the first order; a function is algebraic of the second order if it had a first-order algebraic function as one of its variables, and so on. A function of order μ is of degree m if it has m functions of that order. Such a function can be written in the form

$$v = q_0 + p^{1/n} + q_2 p^{2/n} + \cdots + q_{n-1} p^{n-1/n}.$$

The first theorem follows (Abel 1826, p. 88):

> Every algebraic function of order μ and degree m can be written in the above form, where n is prime, the qs are algebraic functions of order μ and degree at most $m - 1$, p is an algebraic function of order $\mu - 1$ and $p^{1/m}$ cannot be written as a rational function of p and the qs.

This theorem says that the solution to a polynomial equation, written as a function of the roots, will have the above form.

Abel then summarised his paper at some length until he reached the theorem quoted above (Abel's theorem on the form of a solution), which he stated again. Then, with a nod to Cauchy's work, he obtained his third theorem (Abel 1826, p. 90):

> [Abel's third theorem] If a rational function of $x_1, x_2, \ldots$ takes m different values one can always find an equation of degree m whose coefficients are symmetric functions of the $x_1, x_2, \ldots$ that has these m values as its roots, but it is impossible to find an equation of lower degree having one or more of these values as its roots.

On the basis of these general results he now redescribed his proof that the quintic is not solvable by radicals. Certainly there must be a first-order algebraic function involved, say $v = R^{1/m}$, where R is a rational function of the coefficients of the given quintic equation and m is a prime number. Because v takes m values as the roots of the quintic are permuted, it must be that m divides 5! and so either $m = 2, 3$ or 5. Cauchy's work ruled out the case $m = 3$. Abel then excluded each of these cases, as in the memoir itself. He concluded that the general quintic equation was not solvable by radicals, and that therefore the general equation of any degree higher than 5 was not solvable by radicals either.

It is a great moment in the history of mathematics, but there are reasons for disquiet. The paper is genuinely not easy to understand; not when it was published, and not now because it does not translate nicely into modern 'Galois theory'.[6] It is probably clear that it does not imply that the quadratic, cubic, and quartic equations

---

[6]But see Brown, 'Abel and the insolvability of the quintic', http://www.math.caltech.edu/~jimlb/abel.pdf, (Pesic 2003), and (Rosen 1995).

are not solvable by radicals, but why does it not prove that *no* quintic equation is solvable by radicals? What is meant by the phrase the 'general quintic equation'?

Kiernan's comments here are also illuminating. He remarked that

> By a lengthy computational argument, Abel concludes that every expression, rational in five quantities and assuming five distinct values, must be of the form

$$r_0 + r_1 x + r_2 x^2 + r_3 x^3 + r_4 x^4$$

> where r_0, r_1, r_2, r_3, r_4 are expressions symmetric in the five quantities, and where x is some one of the five quantities. This passage, fully one-fourth of the entire article in length, has been the most controversial part of the presentation. The results are established by direct computation, [...] Sylow's comment on Abel's proof is quite appropriate: "It must be admitted that it could be shorter and clearer; but its validity is free of any serious objection."[7]

Among Abel's other remarks about equations that are solvable by radicals is his short paper (Abel 1829) in which he established that if the roots of an irreducible equation are such that all are rationally known when any one of them, say x, is, and if θx and $\theta_1 x$ are any two of these roots and $\theta\theta_1 x = \theta_1\theta x$ then the equation is solvable by radicals. Moreover, if the degree of the equation is $p_1^{n_1} p_2^{n_2} \ldots p_k^{n_k}$ where $p_1, p_2, \ldots, p_n$ are distinct primes, then the solution of the equation reduces to the solution of n_1 equations of degree p_1, n_2 equations of degree p_2, ... and n_k equations of degree p_k. In modern terms, this is the claim that an equation whose Galois group is commutative then the equation is solvable by radicals, and it is the paper from which the term 'Abelian' for a commutative group is derived, following (Kronecker 1853).

9.4 Wantzel on Two Classical Problems

We now look at the discovery of the (nowadays) celebrated impossibility theorems that follow from Galois theory: it is impossible by straightedge and circle to trisect an arbitrary angle or to find the side of a cube double the volume of a given cube. These are two of the so-called three classical problems of Greek geometry:

1. the duplication of a cube: finding a cube equal in volume to twice a given cube by straightedge and circle alone (solve $x^3 = 2$);
2. trisect an angle by straightedge and circle alone (solve $4x^3 - 3x - s = 0$);
3. find by straightedge and circle alone a square equal in area to a given circle [i.e., disc] (construct a segment of length $\sqrt{\pi}$).

The last of these is very deep, but the first two are usually taught today as elementary consequences of Galois theory. What does that involve, and how were they discovered?

---

[7]In *Oeuvres d'Abel*, 2, 293.

It turns out that the history of the three classical problems is complicated. It is clear that they were only some of a number of unsolved problems around the time of Plato; another asked for the construction of a regular heptagon, for example. On the other hand, it is not clear what made them difficult. It was known from early on that the cube can be duplicated by a construction involving the intersection of a hyperbola and a parabola. Later, Archimedes gave a construction for trisecting an angle, but it involves sliding a marked straightedge around while ensuring that it passes through a given point. The restriction to the use of a straightedge and circle alone is implied but nowhere absolutely stated; it gained weight from the absence of any other kind of 'instrument' in Euclid's *Elements*, but that may be irrelevant. Finally, no-one disputes (or disputed) that there is an angle one-third the size of a given one, or that there is a cube twice the size of a given one—the only question was how to exhibit the answer by a reliable procedure. Squaring the circle was regarded as impossible from very early on, and remains to this day a synonym for impossibility.

Nonetheless, the idea spread that these three problems could not be solved. Here we shall look only at duplicating the cube and trisecting the angle. It is not difficult to persuade yourself that a straightedge and circle construction can only involve rational operations on known lengths and the extraction of square roots. Descartes spelled out explicitly in his *La Géométrie* of 1637 that straightedge and circle constructions provide solutions to quadratic equations (Fig. 9.2).

To solve the equation $z^2 = az + b^2$, Descartes drew a right-angled triangle NLM with $LM = b$ and $NL = a/2$. He extended the hypotenuse NM to meet the circle

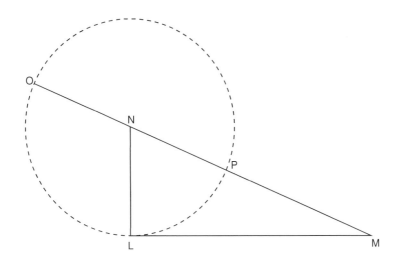

Fig. 9.2 Descartes's solution to a quadratic equation

through L with centre at N at O and P, so $NL = NP = NO$. By the Pythagorean theorem,

$$MN^2 = LN^2 + LM^2 = \frac{1}{4}a^2 + b^2 .$$

So

$$MN = \sqrt{\frac{1}{4}a^2 + b^2} \text{ and } OM = ON + NM = \frac{1}{2}a + \sqrt{\frac{1}{4}a^2 + b^2} .$$

Descartes went on to claim that cubic problems cannot be solved this way. His argument was that circles and lines give us one number between two others, and cubic problems required two. This is either a deep insight or a triviality blind to the complexities of the problem; neither way is it a solution. In the next century Jean-Étienne Montucla, the founder of the history of mathematics, gave a better argument (Montucla 1754, pp. 273–285).[8] He argued that algebraic constructions correspond to finding intersection points of curves. By a traditional result (Bezout's theorem) a curve of degree k and a curve of degree m meet in km points (possibly complex, possibly at infinity, possibly multiple and counted accordingly). Repeated use of quadratics will (might?) solve equations of degree a power of 2, but only those, and therefore cubics cannot be solved.

Our two classical problems ask whether quadratic methods can solve cubic equations (to phrase them in a way current since the sixteenth century). The consensus by 1800, and probably all along, was that the answer was No. That answer is correct, and we can ask: Who first established this result?

The answer is a French mathematician, Pierre Laurent Wantzel, who published his demonstration in 1837, when he was 23, in the journal *Journal des mathématiques*. Wantzel promptly vanished into obscurity, however, from which he has only recently been rescued by the Danish historian of mathematics Jesper Lützen (see (Lützen 2009)).[9] This gives us two questions: What did Wantzel do, and why did it fall dead from the press?

Wantzel's paper was in four parts. First he translated the geometric problem into an algebraic one. Second, he showed that in this way the solution to the geometric problem appears as a root of an equation of degree a power of 2. Third, under certain further assumptions, this equation is irreducible. Fourth, duplicating a cube and trisecting an angle lead in general to irreducible cubic equations—and therefore cannot be solved by a sequence of quadratics.

The first two of these arguments are unproblematic. Wantzel's proof of step three was deficient. He needed the irreducibility of the polynomial P of degree 2^n

---

[8]For a further discussion, see (Lützen 2009).

[9]There was one earlier mention, by Petersen, a Danish mathematician in 1877, but no-one picked it up except Felix Klein, who dropped the reference to Wantzel by name. The first historian of mathematician to mention Wantzel by name was Florian Cajori in 1918.

and rational coefficients because any algebraic number is the root of a polynomial of degree a power of 2—just add the necessary extra roots (but get, of course, a reducible polynomial). Moreover, Wantzel was interested in only one root of his polynomial P of degree 2^n but he has made many choices on the way—he had to show his claim was independent of these choices. He did his best. He succeeded in showing that if P has a root in common with another polynomial, F, with rational coefficients then all of the roots of P are roots of F. But this does not show that P is irreducible, only that it is a power of an irreducible polynomial. However, it turns out that this gap in Wantzel's argument is easy enough to fix (Pedersen and Klein did so).

Mistakes in mathematics make an interesting subject. There are a lot of them, even in published mathematics, certainly in the nineteenth century. They range from the trivial (the reader can readily identify them and correct them) through those that destroy the given proof of a claim but where the claim can be proved in another way, to the irretrievably wrong. They have a historical dimension: 'wrong by the standards of the day' is not the same as 'wrong by our standards'—generally the former implies the latter. Wantzel's argument needs rescue, but good readers could and did supply it.

So why did Wantzel not become famous? He published an original result in a major journal while still a student in a main-stream place. He was even quite well-known, being the first student to come top in the entrance examinations for both the École Polytechnique and the École Normale Supérieure. Lützen has two interesting explanations. The first is that perhaps the result was already thought to be true. Gauss had claimed in the *Disquisitiones Arithmeticae* §366 that the only polygons constructible by straightedge and circle have $2^n p_1 p_2 \dots$ sides, where the p_i are distinct Fermat primes. This implies in particular that the regular 9-gon is not constructible, but its sides subtend an angle of 40° at the centre, which is one-third of the constructible angle of 120°. Therefore angles cannot, in general, be trisected by straight edge and circle.

Gauss wrote

> In general therefore in order to be able to divide the circle geometrically into N parts, N must be 2 or a higher power of 2, or a prime number of the form $2^m + 1$, or the product of several prime numbers of this form, or the product of one or several such primes times 2 or a higher power of 2. In brief, it is required that N involve no odd prime factor that is not of the form $2^m + 1$ nor any prime factor of the form $2^m + 1$ more than once. The following are the 38 values of N below 300: 2, 3, 4, 5, 6, 8, 10, 12, 15, 16, 17, 20, 24, 30, 32, 34, 40, 48, 51, 60, 64. 68, 80, 85, 96, 102, 120, 128, 136, 160, 170, 192, 204, 240, 255, 256, 257, 272.

These are the closing words of the book, so they are oddly prominent—they might be read as the conclusion of the book—and it is true that Gauss's claim was not accompanied by a proof, only a hint that the construction of other regular polygons required that some irreducible equations be reducible. But nor is there any evidence that anyone doubted it, and nor has anyone doubted that Gauss knew how to prove the claim. So to anyone persuaded of Gauss's claim Wantzel was saying nothing new.[10] Such a reader could have found Wantzel's paper little more than a spelling out of ideas and methods published a generation before by Gauss.

Lützen's second argument is that perhaps the result was simply not found to be interesting. Readers could argue that this merely confirmed what they had suspected all along: two of the classical problems are, after all, insoluble. So what? What can be done with an impossibility proof? Nothing. On the other hand, the equations are indisputably solvable—why care about a requirement that they be solved in a particular and frankly old-fashioned way. A number of mathematicians (Gauss himself, Abel, and on another topic Liouville) phrased the problem in a more positive way: Look for a solution of such-and-such a kind, and if we fail prove there is no solution of *that* kind, then look for solutions of another kind.

Lützen points out that Wantzel, however, had a liking for negative results. He was one of the first to publish a simplification of Abel's proof that the quintic cannot be solved by radicals, and his simplification was adopted by Joseph Serret in his *Cours d'algèbre* ((1849) and subsequent editions). Wantzel was also the first to prove that if a cubic equation with rational coefficients and all its roots are real then necessarily any expression for the roots in terms of radicals will involve complex numbers.

Our interest in this result is because it fits in nicely with our understanding of Galois theory, a topic unknown to Wantzel and his readers. More broadly, it fits into a whole range of topics where we are interested in solutions of various kinds and solution methods of various kinds. Impossibility results have become positive results in mathematics—but they were not seen positively in 1837.

Wantzel, it seems from the one article on his life by someone who knew him, did not live up to his early promise. He was unwilling to stick at any one topic for long, was given a heavy teaching load, abused coffee and opium, and died in 1848 aged only 33, his work already forgotten.

9.5   Wantzel on the Irreducible Case of the Cubic

In 1843 Wantzel published an explanation of why the irreducible cubic equation with three real roots cannot be solved by the extraction of real roots alone.

He began with a classification of irrationals or adjoined quantities that was, as he said, close to Abel's. He said that a number obtained by adjoining to the rational numbers a root of an equation of the form $x^{m_1} - a = 0$, where m_1 is prime, is of type 1; a number obtained by adjoining to numbers that are rational or of type 1 a root of an equation of the form $x^{m_2} - a = 0$, where m_2 is prime, is of type 2, and so on.

---

[10]There is also another claimant: Abel. But what he said on the subject was in an unpublished fragment not discovered until 1839, two years after Wantzel published his paper; Wantzel cannot have known of Abel's claim.

He then supposed he had an irreducible cubic equation with three real roots, x_0, x_1, x_2. The root x_0 is of type n say, and satisfies an equation of the form

$$x_0 = A + Bu + \cdots + Mu^{m-1},$$

where $u = \sqrt[m]{a}$, a is of type $n - 1$, and $A, B, \ldots, M$ are of type at most n and if of type n then of lower degree than x_0. Moreover, we can assume without loss of generality that $B \pm 1$.

The expression for x_0 takes m values as the m roots of the equation $u^m - a = 0$ are substituted for u, and all these values of x_0 are roots of the given cubic. So m equals 2 or 3.

If $m = 3$ then the possible values of u are $u, \alpha u, \alpha^2 u$, where $\alpha^3 = 1$. So

$$x_0 = A \pm u + Bu^2,$$

$$x_1 = A \pm \alpha u + B\alpha^2 u^2,$$

$$x_2 = A \pm \alpha^2 u + B\alpha u^2,$$

and therefore

$$x_0 + \alpha x_1 + \alpha^2 x_2 = \pm 3u.$$

But α and α^2 are conjugate imaginaries, and x_0, x_1, x_2 and u are real, so $x_1 = x_2$. But this contradicts the assumption that the cubic equation is irreducible.

Wantzel then argued that if $m = 2$ then $x_0 = A + \sqrt{a}$ and so $x_1 = A - \sqrt{a}$ and $x_0 + x_1 = 2A$. So $2A$ is a root of a new cubic equation, which has as its roots the pairwise sums of the given equation, is also irreducible, and does not have a rational root (but A is of either lower degree or lower type than x_0). The argument that was applied to x_0 applies also to $2A$, and can be repeated to provide a sequence of radicals of decreasing type and degree until it produces a single radical of the form $A' + \sqrt{a'}$, but this satisfies a quadratic equation and therefore cannot satisfy an irreducible cubic equation.

He therefore concluded that

It is impossible to express in real roots alone the roots of an equation of degree three when all its roots are real.

Applied to the irreducible cubic equation $x^3 + ax + b = 0$ with three real roots, x_0, x_1, x_2 Wantzel's argument produces $2A = -x_2$, because $x_0 + x_1 + x_2 = 0$ so the new cubic equation has roots $-x_0, -x_1, -x_2$ and is $x^3 + ax - b = 0$. Repeating this argument returns the original cubic equation $x^3 + ax + b = 0$, so there is no decreasing sequence, as Wantzel required. He would have done better to argue that if the roots are of type m then they are of type $m - 1$ and ultimately are quadratic extensions and cannot be roots of a cubic equation.

However, the insight that each extension being of degree 2 no subfield of degree 3 can be produced is valid. We would say today that a tower of quadratic equations produces a field extension of degree 2^m for some m and so no such tower can contain an element that generates an extension of the rationals of degree 3.

9.6   Exercises

1. Prove that the group S_5 has no subgroups of order 15, 30, or 40.
2. Find an explicit straightedge and circle construction for the regular pentagon.
3. Find $\cos(20°)$ as the solution of a cubic equation.

Questions

1. How convincing do you find the argument that a proof that something cannot be done would not have been regarded as an interesting result in the 1830s?
2. What other negative results of that kind can you find that were known by or derived during the 1830s. Consider, for example, Fourier series representations of a function, the Taylor series representations of a function, non-Euclidean geometry and the parallel postulate.

Appendix A: Abel on the Cases $m = 5$ and $m = 2$

This account summarises (Rosen 1995).

Abel proceeded in two steps: first he showed that a field F that contains all the roots of a polynomial and is contained in a radical tower can be the top of a radical tower—this is the gap in Ruffini's analysis. Second, he showed that for polynomials of degree 5 no such field F is the top of a radical tower.

If $m = 5$ then we can write

$$R^{1/5} = v = r_0 + r_1 x + r_2 x^2 + \cdots + r_4 x^4,$$

where the rs are symmetric functions in the roots $x_1, \ldots, x_5$ and

$$x = s_0 + s_1 R^{1/5} + s_2 R^{2/5} + s_3 R^{3/5} + s_4 R^{4/5}.$$

As before, one deduces that

$$s_1 r^{1/5} = \frac{1}{5}(x_1 + \alpha^4 x_2 + \alpha^3 x_3 + \alpha^2 x_4 + \alpha x_5).$$

But this is impossible, because $s_1 R^{1/5}$ takes only 5 values as the roots are permuted, but the expression on the right-hand side takes 120 values.[11]

The case $m = 2$ is a little longer to treat. We now have

$$R^{1/2} = p + qs, \text{ where } s = \prod_{i<j}(y_i - y_j)$$

and p and q are symmetric functions of the roots. Indeed, we also have $-\sqrt{R} = p - qs$, so by simple algebra, $p = 0$ and $\sqrt{R} = qs$. We may write any expression of the form

$$\left(p_0 + p_1 R^{\frac{1}{2}} + p_2 R^{\frac{1}{2}} + \cdots\right)^{\frac{1}{m}}$$

—anything we get by adjoining $R^{\frac{1}{2}}$ and then passing to the next adjunction—in the form

$$\left(\alpha + \beta\sqrt{s^2}\right)^{\frac{1}{m}},$$

where α and β are symmetric functions. So the solutions will be algebraic expressions in this square root. Suppose

$$r_1 = \left(\alpha + \beta(s^2)^{\frac{1}{2}}\right)^{\frac{1}{m}} \text{ and } r_2 = \left(\alpha - \beta(s^2)^{\frac{1}{2}}\right)^{\frac{1}{m}}.$$

Abel noted that $(\alpha^2 - \beta^2 s^2)$ is symmetric and claimed that $r r_1 = (\alpha^2 - \beta^2 s^2)^{1/m}$ must be symmetric, because if it not then it must be the case that $m = 2$, but then $r_1 r_2$ has 4 values, which is impossible. But if $r r_1$ is symmetric, say $r r_1 = v$, then

$$r_1 + r_2 = \left(\alpha + \beta s^{\frac{1}{2}}\right)^{\frac{1}{m}} + \gamma\left(\alpha - \beta s^{\frac{1}{2}}\right)^{\frac{-1}{m}} = p_0 = R^{1/m} + \gamma R^{-1/m}.$$

In this expression, replace $\mathbb{R}^{1/m}$ by $\alpha^j \mathbb{R}^{1/m}$, $1 \le j \le m-1$, where α is a primitive mth root of unity, and let the value of the expression corresponding to α^j be denoted p_j. Then

$$(p - p_0)(p - p_1)\ldots(p - p_{m-1}) = p^m - A_0 p^{m-1} + A_2 p^{-2} + \cdots.$$

---

[11]In the original version of the paper he did not stop to explain why; the above argument is in the version published in Crelle's *Journal* in 1826.

Then coefficients A_j are rational functions in the coefficients and therefore symmetric functions in the roots. To Abel, at least, the equation was "evidently" irreducible. Therefore there are m values for the p_j and therefore also $m = 5$. It follows by the now-familiar argument, on noting that

$$R^{\frac{1}{2}} = \frac{1}{5}\left(y_1 + \alpha^4 y_2 + \alpha^3 y_3 + \alpha^2 y_4 + \alpha y_5\right) = \left(p + p_1 S^{\frac{1}{2}}\right)^{\frac{1}{5}},$$

that an expression of the form $t_1 R^{1/5}$ on the one hand takes 10 values (5 for the five roots, 2 for the square root) and on the other hand takes 120 values. This contradiction concludes the paper.

It seems to me that Abel did not argue that the best-possible resolvent is not up to the task (say, by showing that it only solves a restricted class of quintic equations), rather, he showed that no suitable resolvent exists, and he did so via the theory of permutations.

I conclude with some remarks about the principal obscurity in Ruffini's work, which Abel dealt with. The element $\alpha = 2\cos(2\pi/7)$ satisfies the polynomial equation

$$x^3 + x^2 - 2x - 1 = 0,$$

which is irreducible over $\mathbb{Q}$, the field of rational numbers. Over the field $\mathbb{Q}(\alpha)$ the polynomial splits completely, but the extension $\mathbb{Q} : \mathbb{Q}(\alpha)$ is not a radical extension because the polynomial is not a pure cubic.

On the other hand, Abel proved that if a field K contains enough roots of unity (think of it as $\mathbb{Q}$ with such roots of unity adjoined) and $f(x)$ is a polynomial with its coefficients in the field K, then the smallest field L that contains all the roots of $f(x)$ will arise in a sequence of radical extensions. In symbols, let

$$\mathbb{Q} < K < \cdots < E_j < E_{j+1} < \cdots < E_n$$

be a chain of radical extensions, and suppose that $E_{n-1} < L < E_n$. Then in fact $L = E_n$.

So in the presence of enough roots of unity a field contained in a chain of radical extensions is a radical extension itself, but without the appropriate roots of unity this is not necessarily the case. In particular, in the presence of enough roots of unity an extension of degree 5 must be obtained as a pure fifth root. This opened the way for Abel to concentrate on the case where the field with the roots of an irreducible quintic adjoined is precisely the top of a chain of radical extensions, and to show that this was impossible.

I omit the proofs, which will be found in (Rosen 1995) in the language of Galois theory.

Appendix B: Cauchy's Theory of Permutation Groups

Fig. 9.3 Augustin Louis
Cauchy (1789–1857)

Cauchy's work was done in two phases, the first between 1812 and 1815 and the second around 1844–45. We have already seen that his work in the first phase, which was quite possibly inspired by the work of Ruffini, helped Abel to prove that the general quintic equation is not solvable by radicals. Now we look at it in a little more detail.

In his paper (1815) *Sur le nombre, etc* (the full title is very long!) Cauchy took up the question of how many different values a rational expression in n variables can take when its variables are permuted in every possible way. He noted that Lagrange, Vandermonde, and above all and more recently his co-religionist Ruffini had shown that it was not always possible to find a polynomial which took a specified number of distinct values (even a number of values that divided $n!$, which is obviously necessary). Ruffini had shown that it is not possible to find a function of $n = 5$ variables that took only 3 or 4 values. Cauchy now produced the more general theorem for S_n that was mentioned above.

In 1844, Cauchy returned to the theme, for reasons it is interesting to speculate about (as we shall below). His treatment now was much more systematic. He spoke of an arrangement of n symbols, and a 'substitution' or permutation as a transition from one arrangement to another. He introduced the permutation notation (x, y, z), the product of two permutations, the idea that forming the product of two permutations is not necessarily commutative, powers of a permutation, the identity permutation (written 1 and called a unity), the degree or order of a permutation (the lowest positive power of it that is equal to 1). He then introduced the cyclic notation, and the idea of similar permutations (ghg^{-1}). The modern word is 'conjugate', but Cauchy reserved that word for another concept, namely the idea of a group defined (generated) by a set of permutations—which he called a system of *conjugate* substitutions. He then proved that

The order of a system of conjugate substitutions in n variables is always a divisor of the
number N of arrangements that one can make with these variables.

Here $N = n!$. Cauchy also proved that the order of any element of this group divides
the order of the group.

Cauchy also smuggled in the concept of a normal subgroup, although without
referring to Galois's proper decompositions, and established a partial converse to
Lagrange's theorem in his (1844, 250), when he showed that if a prime p divides
the order of a group then there is an element of order p in that group and so a
subgroup of order p. Recall that the condition that p be prime is necessary: there is
no subgroup of order 6 in A_4 although $6|12 = |A_4|$. I omit a summary of Cauchy's
proof.

Chapter 10
Galois's Theory

10.1 Introduction

Évariste Galois lived one of the shortest, most dramatic, and ultimately tragic lives in the history of mathematics (Fig. 10.1).[1]

Fig. 10.1 Évariste Galois (1811–1832)

Galois was born on 25 October 1811, in a suburb of Paris, to a prosperous family; his father became mayor of Bourg-la-Reine during Napoleon's brief return from exile in 1814. When he was 12 he went to the prestigious Collège Royal de Louis-le-Grand near the Sorbonne in central Paris, where he was introduced to

---

[1]There are many accounts, not all of which agree on even the most basic facts; this account follows (Rothman 1989). For an analysis of how Galois has been portrayed in fiction, and his transformation into a cultural icon (which was very evident in Paris in 2012) see (Weber and Albrecht 2011).

© Springer Nature Switzerland AG 2018   115
J. Gray, *A History of Abstract Algebra*, Springer Undergraduate Mathematics Series,
https://doi.org/10.1007/978-3-319-94773-0_10

serious mathematics at the age of 15 by his teacher H.J. Vernier. He was completely captivated, and quickly began to read advanced works: Legendre's *Géométrie* and possibly Lagrange's *Traité de la Résolution des équations numériques de tous les degrés* ('The Resolution of Algebraic Equations') in its augmented second edition of 1808. Vernier was understandably delighted with his work, but other teachers reported that he was increasingly withdrawn, "singular", and even "bizarre." He insisted on taking the entrance exam to the nearby École Polytechnique a year early, but he failed, which deepened his dislike of authority. He then entered the advanced class at the Collège de Louis-le-Grand, where he had the good fortune to be taught by L.P.E. Richard, who immediately recognised Galois's brilliance, and called on the École Polytechnique to accept the young man without examination. They declined, but with Richard's encouragement Galois published his first paper (on number theory) in the 1828 issue of Gergonne's *Annales de mathématiques*. In 1829 he wrote his first papers on the solvability of equations of prime degree, and on 25 May and 1 June he submitted them to the Académie des Sciences. Cauchy was appointed to write a report on them.

Then the dangerous world of French politics caught up with him. Galois's father was the innocent victim of a politically inspired plot against him organised by the local priest, and committed suicide on 2 July 1829. Within days of this blow Évariste sat the entrance exam for the École Polytechnique, but he failed again, and now had no choice but to sit for what was then the lesser college, the École Normale Supérieure, which he did in November 1829. This time he passed. Cauchy then failed to report on Galois's work, perhaps because he thought that Galois should re-work his papers as an entry for the Grand Prize in Mathematics organised by the Academy (for which the closing date was March 1, 1830). At all events, Galois did submit just such an entry in February. But in May Fourier, the permanent Secretary of the Academy and chief judge of the competition, died and Galois's paper could not be found. Galois felt convinced that he was the victim of an establishment conspiracy—a reasonable opinion in the circumstances (even if wrong, as it may well have been)—and moved steadily towards the revolutionary left (Fig. 10.2).

Three more articles by Galois soon appeared in Ferussac's *Bulletin*, but the political events of 1830 were more important. The Bourbon monarch Charles X fled from France after 3 days of rioting in Paris, and the Orleanist King Louis-Philippe was installed in his place (Cauchy followed Charles X voluntarily into exile in September). The students of the École Polytechnique played a prominent role in these affairs, but the students of the École Normale, Galois included, were locked in by Guigniault, the head of the school. Galois took up with the revolutionaries Blanqui and Raspail, and by the end of the year Guigniault expelled him.

Galois threw himself into the revolutionary politics that had returned to Paris almost with the fervour of the 1790s. On 9 May 1831 he attended a militantly republican dinner called to celebrate the acquittal of 19 republicans on conspiracy

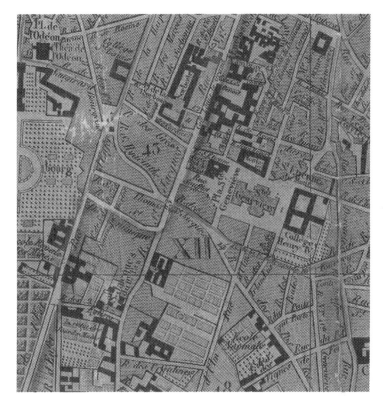

Fig. 10.2 Scholarly Paris in 1830

charges.[2] About two hundred people were there, including the novelist Alexandre Dumas, who later wrote[3]:

> It would be difficult to find in all Paris two hundred persons more hostile to the government than those to be found re-united at five o'clock in the afternoon in the long hall on the ground floor about the garden [of the Vendanges des Bourgogne].

Dumas went on:

> Suddenly, in the midst of a private conversation which I was conducting with the person on my left, the name Louis-Philippe, followed by five or six whistles caught my ear . . . A young man who had raised his glass and held an open dagger in the same hand was trying to make himself heard. He was Évariste Galois, ... All I could perceive was that there was a threat and that the name of Louis-Philippe had been mentioned; the intention was made clear by the open knife.

---

[2] At this time political assemblies were banned, so dinners were a popular, and legal, substitute.
[3] Quoted in (Rothman 1989, pp. 165–166).

Dumas escaped, but Galois was arrested the next day and put on trial on June 15 for threatening the King's life. He was acquitted, and within a month he appeared on Bastille Day dressed in the uniform of the banned Artillery Guard. This was interpreted as an extreme act of defiance—he was carrying several weapons—and he was arrested again, and eventually sentenced, on 23 October, to 6 months in the prison of Sainte-Pélagie. The eminent botanist and fellow republican Raspail, who was to spend 27 months in prison between 1830 and 1836 was among his fellow-prisoners.[4] Raspail recorded a moment in prison on 25 July 1831 when Galois, taunted by his fellow prisoners, had become drunk after emptying a bottle of brandy at a single draught, and exclaimed[5]:

> You do not get drunk, you are serious and a friend of the poor. But what is happening to my body? I have two men inside me, and unfortunately I can guess who is going to overcome the other ... And I tell you I will die in a duel on the occasion of a worthless coquette. Why? Because she will invite me to avenge her honour which another has compromised.

The Academy now rejected another manuscript by Galois, who turned his back on the usual places for a mathematician to publish and arranged to have his work published by Auguste Chevalier, a friend of his who belonged to the proto-socialist Saint-Simonian movement inspired by Claude Henri de Rouvroy, comte de Saint-Simon. On 29 April 1832, he left prison. A month later he wrote letters to friends sketching what he had achieved in mathematics, and on the morning of the 30th of May he went to his fatal duel. Shot in the abdomen, he died in hospital the day after, still only 20 years old. Politics marked his life, although it may not have caused his death as some authors suggest, but his funeral was the occasion for a republican demonstration that sparked off a week of rioting.

Galois's papers, those already in print and those left in manuscript, were eventually published in 1846, as will be described below. They run to 60 pages of text, and in addition to an early published paper on the theory of equations there are papers on continued fractions and the curvature of curves in space. The letter he wrote to Chevalier is reproduced, followed by what is now called the first memoir—*Sur les conditions de résolubilité des équations par radicaux* ('On the conditions for the solvability of equations by radicals')—and the second memoir (presented as a fragment): *Des équations primitives qui sont solubles par radicaux* ('Primitive equations that are solvable by radicals').

We shall be concerned here only with the letter to Chevalier, the first memoir, and a paper on several overlapping topics entitled *Analyse d'un Mémoire sur la résolution algébrique des équations* ('Analysis of a memoir on the algebraic solution of equations') that Galois published in 1830. For the second memoir the reader should consult the careful analysis in (Neumann 2006). So this account marginalises the concept of primitivity, which Jordan was to take up and which remains important

---

[4]Raspail lived to receive the Cross of the Legion of Honour from Louis-Philippe, and was all his life a militant republican. A boulevard and a metro stop in Paris are named after him.

[5]Quoted in (Rothman 1989, p. 94).

to this day, and indeed it will scatter its remarks about (Galois 1830) to later chapters (in particular, Chap. 23).

Galois did not become famous for showing that the quintic equation cannot be solved by radicals. That was already known. It was also known, as Lagrange had demonstrated, that all cyclotomic polynomials of prime degree are solvable by radicals. What Galois did was to give a general theory concerning which equations of whatever degree can be solved algebraically—and, of course, he showed how the quintic equation fitted into his very general account.

As Neumann observes, we cannot be sure what Galois read because he cited very few sources. It would seem that he read Gauss's *Disquisitiones Arithmeticae*, presumably in the French translation, from the occasional references to Gauss and because there is no other relevant work by Gauss. We can presume that he read some classic works under the influence of Louis Richard, his teacher in the final years at school, who appreciated Galois's brilliance, but we cannot be sure which. It is likely that he read several works by Lagrange, but we do not know precisely which. Neumann (p. 393) says of the posthumous third edition of (Lagrange 1808) that it "is the one that Galois would probably have read". If he did read that work, he would have found in Note XIII an account of the various methods for solving equations of degree less than five, but when it came to the quintic equation only the remark that its resolvent was of degree six, and so it was useless to proceed further.

The circumstances in which Galois wrote his memoir, and his early and surely needless death at the age of 20, make this a difficult memoir to understand. Our (historians') task is to understand the memoir as well as it could have been in its day, and to that end any present-day knowledge is welcome. It is not a historian's task to misunderstand it, to make it say what it does not, to supply alien formulations (such as present-day ones when they are significantly different), to cover up gaps by explaining them away. This is because we have not only the mathematics before us, which we do not wish to fail to understand, but also the historical events that, as historians interested in the development of mathematics, we wish to understand.

The historical facts are these. No-one took up Galois's ideas when they were first, and it must be said, obscurely published. They were published in 1846 to only slightly better effect, and then gradually became intelligible through the work of a number of first-rate mathematicians. We shall trace this slow progress over the course of the next several chapters. It cannot therefore be the case that what he wrote was easy by the standards and with the methods of the 1830s.

Could it be that what he wrote is easy by the standards of today? Have mathematicians as it were built the structure that holds his insights together? Can we read his memoir and, given the right background in modern algebra, simply recognise this or that piece of our theory that he, remarkably, foresaw? Such a state of affairs is surely unlikely—but one can imagine that what he wrote acted as a challenge to fill in the gaps. However, the situation, as we shall see, is that what he wrote was reworked, not merely written out (much) more fully. Understanding how this was done is the task of the historian of algebra.

What then are we to make of this, our first meeting with Galois's work? We can take the view that though he had done his best to present the complete resolution of the question: 'When is a polynomial equation solvable by radicals?' what we can see are pieces of this resolution, key pieces presumably, but we don't find them convincing. Either they are not all here, or there is something about them that we don't see. There is a way of thinking about the problem that Galois presented, and which we don't follow. Is that because he was too quick, in which case patient work should enable us to catch up, or is there some deep insight he has failed to present?

10.2   Galois's 1st Memoir

Galois began by explaining that one can always use any rational function of the coefficients of a polynomial equation to form the coefficients of new polynomials, and one may adjoin, as he put it, the nth roots of numbers as needed.[6]

He then defined a *permutation* on n symbols to be any arrangement of them—what we would call an arrangement—and a *substitution* as the passage from one *permutation* to another—what we would call a permutation! To minimise the risk of confusion, I shall use Galois's terms, but in italics. His terminology is the reason the later development of finite group theory concerned itself with *substitution* groups. But it must be said that he used the words inconsistently.

His use of the term 'group', however, calls for further comment. Galois spoke of groups of permutations and groups of substitutions. By a *group of substitutions* he meant a collection of substitutions closed under composition, and since his collections were always finite this makes a group of substitutions a group in the modern sense. By a *group of permutations* he meant a collection of arrangements with the property that the substitutions that turn any given permutation into any other form a *group of substitutions*. Neumann (2011, p. 22) observes that Galois was not only inconsistent in his use of these two terms, although the context usually makes matters clear, he gradually came to attach more theoretical weight to the concept of a *group of substitutions*. Because the concept is emerging here I will also put the word 'group' in italics when discussing Galois's Mémoire.

Galois said that the following results were well known[7]:

1. An irreducible equation cannot have any root in common with a rational equation without dividing it.
2. Given an arbitrary equation which has no equal roots, of which the roots are $a, b, c, \ldots$, one can always form a function V of these roots, such that none of the values that are obtained by permuting the roots in this function in all possible

---

[6]Galois's memoir is available in English in Neumann, *The Mathematical Writings of Évariste Galois* (2011), and also in Edwards, *Galois Theory*.

[7]All these results are taken from (Neumann 2011, pp. 111–113).

ways will be equal. For example, one can take

$$V = Aa + Bb + Cc + \cdots,$$

$A, B, C, \ldots$ being suitably chosen whole numbers.

3. The function V being chosen as is indicated in the preceding article, it will enjoy the property that all the roots of the proposed equation will be rationally expressible as a function of V.

4. (Known, he said, to Abel and stated in his posthumous memoir on elliptic functions): Let us suppose that the equation in V has been formed, and that one of its irreducible factors has been taken so that V is a root of an irreducible equation. Let $V, V', V'', \ldots$ be the roots of this irreducible equation. If $a = f(V)$ is one of the roots of the proposed equation, $f(V')$ will also be one of the roots of the proposed equation.

Let us look briefly at these claims. The first says that if an irreducible polynomial $f(x)$ has a root in common with another polynomial $h(x)$ then $f(x)$ divides $h(x)$. It is proved by applying the Euclidean algorithm to the two polynomials.

The second is proved informally this way. Denote the roots of the polynomial by $x_1, x_2, \ldots, x_n$. There are $n!$ expressions of the form $a_1 x_n + \cdots + a_n x_n$ for each choice of the integer coefficients $a_1, a_2, \ldots, a_n$. It is enough to check that the coefficients can be chosen so that $a_1 x_1 + \cdots + a_n x_n$ never equals one of its permutations (the general case reduces to this one by relabelling the roots). An equality between $a_1 x_1 + a_2 x_2 + \cdots + a_n x_n$ and one of its permutations (say $a_1 x_2 + a_2 x_1 + \cdots + a_n x_n$), is a linear condition on the coefficients $a_1, a_2, \ldots, a_n$. This eliminates a one-dimensional subspace of $\mathbb{Z}^n$, the space of n-tuples of integers, because the roots are assumed to be distinct and not rational. There are at most $n!$ equations of this kind, and it is enough to choose $a_1, a_2, \ldots, a_n$ from $\mathbb{Z}^n$ with these one-dimensional subspaces removed, which is clearly non-empty.

In more modern terms, the third claim says that we have the given equation, $f(x) = 0$, which we take to be irreducible of order n over some fixed extension field K of the rational numbers that contains all the roots of unity we shall want, and having n distinct roots, $x_1, \ldots, x_n$. We have the group S_n, and an element $V = a_1 x_1 + \cdots + a_n x_n$ for suitable integers $a_1, \ldots, a_n$ with the property that V is what we call a *primitive* element of the splitting field for the given equation—that is, we adjoin it and we adjoin all the roots of the given equation, which is to say all the roots are rational functions of this single V. We denote the $n!$ values of V under the action of S_n by $V = v_1, V_2, \ldots, V_{n!}$. Denote the polynomial

$$(X - V_1)(X - V_2) \cdots (X - V_{n!})$$

by $F(X)$ and suppose it factors as $F = G_1 G_2 \ldots G_k$, each factor G_j being irreducible over the field K.

The group of the equation $f(x) = 0$ is obtained as follows. Let G be one of the irreducible factors of F above, of degree d, say. Then the splitting field of $f(x)$ is the simple extension of K obtained by adjoining a root of G. Because G is of order d there are d permutations of the roots that map the roots of G among themselves.

The third claim says that all the roots of the equation are expressible as rational functions of V, and is harder to understand than the first two. Edwards (1984, pp. 43–44) notes that Poisson commented that "The proof of this lemma is insufficient", but that it could be established using some specific ideas introduced by Lagrange. Galois gave an argument to this effect that can be sketched as follows. For, if $V = \phi(x_1, \ldots, x_n)$ and $g \in S_{n-1}$, denote by $g\phi(x_1, \ldots, x_n) = \phi(x_1, g(x_2), \ldots, g(x_n))$—here the subgroup S_{n-1} acts on $x_2, \ldots, x_n$ leaving x_1 fixed. Now consider the expression $\prod_{g \in S_{n-1}} (V - g\phi)$ as a function of x_1, given by the equation $F(V, x_1) = 0$. It is symmetric in the variables $x_2, \ldots, x_n$ and so it can be written as a polynomial in x_1 of degree $(n-1)!$. The root x_1 is a common root of this equation and the given one.[8]

Let us take the case of an irreducible cubic equation, say

$$x^3 + ax^2 + bx + c = 0,$$

with roots u, v, w. We have the equations

$$u + v + w = -a, \ uv + vw + wu = b, \ uvw = -c,$$

to which we now add the equation

$$u + kv + mw = V.$$

Galois asks us to consider the expression $V' = u + kw + mv$, and the product (I've momentarily replaced his v by a more neutral symbol t)

$$(t - V)(t - V') = (t - u + kv + mw)(t - u + kw + mv).$$

This is symmetric in v and w, so expanded out as a polynomial in t it becomes something of the form

$$t^2 - \sigma_1(v, w)t + \sigma_2(v, w).$$

Now plainly the equation $(t - V)(t - V') = 0$ has V as a root, so the equation $t^2 - \sigma_1(v, w)t + \sigma_2(v, w) = 0$ regarded as an equation in u has u as a root. But u is a root of our original cubic equation, so u is a result of the remainder on dividing the cubic equation by this quadratic one, but that remainder is linear, so u is rational in a, b, c, and V. Galois's claim is that this argument works in general.

---

[8]What has to be shown is that x_1 is the only root of the two equations, see (Edwards 1984, pp. 44–45).

Galois gave this argument in support of the last of these claims: form the product $\prod_{g \in S_n}(V - g\phi)$. This expression is divisible by the given equation, so V' is some $g\varphi$. Let $F(V, x_1)$ be as before [so we note that the product is here taken over all $g \in S_{n-1}$]. Now repeat the previous argument by combining the given equation and $F(V', b) = 0$ to deduce $b = f(V')$.

Galois now gave some results of his own, starting with this Theorem.[9]

Theorem (Galois's Proposition I). *Let an equation be given of which the m roots are $a, b, c, \ldots$. There will always be a* group of permutations *of the letters $a, b, c, \ldots$ which will enjoy the following property:*

1. *That every function invariant under the roots invariant under the* substitutions *of this group will be rationally known;*
2. *Conversely, that every function of the roots that is rationally determinable will be invariant under the* substitutions.

He added the following note[10]:

Here we call invariant not only a function of which the form is invariant under the *substitutions* of the roots among themselves, but also one whose numerical value will not change under these *substitutions*. For example, if $Fx = 0$ is an equation, Fx is a function of the roots which is not changed by any *permutation*.

When we say that a function is rationally known, we wish to say that its numerical value is expressible as a rational function of the coefficients of the equation and of some adjoined quantities.

In other words, every equation with roots $x_1, \ldots x_m$ admits a *group G* with the property that every G-invariant function is rationally known and conversely every rationally known function is G-invariant. We must remember that the very concept of a group was unknown at this stage, and Galois was introducing it here in terms of permutations and substitutions.

Before sketching a proof, Galois observed that in the case of what he called *algebraic* equations the *group G* is the group S_n of all *permutations* because the rationally known functions are the symmetric functions.[11] But, in the case of the cyclotomic equation corresponding to an nth root of unity, the *group* is the cyclic group of order $n - 1$ (Galois tacitly assumed that n is prime here). This points, albeit obscurely, to an important advance Galois made over Lagrange. Galois formulated the theory so that it applied not to equations with arbitrary coefficients as parameters, but to all equations with numerical coefficients (as, for example, the cyclotomic equation). So the *group* he was directing attention to is specific to the equation that is to be solved; it might be S_n or some subgroup. This is why he added a footnote at this point to explain that a function is invariant if not only its form is unaltered by the *substitutions* of the roots, but its numerical value does not alter.

---

[9]Taken from (Neumann 2011, pp. 113–115).

[10]Taken from (Neumann 2011, p. 113).

[11]Edwards (1984, p. 54) explains that an "algebraic equation" is one with letters rather than numbers as its coefficients.

Galois also added that he said a function is rationally known if its numerical value is a rational function of the coefficients of the given equation and any quantities so far adjoined. This makes the concept a moving one, dependent on where we are in the process of solving the equation.

The distinction between formal invariants and numerical invariants is worth underlining, if only because later writers were not always clear. Consider the equation $x^4 + 1 = 0$, whose roots are (setting $\varepsilon = e^{i\pi/4}$)

$$x_1 = \varepsilon, x_2 = i\varepsilon, x_3 = -\varepsilon, \text{ and } x_4 = -i\varepsilon.$$

The group for this equation turns out to be

$$\{e, (12)(34), (13)(24), (14)(23)\},$$

and the expression $x_1 x_2^2 + x_3 x_4^4$ is not formally invariant under this group, but its numerical values are (they are zero).[12]

Galois now proved the theorem. He argued that there is a rational function V of the roots of the given equation such that every root of the given equation is a rational function of V. Consider the irreducible equation satisfied by V. Let $V', V'', \ldots V^{(k-1)}$ be the roots of this irreducible equation and let $\phi(V), \phi_1(V), \ldots, \phi_{m-1}(V)$ be the roots of the given equation. Then the k substitutions of the roots obtained by sending V to $V', V'', \ldots$, and V itself, and simultaneously each $\phi_j(V)$ to $\phi_j(V'), \phi_j(V''), \ldots$ is the required *group G*.

Then came another theorem.[13]

Theorem (Galois's Proposition II). *If one adjoins to a given equation the root r of an irreducible auxiliary equation,*

1. *one of two things will happen; either the* group *of the given equation will not be changed, or it will be partitioned into p groups* each belonging respectively to the proposed equation when one adjoins to it each of the roots of the auxiliary equation;
2. *these* groups *will enjoy the remarkable property that one will pass from one to another by operating on all the* permutations *of the first with one and the same substitution of letters.*

This theorem is a casualty of the circumstances in which Galois wrote this memoir. He struck out the statement that the number p is the degree of the auxiliary equation, and failed to state that the partition is into j groups, where j divides p. These groups are what we would call conjugate, and the substitutions will be H and gHg^{-1} in the other. What Galois meant by writing "all the *permutations* are

---

[12]This example comes from (Bolza 1893, p. 99), which is Bolza's review of the English translation of (Netto 1882), see Sect. 15.

[13]Taken from (Neumann 2011, p. 119).

obtained from the first one by operating with one and the same substitution" is that if the *permutations* of one are called H, the *permutations* in another will be called gHg^{-1}. In other words, if g is the substitution that relabels all the permutations (i.e., arrangements) in one group as the permutations in the other, then the corresponding substitutions are related as h and ghg^{-1}. His insight becomes clearer with an example, such as the one provided in Sect. 10.5 at the end of this chapter. Missing entirely at this point in the memoir is any discussion of the *importance* of H being a *normal* subgroup; Galois dealt with that in his letter to Chevalier.

Galois's insight is that a useful adjunction alters the group G of the given equation, which it breaks up into p mutually conjugate subgroups of G, each belonging to the original equation when the roots of the auxiliary equation are adjoined. Moreover, these 'groups' are such that one passes from one to the next by applying the same *substitution* to all the *permutations* of the first—so the groups are conjugate, as we would say. The given equation may, and usually will, remain irreducible even with the new numbers adjoined.

Proof. (Here Galois wrote: "There is something to be completed in this proof. I do not have the time.") Either the original equation is still irreducible after the adjunction, in which case its *group* does not change, or the equation becomes reducible. In this case the equation for V breaks into j factors:

$$f(V,r) \times f(V,r_1) \times \cdots \times f(V,r_{j-1}),$$

where the other values of r are $r_1, \ldots, r_{j-1}$.

Thus the *group* of the given equation also decomposes into *groups*, each containing the same number of elements (*permutations*) because each value of V corresponds to a *permutation*. These *groups* are those of the given equation as one adjoins successively $r, r_1, \ldots, r_{j-1}$.

Since all the values of V are rational functions of one another, if, say, V is a root of $f(V,r) = 0$ and $F(V)$ another, then if V' is a root of $f(V',r') = 0$ then $F(V')$ will be another. Here Galois added[14]

$$f(F(V),r) = a \quad \text{function} \quad \text{divisible} \quad \text{by} \quad f(V,r).$$

Therefore (lemma 1)

$$f(F(V'),r') = a \quad \text{function} \quad \text{divisible} \quad \text{by} \quad f(V',r').$$

Then came this theorem.[15]

---

[14]Taken from (Neumann 2011, p. 119).

[15](Neumann 2011, p. 121), emphasis Galois's.

Theorem (Galois's Proposition III). *If one adjoins to an equation* all *the roots of an auxiliary equation, the* groups *in question in Theorem II will enjoy the additional property that the substitutions are the same in each* group.

Galois could only note "One will find the proof". At this point, he has claimed that all the subgroups obtained in Theorem II are the same, and the reduced group of the equation is what we would call normal in the one before.

And then comes this theorem[16]:

Theorem (Galois's Proposition IV). *If one adjoins to an equation the* numerical *value of a certain function of the roots, the group of the equation will be reduced in such a way as not to have any other permutations than those under which this function is invariant.*

Galois now came to his fundamental question: "In what cases is an equation solvable by simple radicals?" He began by saying

> I shall observe to begin with that, to solve an equation, it is necessary to reduce its group successively to the point where it does not contain more than a single permutation. For, when an equation is solved, an arbitrary function of its roots is known, even when it is not invariant under any permutation.

Certainly, he then argued, an equation solvable by simple radicals is such that the *group* can be diminished to a *group* with a single element by adjoining a suitable sequence of roots of prime degree. Conversely, if the *group* breaks up in this fashion then the equation is solvable by simple radicals. One can suppose the radicals that reduce the *group* are chosen to be of least degree at each stage, and one can assume that such pth roots of unity as are necessary have been adjoined, for they satisfy an irreducible equation of degree less than p. This follows, as Galois said, from Gauss's work on cyclotomic polynomials. Therefore, by Theorems II and III, the *group* of an equation solvable by radicals must decompose, as he put it, into p subgroups obtained one from the other by the same *substitution* and all containing the same *substitutions*. Or, as we would say, there is a normal subgroup of the original group of the equation of index p. Therefore the original group of the equation must decompose into a sequence of smaller groups in this way, each group of prime index in the one before.

Conversely, Galois went on, if the *group* decomposes in this fashion then there is a sequence of adjunctions that brings the decomposition about. Galois sketched a proof, which was obscurely worded (and at one point he wrote the exact opposite of what he meant) and relies on the resolvent V. I omit the proof, which was to be rescued by Jordan (see §518 of his *Traité* below).

Galois then indicated how this analysis held up in the case of quartic equations. The *group* of the equation contains 24 *substitutions* and on adjoining a [suitable] square root decomposes into two *groups* each of 12 *substitutions*. Galois listed the elements of one of these *groups*: it is identical with what we call A_4—the

---

[16](Neumann 2011, p. 121), emphasis Galois's.

other is the coset of A_4 in S_4. This group splits into three *groups*, in line with Galois's second and third theorems on adjoining a radical of the third degree (a cube root?). One of these is the group that came to be called a Klein group, $K_4 = \{e, (12)(34), (13)(24), (14)(23)\}$, and it splits into two 'groups': $\{e, (12)(34)\}$ and $\{(13)(24), (14)(23)\}$—notice that this second 'group' is also just a coset of the first. Adjoining a square root isolates the *group* $\{e, (12)(34)\}$, which is resolved by adjoining a square root.

Galois concluded by turning to the investigation of irreducible equations of prime degree, p. Adjunction must start with a qth root for some prime q. But he did not prove the unsolvability of the quintic by his general theory. Instead he gave an argument which was a lot closer to the hints of Lagrange and, especially, the methods of Abel.

First of all he asked: what is the *group* of an irreducible equation of prime degree p if it is solvable by radicals? He answered his question by showing that such an equation is either unaffected by the adjunction of a radical or it is reduced completely. Therefore, in the reduction process, the last *group* reached before the *group* with only one element must be a *group* of p elements, and by a result of Cauchy's this means the *group* must be the cyclic group of order p. The preceding group to this one must be made up, Galois continued, of a certain number of *groups* having the same *substitutions* as this one. Let me note that this makes sense; Galois was alluding to the corresponding coset decomposition. Galois now labelled the roots $x_0, x_1, \ldots, x_p = x_0$ and considered what must happen to x_k; it must go to $x_{f(k)}$, and so one has

$$f(k + c) = f(k) + C,$$

where C is independent of k. Therefore

$$f(k + mc) = f(k) + mC,$$

and setting $c = 1, k = 0$ deduced that $f(m) = am + b$ and so

$$f(k) = ak + b,$$

a and b being constants. This gave him the form of the next *group* up in the decomposition (let me call it G).

Then Galois showed that the whole *group* of the equation must be the *group* just found, and that a polynomial equation of prime degree is solvable by radicals if and only if its *group* is of the form of the *group* G just found.

Now, this *group* has the property that once two of the roots of the corresponding polynomial equation are known rationally, so are the rest. In other words, it acts doubly transitively on the roots. In fact, as no substitution of the form

$$x_k \mapsto x_{ak+b}$$

other than the identity fixes two elements it follows that once two roots are known the rest are and the *group* of the equation reduces to the *group* with one element.

Galois now wrote down the *group* of the equations of degree 5 that are solvable by radicals, listing all its 20 elements as permutations of 5 symbols.

To sum up: The smallest subgroup of the group of an irreducible equation of prime degree must, by a result of Cauchy's, contain p elements, and the only subgroup of S_p of order p is the cyclic group of order p. Then

- (A): the next group in the decomposition must act in this form (modulo p)

$$x_k \mapsto x_{ak+b}.$$

- (B): indeed, all substitutions of the group of the given equation must be of this form.
- (C): therefore, for the given equation to be solvable by radicals, it is necessary and sufficient that once any two roots are given all the others are rationally known.

This is not the case in S_5, as the subgroup G_{20} of order 20 shows, which is generated by $(1, 2, 3, 4, 5)$ and $(2, 3, 5, 4)$.

10.3   From Galois's Letter to Chevalier

From propositions II and III of the first memoir, we perceive a great difference between adjoining to an equation one of the roots of an auxiliary equation and adjoining all of them. In both cases the group of the equation breaks up by the adjunction in sets such that one passes from one to the other by the same substitution, but the condition that these sets have the same substitutions holds with certainty only in the second case. This is called the *proper decomposition*.

In other words, when a group G contains another H, the group G can be divided into sets, each of which is obtained by multiplying the permutations of H by the same substitution; so that

$$G = H + HS + HS' + \dots.$$

And it can also be divided into sets which contain the same substitutions, so that

$$G = H + TH + T'H + \dots.$$

These two methods of decomposition are usually not identical. When they are identical, the decomposition is *proper*.

It is easy to see that when the group of an equation is not susceptible of any proper decomposition, then, however, the equation be transformed, the groups of the transformed equations will always have the same number of permutations.

On the other hand, when the group of an equation admits a proper decomposition, in which it has been separated into M groups of N permutations, then we can solve the given equation by means of two equations, one having a group of M permutations, the other N.

When therefore we have exhausted in the group of an equation all the possible proper decompositions, we shall arrive at groups which can be transformed, but whose permutations will always be the same in number.

If each of these groups has a prime number of permutations, the equation will be solvable by radicals; otherwise not.

The smallest number of permutations which an indecomposable group can have, when this number is not a prime, is $5 \cdot 4 \cdot 3$.

10.4 Exercises

As we have seen, Galois discussed which quintic equations are solvable by radicals in terms of a group of order 20 that he introduced, and which is connected with equations of the form $x^5 - a = 0$, where a is a rational number that is not a perfect fifth power. These exercises relate to the much simpler, but similar story that concerns cubic equations of the form $x^3 - a = 0$, where a is any rational number that is not a perfect cube. Without loss of generality, we shall assume that $a = 2$.

The roots of the equation $x^3 - 2 = 0$ are $\rho = \sqrt[3]{2}$, $\omega\rho$, and $\omega^2\rho$, where ρ is the real cube root of 2 and ω is a cube root of unity, for definiteness $\omega = \frac{1}{2}(-1 + i\sqrt{3})$. We are interested in the field $K = \mathbb{Q}(\rho, \omega\rho, \omega^2\rho)$ that contains all the roots, and the intermediate field $\mathbb{Q}^* = \mathbb{Q}(\omega)$.

These roots may be permuted in six ways if we keep only the rational numbers fixed. Notice that any automorphism of the field K that fixes $\mathbb{Q}$ maps ρ to one of ρ, $\omega\rho$, and $\omega^2\rho$ because

$$2 = f(2) = f(\rho^3) = (f(\rho))^3,$$

so $f(\rho)$ is a cube root of 2. Similarly, $f(\omega)$ must be a cube root of unity, and therefore one of 1, ω, and ω^2.

The transposition $(\rho, \omega\rho)$ is given as this automorphism of the field K that fixes $\mathbb{Q}$:

$$f(\rho) = \omega\rho, \, f(\omega\rho) = \rho.$$

We find that

$$\rho = f(\omega\rho) = f(\omega)f(\rho) = \omega^n f(\rho) = \omega^{n+1}\rho,$$

for some integer $n = 0, 1$, or 2, so we must have $n = 2$. Therefore $f(\omega) = \omega^2$.

We now check that $f(\omega^2\rho) = \omega^2\rho$, as required.

$$f(\omega^2\rho) = (f(\omega))^2 f(\rho) = \omega^4\omega\rho = \omega^2\rho.$$

1. Define the other transpositions $(\rho, \omega^2\rho)$ and $(\omega\rho, \omega^2\rho)$ explicitly.
2. The 3-cycles are also automorphisms of the field K that fix $\mathbb{Q}$. One is given by

$$f(\rho) = \omega\rho, \ f(\omega\rho) = \omega^2\rho.$$

   Check that $f(\omega) = \omega$ and so $f(\omega^2\rho) = \rho$, as required.
3. Notice that this automorphism and its inverse are automorphisms of the field K that fix $\mathbb{Q}^*$. In terms of the groups of the equation, we have therefore shown that

$$\Gamma(K, \mathbb{Q}) \cong S_3, \ \text{and} \ \Gamma(K, \mathbb{Q}^*) \cong A_3 \cong \mathbb{Z}_3.$$

4. Consider now the integers modulo 3, namely $\{0, 1, 2\}$ and consider the maps

$$j \mapsto aj + b,$$

   where $a \in \{1, 2\}$ and $b \in \{0, 1, 2\}$. They form a group of order 6. The map $z \mapsto z+1$ corresponds to the 3-cycle $(0, 1, 2)$ and so generates a normal subgroup of order 3, and the map $j \mapsto aj$ is the transposition $(1, 2)$. Find the maps of this form corresponding to $(0, 1)$ and $(0, 2)$.

If we now associate 0 with ρ, 1 with $\omega\rho$, and 2 with $\omega^2\rho$ then we obtain a new representation of the group $\Gamma(K, \mathbb{Q})$.

So if we first adjoin ω to $\mathbb{Q}$, to obtain $\mathbb{Q}^*$, and then adjoin any root of the equation $x^3 - 2 = 0$ we get the other roots automatically. This is an example of what was later called a normal extension of $\mathbb{Q}^*$ by K. Notice that the group $\Gamma(K, \mathbb{Q}^*)$ is a normal subgroup of the group $\Gamma(K, \mathbb{Q})$.

However, if we adjoin a root, say ρ, to $\mathbb{Q}$, to obtain $\mathbb{Q}(\rho)$ and then adjoin another root, say $\omega\rho$ to obtain K we find that the group $\Gamma(\mathbb{Q}^*, \mathbb{Q}(\rho))$ is of order 2, and the group $\Gamma(\mathbb{Q}(\rho), \mathbb{Q})$ is of order one! This indicates why non-normal extensions—such as this one—are not productive.

10.5   A Cayley Table of a Normal Subgroup

The first table below is a Cayley table for the group S_3. Each row of it presents an arrangement of the elements of the group, or, in Galois's terminology, a *permutation* of them. The passage from the top row to any other is obtained by multiplying the elements in the top row on the left by the corresponding element on the left, thus effecting what Galois called a *substitution*. What Galois noticed was that in the case of a 'normal' subgroup H of a group G—to use modern terminology—one in which $gH = Hg$ for all $g \in G$, the table forms into blocks that themselves can be regarded as permutations and which act on each other to perform substitutions, as in the second table, which is for the quotient group S_3/A_3.

S_3	e	(123)	(132)	(12)	(13)	(23)
e	e	(123)	(132)	(12)	(13)	(23)
(123)	(123)	(132)	e	(23)	(12)	(13)
(132)	(132)	e	(123)	(13)	(23)	(12)
(12)	(12)	(13)	(23)	e	(123)	(132)
(13)	(13)	(23)	(12)	(123)	e	(132)
(23)	(23)	(12)	(13)	(123)	(132)	e

S_3/A_3	A_3	A_3s
A_3	A_3	A_3s
A_3s	A_3s	A_3

10.6 Galois: Then, and Later

The bicentenary of Galois's birth generated some valuable historical work that was published in an issue of the *Revue d'Histoire des Mathématiques*. In a long article (Boucard 2011) Jenny Boucard argues for Louis Poinsot as a missing link between Gauss and Galois. Poinsot was known for his work on algebra and number theory, and she finds a memoir of his in 1818 on polynomial congruences modulo a prime and a previously unpublished work of his on permutations of 1820 indicative of a coming together of these two fields.

Caroline Ehrhardt, in her (2011), traces Galois's ideas from his day to the modern period in an attempt to understand why they were eventually found to be so exciting and how they transformed algebra in later generations. She notes, as we do below, the difference between the neglect Galois's ideas were met with in Galois's lifetime, and the much more positive response they got after their republication in 1846, and she observes that while Serret promoted Galois's ideas in the 1860s in the spirit of classical algebra and the solution of equations, Jordan offered a radically different interpretation that moved away from the solution of equations and into the theory of groups that Jordan was trying to create. She further traces the implications of this split as far as Picard's republication of Galois's papers in 1896. This theme is pursued at greater length in (Brechenmacher 2011), a paper that looks outside France at networks of texts associated with Kronecker, Klein, and Dickson. Brechenmacher also amplifies a theme in (Goldstein 2011) where the implications of Galois's work for Hermite were traced: the importance—going back indeed to Abel—of Galois's ideas for the theories of elliptic and abelian functions.[17] This issue is pursued in the next chapter and in Chap. 15.

---

[17]Some of these connections are also traced in (Bottazzini and Gray 2013).

Chapter 11
After Galois

11.1 Introduction

Galois's theory was considered very difficult in its day, and was also poorly published. This chapter looks at what had to happen before it could become mainstream mathematics, and how as it did so it changed ideas about what constitutes algebra and started a move to create a theory of groups.

11.2 The Publication of Galois's Work

Galois's work did not pass entirely unnoticed in his lifetime, but what little was said about it was not encouraging. The Academy of Sciences had commissioned a report on Galois's first memoir by Poisson and Lacroix and it was submitted on 4 July 1831 (although signed only by Poisson).[1] In it they observed that in a posthumously published paper Abel had already claimed that

> If three roots of an irreducible equation of prime degree are such that any one of these roots can be expressed rationally in terms of the other two then the equation is solvable by radicals.

They admitted that this differed from Galois's theorem in that it asserted a sufficient, but not a necessary, condition for an equation to be solvable by radicals, but having thus hinted that Galois's work was not completely original they went on to disparage its quality. They argued, reasonably enough, that even if the truth of Galois's

---

[1] See *Procès-verbaux des séances de l'Académie* 6, 660–661, quoted in Taton (1947, pp. 120–122).

J. Gray, *A History of Abstract Algebra*, Springer Undergraduate Mathematics Series,
https://doi.org/10.1007/978-3-319-94773-0_11

principal theorem was admitted it was not possible to use the criterion in practice, for

> a condition of solvability, if it exists, must have an exterior character that one can verify by inspecting the coefficients of the given equation or at most by solving other equations of a degree less than the degree of the given equation.

But, they went on

> Be that as it may, we have made every effort to understand Galois's demonstration. His reasoning is neither sufficiently clear nor sufficiently developed for us to be able to judge its correctness and we will not be in a state to give an idea of it in this report.

However, they said, the author intends to give a complete account, and often the various parts of an extended work shed light on each other and make the whole thing easier to understand than the isolated parts, so they recommended waiting until the author publishes the whole work, when one might be able to reach a definitive opinion, but until then they had to withhold any recommendation to publish it.

Nor did things improve after Galois's death. The by-then elderly Sylvestre Lacroix wrote in the sixth edition of his *Compléments des Element d'Algèbre* (1835, 382) that

> In 1831 a young Frenchman, Évariste Galois, who died the following year, announced in a paper presented to the Academy of Science that, "in order for an irreducible equation of prime degree to be solvable by radicals, it is necessary and sufficient that, once any two of the roots are rationally known, the others can be determined from these by rational operations." But the memoir seemed, to the committee appointed to review it, to be very nearly unintelligible.[2]

In 1843, Joseph Liouville announced to the Paris Academy of Sciences that he had discovered among Galois's unpublished papers a "solution which is both correct and profound" to the problem of determining when an irreducible equation of prime degree is solvable by radicals.[3] Galois had died in 1832, and Liouville became interested in Galois's work when some friends of Galois brought it to his attention; Liouville said that Chevalier had entrusted the manuscripts to him. It has been conjectured that the revival of interest in 1842 stemmed from the striking coincidence that Galois's school teacher, Louis Richard, who had been very impressed by Galois, had acquired a second brilliant student, Charles Hermite, who rapidly attached himself to Liouville, and that this connection suggested an opportunity for the friends of Galois. All of which makes it plausible that these people could have gone to Liouville in the hope that he, through his own ability as a mathematician and as the

---

[2]Quoted in Kiernan (1971, pp. 90–91), who observes that the remark was also picked up in (Galois 1846).

[3]This account follows Lützen (1990, Chapters III, V, and XIV). For this quote, see p. 129.

editor of a major journal, could assist in doing justice to Galois. But it does not mean that any of then had fully read or understood Galois's ideas, and that seems unlikely.

Chevalier also prepared an annotated version of the first of Galois's memoirs, and Liouville prepared it for publication in his journal in December 1843, only to withdraw it at the last minute. His reasons remain unclear; did he find a error, did he decide to do Galois better justice? Whatever the reason, in 1846 he published a longer account, including a fragment of another paper, the last letter to Chevalier, and all of Galois's previously published papers. The legend of Galois was born.

In 1843, when addressing the Academy, Liouville had admitted that Galois's memoir was difficult and had promised to write a commentary elucidating it, but in 1846 there was no commentary. Most likely, he had found it too hard to fit in the necessary study alongside all his other commitments. But he did offer a short course of private lectures on Galois's ideas—the date is uncertain but the winter of 1843–1844 is likely—which Serret attended. Unpublished notes by Liouville from this period show that he had filled in the largest gaps in Galois's sketchy exposition, but it seems he never pulled them together into a systematic exposition. We know he intended to, because Serret said so in his *Cours d'Algèbre Supèrieure* of 1849. Serret was to go on to be a prominent exponent of Galois's ideas in successive editions of this book, but not in 1849 (Serret, by the way, was another student of Richard's). A detailed analysis of Liouville's unpublished notes conducted by Lützen suggests that the biggest problem Liouville had was adapting to the fundamentally group-theoretic way of thinking in Galois's work; Liouville adhered to Cauchy's way of thinking about rational functions and their values. This makes them closer to Galois's work than the twentieth century reworking of his theory, but muffles the insight all the same.

In a sense, therefore, Liouville's influence on the rediscovery of Galois's ideas was indirect. He interested the young Joseph Bertrand in it, and when in 1845 Bertrand published a memoir on the classic topic of the number of values a function of n variables can take as its variables are permuted, the referee was Cauchy (reasonably enough). This fact alone would help explain why Cauchy returned to the subject, but the announcement at the Academy in 1843 might well have done the trick.

In one way, though, Liouville retarded the cause. By announcing his intention to publish a commentary but never writing one Liouville made other people wait and hold back from research they otherwise might have done. Serret was to wait 20 years before finally publishing his own ideas, in the third edition of his *Cours d'algèbre supérieure*, 1866. In Italy Enrico Betti eventually wrote his own commentary in 1851, but he still urged Liouville to publish his own.

11.3   Serret's *Cours d'Algèbre Supérieure*

The third edition of Serret's *Cours* has been described (Wussing 1984, p. 129) as "The first textbook-like presentation of Galois theory and, in particular, the earliest algebra-oriented presentation of group theory".[4] As Wussing (1984, p. 130) went on to note, Serret, in all editions of his *Cours*, saw algebra as the analysis of equations. In 1854 all he did with Galois's work was to summarise it in a few lines (on p. 4). He asked the question: When is an equation solvable by radicals?, and replied with references to Gallois (as Serret spelled the name).

> This difficult question has been completely answered, at least for irreducible equations of prime degree, by Evariste Gallois, former student of the École Normale and one of the deepest geometers that France has produced. In a memoir presented to the Academy of Sciences in 1831 and published in 1846 thanks to the efforts of Liouville, Gallois in effect proved this beautiful theorem: For an irreducible equation of prime degree to be solvable by radicals it is necessary and sufficient that, given any two of the roots, the others are rationally deducible.

He went on to say that a polynomial equation of degree greater than 4 is not in general solvable by radicals: "This important proposition, stated by Ruffini, was put beyond doubt by the recent work of Abel." In 1866 Serret merely added a few more lines to note the work of Hermite, Kronecker, Betti and other (unnamed) authors (and spelled 'Galois' correctly).

As Wussing went on to say (1984, 131), this

> ...indicates a certain precariousness in the evaluation of Galois's contribution by members of the French school in the early sixties. Galois's work is regarded as being of greatest significance in the theory of solution of algebraic equations, but the fundamental role of the group concept in that theory and elsewhere in mathematics is not yet realized.

Wussing pointed to Serret's failure to bring out the concepts of group and normal subgroup, and to describe the new method. The one theorem mentioned "is stated in the framework and language of the traditional, Abel-inspired, solvability theory". That said, as Wussing remarked, Serret recognised the significance of permutation theory for Galois's solvability theory, and did discuss Cauchy's permutation theory.

Section IV, Substitutions

1. General properties of substitutions,
2. Properties of conjugate systems of substitutions,
3. The indices of conjugate systems,
4. Some special cases in the theory of substitutions,
5. Applications of the theory of substitutions.

---

[4]Wussing's book remains an important influence on the history of abstract group theory and on the tendency of mathematics in the nineteeth century towards increasing abstraction.

Section V, The algebraic resolution of equations

1. The equations of the third and fourth degrees. General considerations about the algebraic resolution of equations,
2. The impossibility of the algebraic resolution of general equations of degree greater than four,
3. Abelian equations,
4. On a class of equations of the ninth degree solvable algebraically.

The impossibility proof that Serret presented was not Galois's but Abel's. He followed it with Wantzel's shorter and simpler proof of the same result, which he thought to present verbatim but from which he omitted some results about permutations that he had proved earlier in the book, and which is clearer because it is more explicit about the permutations of the roots that are involved. He then proved that an equation is solvable by radicals if all of its roots are known rationally when one of them, say x_1, is known and the others are of the form

$$x_1, x_2 = \theta(x_1), x_3 = \theta(x_2), \ldots.$$

A few pages later Serret looked at the cyclotomic equation and summarised Gauss's results. He dealt in detail with the constructibility by straight-edge and circle of the regular 17-gon.

Wussing then showed how Serret took over Cauchy's "system of conjugate substitutions", amalgamated it with Galois's closed set of substitutions, and came to speak of "substitutions of a group of permutations", although not of a "group of substitutions". This indicates, he suggests, how ambiguous Serret was about going all the way with Galois, and how attracted he was to Cauchy's ideas. Wussing (1984, p. 134) observes that at one point (Serret 1866, 413–420)

> Serret invokes a theorem of Galois, according to which it is possible to form a function V of the roots of an equation with distinct roots which takes on different values under all $n!$ permutations of the roots, such that each of the roots can be represented as a rational function of V. Serret's proof is a direct repetition of Galois's proof as published by Liouville in 1846

which is an oddly traditional, almost Lagrangean, point to pick.[5]

Serret turned to Galois's own contributions in the last chapter, "On the equations that are solvable algebraically", of Section V, when he worked through Galois's paper *Sur les conditions de resolubilité des équations par radicaux* ('On the conditions for the solvability of an equation by radicals') (first published 1846).

---

[5]Later, Serret was to be the main editor of the *Oeuvres* of Lagrange.

Serret stuck closely to what Galois had done and how he had done it, noting that
(*Cours*, vol. 2, 607) "I have followed the order of the propositions adopted by Galois,
but, very frequently, I had to close the gaps in the proofs", and that where, as in the
theorem about the existence of the function V just mentioned, Galois had employed
groups of permutations he (Serret) preferred to stick to substitutions. "After all",
Serret wrote (*Cours*, vol. 2, 608–609), "it is just a simple change in the form of the
statements of the theorems, dictated by the fact that there is no reason to consider
permutations except from the point of view of substitutions by which one passes
from one to the other".

Wussing, after quoting this passage, remarked:

> Nothing need be added to these words. There is probably no better way to describe the
> coexistence and the now (1866) completed amalgamation of the approaches of Cauchy and
> Galois.

I would only add that Serret did prove (§§576, 577) that an irreducible polynomial
equation of prime degree is solvable by radicals if and only if its group is of the form
Galois indicated: the roots are permuted by a group of the form $z \mapsto az + b$, where
a and b are residues mod p and a is non-zero. He then deduced from this that such
an equation is solvable by radicals if and only if all of its roots are known rationally
when any two of them are given. We can think of this result as saying that the group
is solvable if and only if it has a representation of a certain kind—but Serret did not
use such language.

Serret then ended his book with a translation into French of a memoir by
Kronecker published in 1853 that we shall look at next.

11.4   Galois Theory in Germany: Kronecker and Dedekind

German accounts of Galois theory were to emphasize the idea of successive
adjunction of elements and the formation of (intermediate) fields, but before they
reached that level of sophistication they considered the idea of a group. Here we
look at two complementary accounts, by Kronecker and Dedekind, who were to
come to embody very different attitudes to the aims and nature of algebra.

Kronecker

Fig. 11.1 Leopold
Kronecker (1823–1891).
Photo courtesy of the
Archives of the
Mathematisches
Forschungsinstitut
Oberwolfach

The German mathematician Leopold Kronecker learned the new theory of
solvability of polynomial equations from Hermite and others during his stay in Paris
in 1853, when he was 30, and became interested in the study of solvable equations.
While there he wrote one of his earliest papers, in which he expressed himself with
remarkable force. He claimed the previous studies into solvable equations of prime
degree by Abel and Galois[6]

> have essentially given as the result two criteria with which one can decide if a given equation
> is solvable or not. In themselves these criteria shed not the least light on the nature of
> solvable equations *themselves*. Indeed, strictly speaking one can not know at all if (aside
> from what Abel considered in Crelle's *Journal* vol. 4 and the simplest cases connected
> with the binomial equations) there are any equations that satisfy the stated solvability
> conditions. Still less can one *construct* such equations, and one is also not led through
> other mathematical work to such equations. ... And so the solvable equations themselves
> have remained up to now in a certain darkness ...

---

[6]See Kronecker (1853, 3) in *Werke* IV.

Accordingly, Kronecker set himself the task of finding (1853, 4)

> the most general algebraic function involving arbitrary quantities $A, B, C, \ldots$ that satisfies an equation of given degree whose coefficients are *rational* functions of these quantities.

In particular (1853, 8), to find the most general form of an algebraic function of A, B, C, etc. which, on varying the root signs that it contains, gives rise to symmetric and cyclic rational functions of A, B, C, etc.

His conclusion (1853, 10–11) was that the roots of any Abelian equation of any degree are rational expressions in the roots of unity—a celebrated result today known as the Kronecker–Weber theorem, and first proved by Weber in 1886.

He returned to the theme in a second paper with the same title (1856) and gave an expression on p. 36 that characterised the roots of an irreducible polynomial equation. This is explicit, if unwieldy, and it seems to have been regarded by many subsequent writers as either the last word on the subject, or a dead end because, for whatever reason, they do not make use of it. It is, of course, discussed by Netto in his book (1882)—see Chap. 13—who followed it with a group-theoretic account in Chap. 14. However, Kronecker gave no proofs of these results in his papers; he did not even comment when Weber published a proof of the Kronecker–Weber theorem. Edwards argues that Kronecker based his claim on papers that he had written but for some reason never got round to publishing—such papers would now be irretrievably lost. A proof of the formula was, however, given by Netto in his (1882), although Edwards finds it incomplete.[7]

In fact, these early papers by Kronecker are indicative of his life-long interest in the explicit and constructible in mathematics. He sought to construct all the equations which are solvable by radicals, as for instance, the cyclotomic equations $\frac{\alpha^p-1}{\alpha-1} = 0$, p a prime. Those equations in particular he called 'Abelian' in his (1856), their characteristic feature being that all the roots are obtained from one of the roots by successive repetition of a rational function—they are

$$x_1, x_2 = \theta(x_1), x_3 = \theta(x_2), \ldots.$$

It seems that this is the origin of the designation Abelian for a commutative group: the group of a cyclotomic equation permutes the roots cyclically.

In one of his many later works on number theory (Kronecker 1870), Kronecker gave an abstract definition of a finite commutative group, which he put in context as follows[8]:

> In articles 305 and 306 of the *Disquisitiones Arithmeticae*, Gauss based an arrangement of the different classes of quadratic forms on the theory of composition, and Herr Schering has recently dedicated a paper to the further development of the subject in a paper .... The very simple principles on which Gauss's method rests are applied not only in the places noted but also frequently elsewhere—and indeed already in the elementary parts of number theory.

---

[7] Kronecker's papers are discussed and corrected in Edwards (2009), where further historical information can be found. The discussion in Wussing (1984, pp. 120–123) is also informative.

[8] The passage is quoted in Wussing (1984, p. 65) but this is a new translation.

This shows, and it is otherwise easy to be convinced, that these principles belong to a more general and more abstract sphere of ideas. It is therefore proper to free their development from all inessential restrictions, so that it is unnecessary to repeat the same argument when applying it in different cases. This advantage is valid also for the development itself. Also, when stated with all admissible generality, the presentation gains in simplicity and, since only the truly essential features are thrown into relief, in transparency.

Kronecker then presented this definition:

Let θ', θ'', θ''', ... be a finite number of elements such that from any two of them a third can be derived by means of a definite procedure. Thus, if f denotes the result of this procedure for two elements θ' and θ'', which may be identical with each other, then there exists a θ''' equal to $f(\theta', \theta'')$. Furthermore we require that

$$f(\theta', \theta'') = f(\theta'', \theta')$$

$$f(\theta', f(\theta'', \theta''')) = f(f(\theta', \theta''), \theta'''),$$

and if θ'' is different from θ''' then $f(\theta', \theta'')$ is different from $f(\theta', \theta''')$.
Once this is assumed, we can replace the operation denoted by $f(\theta', \theta'')$ by multiplication $\theta' \cdot \theta''$, provided that instead of equality we employ equivalence. Thus using the usual equivalence symbol "$\sim$" we define the equivalence

$$\theta' \cdot \theta'' \sim \theta'''$$

by means of the equation

$$f(\theta', \theta'') = \theta'''.$$

Wussing picked this up because the group elements are abstract and not necessarily presented as permutations, but Kronecker was always concerned to use the group-idea to advance other domains of mathematics, chiefly number theory, and concentrated therefore on the study of the roots of equations, regarding them as given by some construction.

Dedekind

Richard Dedekind was a friend of Riemann's from his student days although, being a pianist, he gravitated to the circles around Dirichlet, who was connected to the Mendelssohns. He came to think of himself as the tortoise to Riemann's hare, and blessed with a long life became a great example of what can be done by constantly thinking upon things. Even so, historians of mathematics were surprised when, in 1976, Walther Purkert, a former student of Wussing's, showed that Dedekind had lectured on Galois theory at Göttingen in 1856–1858. It seems these may be the first lectures anywhere on the subject. As Purkert showed in his (1976) Dedekind developed the idea of an abstract group in the context of Galois theory, although he published nothing on it until 1894, in the famous 11th supplement to

Dirichlet's lectures on number theory (see the 4th edition).[9] Dedekind's manuscript, as described by Purkert, goes far beyond the limited ideas about groups published by his *Werke*, II, paper LXI. In it, Dedekind described the idea of permutations of a finite set of objects is generalised to that of a finite abstract group. For permutations, he showed

Theorem 11.1. *If* $\theta\theta' = \phi$, $\theta'\theta'' = \psi$, *then* $\phi\theta'' = \theta\psi$, *or more briefly,* $(\theta\theta')\theta'' = \theta(\theta'\theta'')$.

Theorem 11.2. *From any two of the three equations* $\phi = \theta$, $\phi' = \theta'$, $\phi\phi' = \theta\theta'$, *the third always follows.*

The subsequent mathematical arguments, however,

> are to be considered valid for any finite domain of elements, things, ideas, $\theta, \theta', \theta'', \ldots$ having a composition $\theta\theta'$ of θ and θ' defined in any way, so that $\theta\theta'$ is again a number of the domain and the manner of the composition corresponds to that described in the two fundamental theorems. (Purkert 1976, p. 4).

The ideas of subgroup (*Divisor*) and coset decomposition of a group G are defined, and a normal subgroup (*eigentlicher Divisor*) is defined as one, K, satisfying

$$K = \theta_1^{-1}K\theta_1 = \theta_2^{-1}K\theta_2 = \cdots = \theta_h^{-1}K\theta_h,$$

where the θs are coset representatives for K (cf. Galois: 'propre'). Dedekind showed that the cosets of a normal subgroup themselves formed a group, in which K played the role of the identity element. It seems that he went on to apply his theory to the study of polynomials, however, the manuscript is clearly incomplete, and a 'field-theoretic' part is missing. Nonetheless, a considerable amount of the theory of polynomial equations survives, and is analysed by Purkert.

---

[9] We shall look at a different aspect of this supplement in Chap. 19.

Chapter 12
Revision and First Assignment

It is time to revise the topics so far and discuss the first assignment. Rather than boil the previous eleven chapters down to a misleading 'essence', it is better to raise some questions and let you answer them yourselves, to your own satisfaction.

The first task is to start coming to grips with Gauss's work on number theory. As the extract from the book by Goldstein, Schappacher, and Schwermer explains very clearly, the book redefined—re-created, if you like—the subject of number theory for the ninetieth century, which indicates how wide a framework must be to capture fully what Gauss did.

The mathematics in the *Disquisitiones Arithmeticae* provides evidence with which to test claims about Gauss. Modular arithmetic is not difficult, but Gauss's presentation was unusual in being systematic, and we can wonder why he provided it. The subject is crucial to his account of both quadratic reciprocity and his theory of quadratic form s; clearly he wanted an effective notation for it. As we noted, Sophie Germain was to write to him in 1819 to thank him for introducing the idea of equivalence to replace the previous use of equality; this small step made the theory easier by removing an otherwise potentially confusing abuse of language.

Where does the question of reduction modulo a prime p first enter the subject? How often did Lagrange, or Gauss, need to find a square root modulo p, as opposed to knowing only whether a number was a square modulo p?

The theorem of quadratic reciprocity is special, as even the thorough build-up through special cases demonstrates; see, for example, those questions about primes p for which, say $+2$ or -2 is a square mod p. But why is it so special? Why did Gauss call it the 'golden theorem'?

Chapter 4 introduces the one truly difficult part of Gauss's *Disquisitiones Arithmeticae* which is the theory of composition of quadratic form s (we saw that even Kummer found this hard). Key words here are 'fixed relationship' and 'character', and one jewel to appreciate is the table about forms of discriminant -161. Before climbing any mountain, you need to be confident that the view from the top is worth it (not to mention the health benefits). In the present case, we should

© Springer Nature Switzerland AG 2018
J. Gray, *A History of Abstract Algebra*, Springer Undergraduate Mathematics Series,
https://doi.org/10.1007/978-3-319-94773-0_12

ask: why was Gauss's *Disquisitiones Arithmeticae* so important? What was good about it? What was difficult about it? What impact did it have?

If you take a few minutes you will be able to show that the 16 equivalence classes of positive forms of discriminant −161 are the only ones, and they have the characters stated. You will not be able to show that in general only half the possible characters arise, or prove theorems about composition. But this puts you in the position of anyone reading the *Disquisitiones Arithmeticae*, and indeed of Gauss himself at one stage in the course of doing the work. It is impressive to have spotted the pattern that the table illustrates, and even more so to have been able to prove it exists in general.

Cyclotomy was the topic that first drew people's attention—I would suppose from a mixture of interest and relative ease of access. Can you be more precise? If you explore that question in detail, you might become a little unhappy with the usual story about complex numbers in which Wessel and Argand play major roles.

The raw ingredient in the theory of cyclotomy, the clever idea that makes it work, is the existence of primitive elements modulo a prime. Using this concept, Gauss exhibited a whole class of polynomial equations and showed how to solve them by solving a succession of much simpler equations. What is the big idea: that these equations can be solved, or that they can be solved systematically?

Gauss frequently insisted that one of the merits of number theory was that simple conjectures had deep proofs and that these proofs revealed hidden connections. Chapter 6 offered some of the evidence: the connection between the theorem of quadratic reciprocity and composition of forms; and the connection between the theorem of quadratic reciprocity and cyclotomy (which is hard, and a bit mysterious). The least that Gauss's proofs establish is that the connection is genuine—can you say anything more?

There is surely no dispute at this stage that what Gauss did was very difficult. I might be prepared to argue that it was unusually difficult, and that the difficulties were of a particularly conceptual kind—indeed, that that is one of the ways the *Disquisitiones Arithmeticae* had the impact it eventually did. Be that as it may, Dirichlet's lectures were the vehicle by which people learned Gaussian number theory. How do they compare with Gauss's *Disquisitiones Arithmeticae*? In his (1851) Dirichlet published his simplified theory of composition of forms. What was simpler about it? Is it a step forwards, sideways, or backwards?

Finally, on the basis of what you know, how does Gauss's theory of forms surpass Legendre's?

Chapter 8 introduces the other principal topic in this book: the solution of polynomial equations by radicals, and its transformation into Galois theory. Implicit in this presentation is the idea that Galois's work may best be seen as redefining and recreating the field of the theory of equations very much as Gauss rewrote number theory.

The solution of cubic and quartic equations by radicals was a considerable breakthrough that was not followed by progress with the quintic, which is why Lagrange investigated both why equations of lower degree were solvable by radicals and why progress halted with the failure to solve equations of degree 5. What is a Lagrangian resolvent? Why are they useful?

Chapter 9 is more historical: it sets the scene with what was known before Galois and what was known immediately after, so that we can understand the importance of what Galois did. Ruffini almost succeeded, and Abel succeeded completely, in showing that the general quintic equation is not solvable by radicals. A little later Pierre Wantzel showed that two of the so-called three classical problems are not solvable by straight edge and circle. There are many reasons why one mathematician does what another did not, but one concern stands out here, which is tied to the idea of symmetry. What did Ruffini and Abel bring to the problem of the quintic that Lagrange did not, and how did it help?

Chapter 10 is (with Chap. 6) the only really hard lecture in the book. Galois's first Mémoire was not understood by his contemporaries, and arguably not for a generation. Our task as historians is to understand how that happened, not to make his badly written, conceptually profound Mémoire easy to understand. But it is possible to see tantalising glimmers in it. For example, can you see a connection to what went on before in the solution of the quartic equation? Can you line up what Ferrari did in solving the quartic with Lagrange's comments and with the remarks about the solution that Galois made?

It is unlikely that anyone promptly understood all that Galois had to say about quintic equations—but what did he say about which ones are solvable by radicals? What did he mean by the terms 'successive adjunction' and 'rationally known'?

I do wish to insist on the sheer incomprehensibility of Galois's ideas, at least in their day. The wrong way to tackle this painful difficulty in a course on the history of mathematics would be to sort out the mathematics. That will happen, in the proper historical order. Our task here is to analyse the actual historical situation.

We can begin with the texts. We are not dealing with fully written out, published papers, but with two memoirs that disappeared for a time, a letter written hastily to a non-mathematician, a few short, published papers, and manuscripts. (Indeed, in this book, we are dealing with only one memoir, one published paper, and the letter.) As we shall see later, nothing was to be published until 1846, 14 years after Galois's death. These memoirs, branded unclear by the influential Poisson, were truly hard to understand, perhaps because he had not acquired the skill of explaining highly original ideas to a bemused, if expert, audience.

Galois himself was uncompromising on the subject. In the preface to the Mémoires he wrote[1]:

> Since the beginning of this century algorithmics had attained such a degree of complication that any progress had become impossible by these means, without the elegance with which modem geometers have believed they should imprint their research, and by means of which the mind promptly and with a single glance grasps a large number of operations. [...] I believe that the simplifications produced by elegance of calculations (intellectual simplifications, of course; there are no material ones) have their limits; I believe that the time will come when the algebraic transformations foreseen by the speculations of analysts

---

[1] Adapted from Neumann (2011, pp. 251–253).

will find neither the time nor the place for their realisation; at which point one will have to
be content with having foreseen them.

Jump with both feet on calculations, put operations into groups, class them according to
their difficulty and not according to their form; that is, according to me, the mission of
future geometers, that is the path that I have entered in this work.

The romance of Galois's life (which is how it is often, if naively, seen) cannot
obscure the fact that he did not prepare his desk drawer for posterity. It is easier to
imagine, in fact, that all could have been lost.

Next, the problem. It is worth thinking about how you would *prove* that something cannot be done. It is necessary to find some way of looking at all attempts,
those known and those yet to be tried, and showing that there is some intrinsic
element of the task that they must all fail. It is not just that the mathematician must
contemplate a radically different task from that of applying a new solution method,
it is also that new tools must be devised that characterise the allowable methods
and exhibit their intrinsic weakness. It is a curious fact that others were finding
impossibility results in mathematics: Wantzel in algebra, Bolyai and Lobachevskii
with their discovery of non-Euclidean geometry (which was rejected for even longer,
until the 1850s and 1860s). Had it been found possible to solve the quintic by
radicals—had Galois shown it by his new methods—his work would surely still
have been difficult to follow, but to use new methods to show a new kind of result
was an even greater challenge to his eventual readers.

Chapter 11 traces the gradual recognition of the importance of Galois's work in
France and Germany. Do you think any of these people fully grasped it? What does
that say about how mathematics evolves?

Mathematics most likely looks difficult most of the time, although I hope it
also looks interesting and rewarding. One specific problem facing a historian of
mathematics is that the subject is disproportionately full of just those few people
whom we like to think did not find mathematics difficult, and this encourages a
lazy habit of supposing that it was indeed easy for them. We look at their output,
at their youth, and often the absence of any formal training, and find ourselves day-
dreaming that they could have tossed their papers off in a few afternoons and that
the *Disquisitiones Arithmeticae*, a book of some 400 pages, only took 4 years to
write because it was long and written in Latin, which takes extra time.

This is silly. It is true, to borrow a phrase once used of Alan Turing, that Gauss,
Abel, and Galois each reached a place where things that looked difficult to other
people looked easy to them, but one cannot suppose that getting there was easy.
Indeed, I would suggest that one of the reasons for the growing prestige of number
theory in Germany was the recognition that it was hard, and people who succeeded
at it had joined an elite group of people who could do difficult things, with all the
effort and psychological pressure that that involves.

At all events, we have good evidence that Gauss found this work difficult—he
tells us so, when he speaks of many fruitless attempts to find proofs—and some
evidence that Abel did—witness his two attempts to explain his account of the
quintic. And we have an abundance of evidence that people who came to read their
work found it difficult. Dirichlet is an excellent example, Kummer another, and then

there is the sheer failure of anyone to master Galois's work for two decades after 1846.

With these considerations in mind, here is what became my standard first question on the course, in each of its two variants. The question was to be answered in a page, so as to force students to decide what was really important.

Question 1a Imagine you are young British professor finishing a year or two studying mathematics abroad. You are writing a letter advising a very good student who will take your place out here. Describe what has been taken to be important and why in the work of either Gauss (c. 1810) or Galois (c. 1846).

Question 1b Imagine you are British student writing to your former professor about the year you have just spent studying mathematics abroad, either in Brunswick and Göttingen under Gauss around 1810 or in Paris around 1846. What would you tell him about contemporary developments in mathematics? Your answer should describe what has been taken to be important and why in the work of either Gauss or Galois.

I set the date for Gauss at 1810 to allow for the French appreciation of the *Disquisitiones Arithmeticae* to be clear. I set the Galois date at 1846 so that the student (or the professor) could have sat in on Liouville's discussions and might have read his version of Galois's memoirs.

In either version, the question helped put over the point that the course was a history course, in which the views of various mathematicians are at stake, and that these people could no more see their future than we can see ours. It is therefore a mistake to write a history of the period, with all the later insights that have accrued, rather than a viewpoint 'from' the period, and a low mark for such an essay therefore helped students to see the difference, and so to appreciate the historian's task a little more clearly.

Chapter 13
Jordan's *Traité*

13.1 Introduction

The book that established group theory as a subject in its own right in mathematics
was the French mathematician Jordan's *Traité des Substitutions et des Équations
Algébriques* of 1870. Here we look at what that book contains, and how it defined
the subject later known as group theory.

Fig. 13.1 Camille Jordan
(1838–1922)

Although there was a great deal more to Jordan's *Traité* than a discussion of
Galois theory, it will help to stay close to the study of quartic and quintic equations.
Jordan's idea was to associate to every polynomial equation a group, which he called
the *group* of the equation, and to show that questions about solving a polynomial
equation of any degree could be rephrased as questions about its group and tackled
there. Jordan not only gave a novel backbone and a structure to Galois's obscure
remarks, which enabled them to be understood much more profoundly than before,
he endeavoured to move the subject away from the theory of equations and towards a

© Springer Nature Switzerland AG 2018   149
J. Gray, *A History of Abstract Algebra*, Springer Undergraduate Mathematics Series,
https://doi.org/10.1007/978-3-319-94773-0_13

more abstract theory of (substitution) groups. In this he was helped by the existence of a difficulty in the theory that is seldom addressed explicitly: it is not easy to find the 'Galois group' of a given equation. 'Galois theory', whatever that is taken to mean, may be interesting, but it is unlikely to help anyone whose daily business is with solving polynomial equations. We must leave equation solving behind when we follow Jordan, and learn to ask other questions about equations.

13.2   Early Group Theory: Introduction

As part of his creation of Galois theory Jordan employed the concepts of transitive group actions and composition series. This section revises enough group theory to understand what Jordan presented in his *Traité* in 1870. We need two ideas that we have not so far met: the knowledge of which subgroups of S_4 or S_5 act transitively, and of the composition series of a group. For more detail on this, see Appendix G.

Transitive Subgroups

Jordan said a group was transitive (p. 29) if "when applying all its substitutions successively one can move one of its letters to the place of an arbitrary other one; more generally, it will be n-fold transitive the substitutions permit one to move n given letters $a, b, c, \ldots$ to the places formerly occupied by n other arbitrary letters $a', b', c', \ldots$". In our words, a group G acting on a set S does so transitively if for any two elements $a, a' \in S$ there is a $g \in G$ such that $ga = a'$, and it acts n-fold transitively if for any two n-tuples of elements of S, say $a_1, a_2, \ldots, a_n$ and $a_1', a_2', \ldots, a_n'$ there is a $g \in G$ such that $ga_1 = a_1', ga_2 = a_2', \ldots, ga_n = a_n'$.

Jordan saw that the group of a polynomial equation acts transitively on the roots if and only if the equation is irreducible. Accordingly, he regarded solving a polynomial equation as a process of the successive adjunction of numbers, the idea being that worthwhile adjunctions reduce the size of the group and that every time a group is reduced to one that acts intransitively on the roots the polynomial equation factorises. Eventually the group is reduced to the identity, the equation is completely factorised, and all the roots are therefore known.

This programme requires that the transitive subgroups of S_4 and S_5 are known. Jordan knew that if a group acts transitively on a set of n elements then it has an orbit of order n and therefore n divides the order of the group. In the case of subgroups of S_5 this tells us that a transitive subgroup contains a 5-cycle. But in the case of S_4 it tells us that the transitive subgroups have order a multiple of 4, but this allows for the

possibility that we might be looking at the Klein group, which does act transitively.[1]
We deduce that

- The transitive subgroups of S_4 are: S_4, A_4, $S(square)$, and K the Klein group.
- The transitive subgroups of S_5 are: S_5, A_5, $G(20)$, $S(pentagon)$, and C_5, the cyclic group of order 5.

Composition Series

Jordan then turned to 'normal' subgroups and composition factors. He was interested in a subgroup H of a group G that commuted with all the elements of G—a relationship he called '*permutable*', meaning $gH = Hg, \forall g \in G$. This in turn means that for every $h \in H$ and every $g \in G$ there is an $h' \in H$ such that $gh = h'g$. He said G was composite if it had a subgroup of this kind (other than itself and the identity), otherwise he said it was simple, and he observed that every group admits a chain of subgroups each one 'permuting', if I may invent that word, with the group before. He did not introduce a special word for a subgroup that 'permutes' with a larger group; the word 'normal' is used today and it would be acceptable to use it here in quotes. In a chain of 'normal' subgroups, each one 'normal' in the one before, say

$$G = G_0 > G_1 > \cdots > G_j > G_{j+1} > \cdots > G_n = \{e\},$$

Jordan called the orders of the quotients $\frac{|G_j|}{|G_{j+1}|}$ the composition factors, and (Book II, Chapter 1, Section IV) he proved the theorem that the composition factors are unique up to order provided the chain cannot be refined (such a chain is called *maximal*). Notice that for Jordan the numbers in the composition series for a group are obtained by dividing one number by another; they are not the size of the corresponding quotient group, $\frac{G_j}{G_{j+1}}$. Jordan was to equivocate about quotient groups for a further few years, until 1873.

The composition series of S_4 is obtained from the maximal chain

$$S_4 \triangleright A_4 \triangleright K \triangleright \mathbb{Z}_2 \triangleright \{e\},$$

where K is the Klein group, and is therefore $2, 3, 2, 2$.

The composition series of S_5 is obtained from the maximal chain

$$S_5 \triangleright A_5 \triangleright \{e\},$$

because the group A_5 is simple, and is therefore $2, 60$.

---

[1] This group only acquired its name after Klein's work in the late 1870s, but it was known earlier.

13.3   Jordan's *Traité*

The publication of Galois's work in Liouville's *Journal* was a challenge to all
mathematicians to understand it, extend it, and apply it. Ultimately, it stimulated
the emerging generation of mathematicians, as Wussing (1984) has described. He
noted an initial period in which Betti, Kronecker, Cayley, Serret, and some others
filled in holes in Galois's presentation of the idea of a group. These modest yet
difficult pieces of work established the connection between group theory and the
solvability of equations by radicals, and then explored the solution of equations by
other means than radicals (generalising the way the trigonometric formulae can be
used to solve some cubic equations). The implicit idea of a group was expressed in
terms of permutation of a finite set of objects, amalgamating Cauchy's presentation
of the theory of permutation groups in 1844–1846 and Galois's terminology. The
elements are often called substitutions or operations, they come with a set of objects
which they permute, there is no sense of a group as something other than a certain
type of collections of 'substitutions'.

The crucial presentations of the idea of permutation groups were made by Jordan
in his *Commentaire sur Galois* (1869) and his *Traité des Substitutions et des
Équations Algébriques* (1870).[2] Jordan's systematic theory of permutation groups
was much more abstract; he spoke of abstract properties such as commutativity, conjugacy, centralizers, transitivity, 'normal' subgroups (and, one might say obliquely,
of quotient groups), group homomorphisms and isomorphisms. So much so that
one can argue (contra Wussing) that Jordan came close to possessing the idea of
an abstract group. Jordan said (*Traité*, p. 22, quoted in Wussing, p. 104) "One will
say that a system of substitutions forms a group (or a *faisceau*) if the product of
two arbitrary substitutions of the system belong to the system itself". He spoke
(*Traité*, p. 56, Wussing, p. 105) of isomorphisms (which he called an *isomorphisme
holoédrique*) between groups as one-to-one correspondences between substitutions
which respect products. It could be that the use of words like 'substitution' and the
permutation notation were well-adapted to their purpose, but should not to be taken
too literally.

A further indication of the high level of abstraction at which Jordan worked is his
use of technical concepts of increasing power—he was one of the first to use Sylow
theory in the treatment of finite groups. Another is the wide range of situations in
the *Traité* in which groups could be found permuting geometrical objects: the 27
lines on a cubic surface, the 28 bitangents to a quartic, the symmetry groups of the
configuration of the nine inflection points on a cubic and of Kummer's quartic with
sixteen nodal points. Powerful abstract theory and a skilled recognition of groups
'in nature' suggests that Jordan had an implicit understanding of the group idea
that he presented in the language of permutation groups only for the convenience

---

[2]I pass over Jordan's first account of Galois's work, his (1865), described by Wussing (1984, p.
136) as being "essentially of the gap-filling variety, while the second (Jordan 1869) is unmistakably
concerned with making precise the group-theoretic component of Galois theory".

of his audience. This is not to deny the role of permutation theoretic ideas in Jordan's work, indicated by the emphasis on transitivity and degree (= the number of elements in the set being permuted), but rather to indicate that ideas of composition and action (as for example change of basis in linear problems) were prominent, and could be seized upon by other mathematicians.

A little more detail is instructive, but Jordan's *Traité* is a work of some 660 pages, and it can only be summarised selectively here.[3] It is in four books. The short Livre I picks up on the topic of Galois (1830) and is about congruences modulo a prime or prime power (what we today would call finite fields). The material on modular arithmetic is needed because the *Traité* is entirely about finite groups (often described as matrix groups with entries modulo a prime number).

Livre II is about substitutions (Galois's word for what later became permutations). It ranges widely, bringing in work by Lagrange, Cauchy, Mathieu, Kirkman, Bertrand, and Serret. Its themes are transitivity, simple and multiple, primitivity (the topic of Galois's second memoir) and composition factors. It also gives a (flawed) proof that the alternating group A_n is simple when $n \neq 4$. Here Jordan presented the opening propositions in the theory of groups: Lagrange's theorem and Cauchy's theorem. Lagrange's theorem is stated in the form it retains to this day: the order of a subgroup divides the order of the group, and the proof is the coset decomposition argument. Cauchy's theorem (§40) is the partial converse: if a prime p divides the order of a group then there is an element of order p in the group. This result is deeper: I omit the proof.

Then comes an account of transitivity, and the 'orbit-stabiliser' theorem: Jordan proved (§44) that given a set of objects permuted by a group G, if these elements can be sent to m different systems of places, and n is the order of the subgroup that leaves these elements fixed, then the order of the group is mn. Jordan was very interested in how transitive a group could be, and in §47 recorded the 'remarkable' example of the Mathieu group on 12 letters that is 5-fold transitive.

Then we get some 160 pages on what we might call finite linear groups, groups whose elements are matrices with entries in a finite field. They come in various types, and the names have not always retained the meaning Jordan gave them: primary, orthogonal, abelian, hypoabelian. A typical theme is the 'normal' subgroups and composition factors of the different groups. He was also interested in the concept of primitivity, which he defined negatively: a group is non-primitive (better, imprimitive) if the elements can be divided into blocks containing the same number of elements and the group maps blocks to blocks.

Livre III, 'The irrationals', is about the behaviour of a given irreducible equation under successive adjunction of irrationals (roots of other polynomial equations). It deals with the solution of the quartic by radicals and the insolubility of the general quintic. Then we get a treatment of equations with particular kinds of group: abelian groups and what Jordan called 'Galois's equation' $x^p = A$. Then we get the geometrical examples mentioned above in which discoveries by Hesse, Clebsch,

---

[3]For a translation of the book's preface, see Appendix D.

Kummer, and Cayley (the 27 lines on a cubic surface) are shown to have interesting group-theoretic interpretations. Then comes material from elliptic function theory: the modular equation and the discovery in 1858 of the solution of the quintic equation according to Hermite and Kronecker. Jordan reworked their contributions in his own way and then extended their work to show that all polynomial equations can be solved by a similar use of hyper-elliptic functions.

The book ends with Livre IV, of almost 300 pages, entitled 'Solution by radicals'. Jordan quickly rehearsed the argument that an equation is solvable by radicals if and only if the composition factors of its group are all prime, and called such a group solvable. This led him to proclaim three problems, of which for brevity I give the first only: construct explicitly for every degree the general [i.e. maximal] solvable transitive groups. The theorems concern maximal solvable groups of transitive groups, as Dieudonné (1962) explained.

Jordan used his theory of composition factors to suggest that a massive process of induction might suffice to find all the permutation groups. After all, the maximal normal subgroup of a given group is smaller than the given group, and its index is determined by the given group. So it might be possible to construct the given group from its composition series if all groups of order less than the given group are known. In the event this ambitious programme failed—groups are vastly more varied than can be handled this way—but it has the germ of a process that was later made more precise by Hölder and became the search for all simple groups.

13.4   Jordan's Galois Theory

We now look at Book III Chapter 1 of the *Traité* in more detail. For a partial translation, see Appendix D.

In Chapter 1 Jordan followed Galois closely, but with the new resources of an explicit theory of groups. He began by considering the group of substitutions of the roots of a polynomial equation with distinct roots, whose coefficients are rational functions of the rational numbers and some auxiliary quantities. This is called the group of the equation, and it can alter with further adjunctions of auxiliary quantities. Jordan showed that any function of the roots that is not altered in numerical value by any substitution in the group is rationally expressible in terms of the coefficients, and conversely.

He showed that irreducible equations have transitive groups and that the converse also holds. In particular, if an equation of degree n is such that all its roots are a rational function of any one of them then the group is of order n.

Then he investigated what happens to the group G of an equation if a rational function φ of the roots is adjoined, and showed that the group reduces to the subgroup H of substitutions that do not alter the numerical value of φ. An analogous result holds if several quantities are adjoined simultaneously. Conversely, if rational functions φ_1 and φ_2 are invariant under the same group then each is a rational function of the other. Jordan also showed that if φ_1 takes the n values $\varphi_1, \varphi_2, \ldots, \varphi_n$

under the substitutions of G then

$$n = \frac{|G|}{|H|}.$$

If one recalls that Jordan did not possess the concept of a quotient group when writing the *Traité* then this result can be given an added degree of significance, because his proof essentially identifies the group of the equation

$$(Y - \varphi_1)(Y - \varphi_2)\ldots(Y - \varphi_n) = 0 \tag{13.1}$$

as the quotient group G/H. His proof also shows that this equation is irreducible.

From this point on it is helpful to think of Jordan's Galois theory as being concerned with the given equation and a succession of auxiliary equations.

Next Jordan showed that if the result of adjoining φ_1 is to reduce the group G to the group H_1 and the result of adjoining φ_2 is to reduce the group G to the group H_2 and if the substitution b transforms φ_1 into φ_2 then the groups H_1 and H_2 are conjugate and indeed $H_2 = bH_1b^{-1}$.

When the process of adjunction is complete and the roots are known, every rational combination of them is also known, so one wants to have not only φ_1 say but also $\varphi_2, \ldots, \varphi_n$. It can happen that adjoining φ_1 necessarily adjoins the others, but it need not, and if one adjoins all the $\varphi_1, \varphi_2, \ldots, \varphi_n$ at once the group of the equation reduces to K, the largest 'normal' subgroup of the groups $H_1, H_2, \ldots$. In this case the order of the group (13.1) is v and the order of the group K is $|G|/v$.

The solution of a polynomial equation was described by Jordan in terms of a sequence of adjunctions. An adjunction may or may not change the group of the equation, and even if it does reduce the order of the group it may not result in a factorisation of the equation—that will happen if and only if the reduced group does not act transitively on the roots.

Then we come to Jordan's Theorem XIII, which says: Let $F(x) = 0$ and $F'(x) = 0$ be two equations whose groups, G and G' respectively have N and N' elements [substitutions]. If the resolution of the second equation reduces the group of the first to a group H having at most N/v elements then the resolution of the first equation reduces the group of the second to a group H' having at most N'/v. Moreover, the two equations are composed with the same auxiliary equation $f(u) = 0$ of degree v and whose group has v elements.

His Theorem XV says: No irreducible equation of prime degree p can be solved by means of auxiliary equations of lower degree. His proof runs as follows: Because the equation is irreducible, its group is transitive and therefore of order divisible by p. By Theorem XIII it remains divisible by p provided one does not use an auxiliary equation of order divisible by p. But the order of the group of an auxiliary equation of degree $q < p$ is a divisor of $q!$ and so prime to p. Therefore the order of the group of the given equation remains divisible by p provided one only employs similar equations.

Finally, Jordan's Theorem XVI says: The general equation of degree n cannot be solved by means of equations of lower degrees unless $n = 4$. The proof goes like this. The group of the equation has order $n!$—that's the meaning of 'general'— and it can be reduced to the alternating group of order $\frac{n!}{2}$ by resolving an equation of degree 2. But this new group is simple (§85). Therefore the equation cannot be solved by means of an auxiliary equation unless the order of its group is at least equal to $\frac{n!}{2}$, but if q is the degree of this auxiliary equation, its order divides $q!$. Therefore q cannot be less than n.

This concludes Jordan's account of the solution of a polynomial equation by auxiliary irrationals, which by definition are the roots of other polynomial equations. His result is a generalisation of Galois's on the solution of equations by radicals. It remained for Jordan to complete his account of the solution by radicals, which is the case where the auxiliary equations are of a special kind, $x^p - a = 0$, corresponding to the adjunction of a pth root of a known quantity.

He did this in Chap. 2, where he dealt with abelian equations and especially those of prime degree, before turning to Galois equations, their groups, and the equation $x^p = A$. Kronecker had earlier introduced the term 'abelian equation' to refer to those equations whose corresponding group is cyclic; Jordan proposed extending the term to equations whose groups are commutative, and the term now adheres to the groups.

At stake is an irreducible equation $F(x) = 0$ of degree p a prime, with the further property that all its roots can be expressed rationally in terms of any two of them, say x_0 and x_1. (Such an equation is $x^p = A$, as we can see, the roots being some quantity α multiplied by the powers of a primitive pth root of unity.) The task is to find the group of this equation.

Jordan noted that the order of the group cannot exceed $p(p - 1)$, which is the number of choices one has for the images of x_0 and x_1 under permutations permuting the roots.

On the other hand it is transitive and therefore its order is divisible by p, and therefore it has an element of order p, which he wrote in the form $\phi(z) = z + 1$— tacitly working mod p. Jordan then showed that the elements of order p are precisely the powers of this one (else the group would be too big). But conjugates of an element of order p are also of order p, so the group consists precisely of elements of the form $z \mapsto az + b$.

Jordan next showed that the given equation can be solved by successively adjoining the roots of two abelian equations, one of degree p the other of degree $p - 1$. This follows from the observation that one first adjoins a rational function of the roots that is invariant under ϕ.

The equation $x^p = A$ is of the kind under study, and its group must have order $p(p - 1)$, because the quotient x_1/x_0 of two roots x_1 and x_0 satisfies the irreducible equation

$$\frac{x^p - 1}{x - 1} = 0$$

of degree $p - 1$.

13.5 The Cubic and Quartic Equations

It is helpful to look again at the account of the solution of low degree equations in the spirit of successive adjunctions. It can also be productive to treat what follows as a series of exercises and derive for yourself the various results.

The Cubic Equation

Let $x^3 + px + q = 0$ be an irreducible equation with roots α, β, γ and discriminant

$$\Delta = 6\sqrt{-3}\sqrt{\left(\frac{q}{2}\right)^2 + \left(\frac{p}{3}\right)^3}.$$

Then

$$\beta + \gamma = -\alpha$$

and

$$\beta\gamma = -q/\alpha$$

and a simple calculation shows that

$$\Delta = (\gamma - \beta)(\alpha^2 - \alpha(\gamma + \beta) + \beta\gamma) = (\gamma - \beta)(2\alpha^2 - q/\alpha),$$

so from the expressions for $\beta + \gamma$ and $\gamma - \beta$ we deduce that

$$\gamma = \frac{1}{2}\left(\frac{\Delta}{2\alpha^2 - q/\alpha} - \alpha\right), \quad \beta = \frac{1}{2}\left(-\alpha - \frac{\Delta}{2\alpha^2 - q/\alpha}\right).$$

So if Δ and α are known then so are β and γ.

In terms of Galois groups, consider

$$\mathbb{Q} \hookrightarrow \mathbb{Q}(\Delta) \hookrightarrow \mathbb{Q}(\Delta, \alpha) \hookrightarrow \mathbb{Q}(\Delta, \alpha, \beta, \gamma).$$

The permutations of the roots that fix every 'number' in $\mathbb{Q}(\Delta)$ are precisely the even permutations (they form the group A_3). We have just seen that the collections of 'known numbers' $\mathbb{Q}(\Delta, \alpha)$ and $\mathbb{Q}(\Delta, \alpha, \beta, \gamma)$ are the same. In the context of groups we see this way: if $\mathbb{Q}(\Delta, \alpha)$ was a proper subset of $\mathbb{Q}(\Delta, \alpha, \beta, \gamma)$ then the permutation (β, γ) would be available to form a non-trivial subgroup of order 2 of the group of permutations $\mathbb{Q}(\Delta, \alpha, \beta, \gamma)$ that fix $\mathbb{Q}(\Delta, \alpha)$—but this would imply (by Lagrange's theorem)—that 2 divides 3, which is absurd.

What a Difference a Discriminant Makes

Consider the equation $x^3 - 2 = 0$ with roots ρ, where $\rho^3 = 2$, $\rho\omega$ and $\rho\omega^2$, where $\omega^2 + \omega + 1 = 0$, so ω is a complex cube root of unity. The discriminant of this equation is $\Delta = -6(1 + 2\omega)$, and we note that

$$\Delta.(-\Delta) = -36(1 + 2\omega)^2 = -36(4 + 4\omega + 4\omega^2 - 3) = 108.$$

The equation $\Delta = -6(1 + 2\omega)$ tells us that $\mathbb{Q}(\Delta) = \mathbb{Q}(\omega)$. Let me write $\mathbb{Q}^*$ here and elsewhere to mean $\mathbb{Q}$ with suitable cube roots adjoined—in this case a cube root of unity. So $\mathbb{Q}^* = \mathbb{Q}(\Delta)$.

Now, if we adjoin ρ to $\mathbb{Q}$ we do not adjoin either $\rho\omega$ or $\rho\omega^2$ because ρ is real and the others are not. (Nor indeed does adjoining just $\rho\omega$ adjoin either ρ or $\rho\omega^2$.) So adjoining one root gives us three conjugate extensions of $\mathbb{Q}$. For example, $\mathbb{Q}(\rho)$ and $\mathbb{Q}(\rho\omega)$ are conjugate by any permutation of the roots that exchanges ρ and $\rho\omega$. There are three corresponding Galois groups:

1. The permutations of the roots that also fix ρ;
2. The permutations of the roots that also fix $\rho\omega$;
3. The permutations of the roots that also fix $\rho\omega^2$.

They are conjugate, but not normal, there is no equation with rational coefficients that has only, say ρ, as a root and accordingly no corresponding Galois group. If we adjoin all three roots at once the group of permutations of the roots that fixes all the 'known numbers' (i.e. the rationals with the roots adjoined) is, trivially, the identity group, but it is indeed the largest 'normal' subgroup of each of the above three groups, as Jordan's theory requires.

Things are very different if we first adjoin Δ to $\mathbb{Q}$ to get $\mathbb{Q}^*$, and then adjoin, say ρ. Now the other roots come in automatically, being rational combinations of ρ and Δ, i.e. of ρ and ω. We have

$$\mathbb{Q}(\Delta) \hookrightarrow \mathbb{Q}(\Delta, \rho) = \mathbb{Q}(\Delta, \rho\omega) = \mathbb{Q}(\Delta, \rho\omega^2) = \mathbb{Q}(\rho, \rho\omega, \rho\omega^2).$$

The groups of permutations of the roots that fix Δ and also fix ρ or $\rho\omega$ or $\rho\omega^2$ are the same (not merely conjugate); they are all adjoined at once by an equation whose group is A_3, and we can now see that the equation for Δ appears as the equation for the quotient group S_3/A_3.

The Quartic Equation

Let us take $x^4 + bx^3 + cx^2 + dx + e = 0$ with distinct roots $\alpha_1, \alpha_2, \alpha_3, \alpha_4$. The discriminant is again denoted Δ, and two cases arise according as we adjoin Δ to $\mathbb{Q}$ or not. (Of course, it can happen that Δ is rational anyway.) In either case, define

$$\tau_2 = \alpha_1\alpha_2 + \alpha_3\alpha_4, \quad \tau_3 = \alpha_1\alpha_3 + \alpha_2\alpha_4, \quad \tau_4 = \alpha_1\alpha_4 + \alpha_2\alpha_3.$$

The expressions

$$\tau_2 + \tau_3 + \tau_4, \ \tau_2\tau_3 + \tau_3\tau_4 + \tau_4\tau_2, \ \tau_2\tau_3\tau_4$$

are symmetric (unaltered by any permutation of the roots). Indeed, with a little work we find that

$$\tau_2 + \tau_3 + \tau_4 = c, \ \tau_2\tau_3 + \tau_3\tau_4 + \tau_4\tau_2 = bd - 4e, \ \tau_2\tau_3\tau_4 = b^2e + d^2 - 4ce,$$

so we could, if we wished, write down the cubic equation of which they are the roots and whose coefficients are fixed by all permutations of the roots and are therefore rational in b, c, d, e.

If we do adjoin Δ to $\mathbb{Q}$, and work over $\mathbb{Q}(\Delta)$, then the group of permutations of the roots that also fix Δ is at most A_4, the even permutations on the four roots. The even permutations of the roots that fix $\tau_2, \tau_3,$ and τ_4 form the group

$$K = \{1, (\alpha_1, \alpha_2)(\alpha_3, \alpha_4), (\alpha_1, \alpha_3)(\alpha_2, \alpha_4), (\alpha_1, \alpha_4)(\alpha_2, \alpha_3)\}.$$

This group is a normal subgroup of A_4, with quotient A_3. Can we adjoin say τ_2 to $\mathbb{Q}(\Delta)$ without adjoining the others? To answer this question, we look at the stabiliser of τ_2 in A_4, which we denote St_2. We find that $St_2 = K$. This means that the groups St_2, St_3, St_4 are all equal, and normal, of course, in A_4, and that any even permutation of the roots that also fixes τ_2 fixes the others, so indeed τ_3 and τ_4 are rationally known when τ_2 and Δ are.

Let me now stress that if the Galois group of an equation is A_3 then the effect of adjoining any rational expression in the roots either changes nothing or brings in all the roots of the equation.

Matters are different if we have not adjoined Δ to $\mathbb{Q}$. Now the group of permutations of the roots is S_4 and K is not a normal subgroup of S_4. We find that in S_4

$$St_2 = K \cup K(\alpha_1, \alpha_2),$$

with similar expressions for St_3 and St_4. These three groups are conjugate, for example

$$(23)St_2(23) = St_3.$$

Comfortingly, the conjugating elements are odd, and cannot be in A_4.

This means that adjoining say τ_2 to $\mathbb{Q}$ does not adjoin τ_3; τ_3 is not a rational function of τ_2. If we adjoin $\tau_2, \tau_3,$ and τ_4 simultaneously, we find that the group of permutations of the roots that fix the τs is indeed K, the largest normal subgroup of S_4 contained in $St_2, St_3,$ and St_4, i.e. in $St_2 \cap St_3 \cap St_4$. This is what Jordan proved.

Consider now the system of adjunctions:

$$\mathbb{Q} \hookrightarrow \mathbb{Q}(\Delta) \hookrightarrow \mathbb{Q}(\Delta, \tau_2) \hookrightarrow \mathbb{Q}(\Delta, \tau_2, \kappa) \hookrightarrow \mathbb{Q}(\alpha_1, \alpha_2, \alpha_3, \alpha_4),$$

where $\kappa = \alpha_1 \alpha_2$. The new quantity κ and $\lambda = \alpha_3 \alpha_4$ are the roots of the equation

$$y^2 - \tau_2 y + e = 0.$$

Finally, α_1 and α_2 are the roots of the equation

$$y^2 - \mu y + \kappa = 0,$$

where $\mu = \alpha_1 + \alpha_2$ stands for a rational expression in $\Delta, \tau_2, \tau_3, \tau_4$ and κ that is found as follows. We have the equations

$$d = -\lambda(\alpha_1 + \alpha_2) - \kappa(\alpha_3 + \alpha_4),$$

$$b = -(\alpha_1 + \alpha_2) - (\alpha_3 + \alpha_4),$$

and solving them for $\mu = \alpha_1 + \alpha_2$ yields

$$\mu = \frac{\kappa b - \lambda}{\lambda - \kappa}.$$

To summarise: we start with a Galois group isomorphic to S_4 and by adjoining Δ reduce the group to A_4. We adjoin say τ_2—recall that τ_3 and τ_4 come automatically—and the Galois group becomes K. Up to this point the quartic equation does not factorise, because the group K acts transitively on the roots. But if we adjoin κ the Galois group reduces to $\mathbb{Z}_2$ which is not transitive and the equation factorises into a product of two quadratics. Finally, we solve the quadratics, the Galois group reduces to the identity and the original quartic is solved.

It is instructive to see where the auxiliary irrationals $\Delta, \tau_2, \tau_3, \tau_4$ and κ come from. Δ is the root of a quadratic, its Galois group is isomorphic to $\mathbb{Z}_2$ that allows us to switch Δ and $-\Delta$ while not altering the rationals (assuming, here, that Δ is not itself rational). Now we adjoin the τs—adjoin any one and we get the others—and they are the roots of a cubic with coefficients in $\mathbb{Q}(\Delta)$ and as it happens the coefficients are in $\mathbb{Q}$, as we saw. This cubic has group A_3 precisely because adjoining one τ adjoins the other, and indeed A_3 is isomorphic to the quotient group A_4/K. The same story holds for adjoining κ and then again for the adjunction of α.

Solvability by Radicals

It remains to connect this story to solvability by radicals. Solvability by radicals requires that each adjunction is a pure radical, something of the form $\sqrt[n]{\sigma}$, where σ is a quantity in the collection of 'numbers' we are trying to enlarge. It is enough to assume n is prime, say p. Adjoining a pure pth root shows up in the Galois group, as follows (where for simplicity I shall assume σ is in $\mathbb{Q}$).

Let the roots of $x^p - \sigma = 0$ be ρ where $\rho^p = \sigma$ and ρ is real, and $\rho\omega^k$, $k = 1, 2, \ldots p-1$, where ω is a primitive complex kth root of unity, or, more simply, the roots are of the form $\rho\omega^k$, $k = 0, 1, 2, \ldots p - 1$. Any allowable permutation of these must be given in the form $k \mapsto ak + b$ where a and b are taken mod p and a is not zero (as Galois proved), because the new expressions are of the form $\rho^j\omega^k = \rho^{j-k}\omega^k$, where j is nonzero mod p and k is any number mod p, and once we know the images of ρ and $\rho\omega$ under a permutation of the roots we know the image of $\rho^j\omega^k$. It helps to notice that there are $p(p - 1)$ expressions of the form $\rho^j\omega^k$.

So for an equation to be solvable by radicals the successive adjunctions must correspond to quotient groups, as above, that are subgroups of the above group of order $p(p - 1)$ for some prime p. In this way we can get to a group of order 20, but not beyond, so only equations with Galois groups that are this group or one of its subgroups are solvable by radicals.

One final question: can you now tackle the first cases of the inverse problem, and find for every subgroup of S_4 a polynomial with that Galois group?

Chapter 14
The Galois Theory of Hermite, Jordan and Klein

14.1 Introduction

In this chapter and the next we look at how Galois theory became established. This will involve us in looking at an alternative approach to the question of solving quintic equations that is associated with the influential figures of Hermite, Kronecker, and Brioschi. Then we consider the alternative promoted by Jordan and by Felix Klein, whose influence on the development of Galois theory in the late nineteenth century is often neglected, but was in fact decisive.

14.2 How to Solve the Quintic Equation

In a celebrated paper of 1858 Hermite showed how to solve the quintic equation. Not, of course, by radicals, but by the use of functions. He drew an analogy with cubic equations. The cubic equation with three real roots can always be solved by trigonometric functions. It is enough to show that any cubic equation can be put in the form of the equation for $\sin 3\theta$ in terms of $\sin \theta$, in which case the roots of the cubic equation are the values of $\sin \theta$ that correspond to the given value of 3θ.

In more detail, we already know that the general cubic equation with real coefficients can be written in the form

$$x^3 + px + q = 0. \tag{14.1}$$

© Springer Nature Switzerland AG 2018
J. Gray, *A History of Abstract Algebra*, Springer Undergraduate Mathematics Series,
https://doi.org/10.1007/978-3-319-94773-0_14

Fig. 14.1 Charles Hermite
(1822–1901). Photo courtesy
of the Archives of the
Mathematisches
Forschungsinstitut
Oberwolfach

The equation for the trisection of an angle is

$$s^3 - 3s/4 + S/4 = 0,$$

where $s = \sin\theta$ and $S = \sin 3\theta$. We suppose that the value of S is given and use the equation to find the values of s. Notice that we may replace s by λs and obtain

$$(\lambda s)^3 - 3\lambda s/4 + S/4 = 0,$$

or

$$s^3 - \frac{3s}{4\lambda^2} + \frac{S}{4\lambda^3} = 0.$$

This equation agrees with Eq. (14.1) if

$$p = \frac{-3}{4\lambda^2}, \quad \text{and} \quad q = \frac{S}{4\lambda^3}.$$

These equations imply that

$$\frac{\lambda}{S} = \frac{-p}{3q},$$

and so

$$S = \frac{-4p^3}{27q^2}.$$

The condition that S is the sine of an angle is that $|S| \leq 1$, which is equivalent to the condition $\left(\frac{q^2}{2}\right) + \left(\frac{p}{3}\right)^3 > 0$ for three real roots that we obtained earlier.

If we allow the sine function to be defined by its power series for complex values and use $\sin(ix) = i \sinh x$ then the above argument also deals with cubic equations with only one real root.

Hermite argued that it was already known that any quintic equation can be written in the two-parameter form

$$x^5 + ax^2 + bx + c = 0,$$

and indeed in the one-parameter form

$$x^5 + bx + c = 0,$$

which is further reducible to

$$x^5 + x + c = 0 \tag{14.2}$$

by replacing x with λx. This was the combined result of the German mathematician Ehrenfried Walther von Tschirnhaus in 1683 and the Swedish mathematician Erland Samuel Bring in 1786, and so Eq. (14.2) is sometimes called Bring's quintic equation. Their methods were entirely algebraic, and required no more than the solution of equations of degree 4.

What Hermite was the first to appreciate was that the equation $x^5 + x + c = 0$ had occurred in the theory of elliptic function s. Certain functions $f(z)$ that arise in this theory satisfy an equation relating $f(5x)$ and $f(z)$, and so, precisely as with the cubic equation and the sine function, the quintic equation was solved by reducing it to the theory of this function f.

The precise derivation of the function called f above was difficult, and other mathematicians, Brioschi and Kronecker, joined in with other approaches, but all agreed that this rich and unexpected connection between the theory of polynomial equations and elliptic function s was important. Discovering it, and establishing a clear route to it, was part of what mathematics was about, in their view.

In what follows I shall sketch how the function f was defined and discovered. It is not possible to do this without assuming some acquaintance with elliptic function theory, but what follows is stripped of as many inessentials as possible.

Jacobi's Equation

Hermite's work drew on some of the many intriguing formulae that Jacobi had discovered back in 1828 and 1829 when he and Abel were discovering and publishing the theory of elliptic function s.

In his (1858) Hermite described what Jacobi had done in this way. He took K and K' as the periods of the elliptic integral

$$\int \frac{d\varphi}{\sqrt{1 - k^2 \sin^2 \varphi}},$$

with modulus k, that is

$$K = \int_0^{\pi/2} \frac{d\varphi}{\sqrt{1 - k^2 \sin^2 \varphi}}, \quad K' = \int_0^{\pi/2} \frac{d\varphi}{\sqrt{1 - k'^2 \sin^2 \varphi}},$$

where $k'^2 = 1 - k^2$, and defined $\omega = iK'/K$. Here, K and K' are thought of as functions of k.

Jacobi had drawn attention to the fourth root of the modulus and its complement, so Hermite, after defining these fourth roots

$$u = \varphi(\omega) = \sqrt[4]{k}, \quad v = \psi(\omega) = \sqrt[4]{k'},$$

explained that the modular equation relates u and v by an equation of degree $n + 1$. To study it further, Hermite introduced the very simple transformation properties relating $\varphi(\omega)$ and $\psi(\omega)$ to

$$\varphi\left(\frac{c + d\omega}{a + b\omega}\right) \text{ and } \psi\left(\frac{c + d\omega}{a + b\omega}\right),$$

where a, b, c, d are integers and $ad - bc = 1$. Indeed, he said, it is enough to consider a, b, c, d integers modulo 2, when there are precisely six transformations, which he exhibited explicitly.

Galois, in his letter to Chevalier, had drawn attention to the striking fact that the modular equation of degree 6 (as well as those of degrees 8 and 12 but no others) can be reduced to an equation one degree less.

Of course, this does not mean that the six roots of this equation can be made to satisfy an equation of degree 5. Rather, Galois had remarked that the 'group' of the modular equation of degree $p + 1$ is composed of 'substitutions' of the form

$$\frac{k}{l} \to \frac{ak + bl}{ck + dl},$$

in which k/l can take the $p+1$ values $\infty, 0, 1, \ldots, p-1$, and $ad-bc$ is a square. Reduction will happen if this 'group' decomposes into p 'subgroups' of $\frac{1}{2}(p-1)(p+1)$ 'substitutions', as it does when $p=5$. We would say today that the Galois group in this case acts by conjugation on these five subgroups, so it is A_5, the Galois group of an equation of degree 5.

This reduction in degree struck Hermite as very important, and he thought that, although only fragments of Galois's work on the question remained, it would not be difficult to rediscover the proof by following the path that he had opened. But, Hermite said, in that way one is only assured of the possibility of the reduction, and an important gap remained to be filled before the problem could be fully solved. For Hermite, this reduction meant that combinations of the roots of the modular equation can be found that satisfy a polynomial equation of degree 5, and that is what he set out to find.

The relevant modular equation here is the one for $n=5$, which is

$$u^6 - v^6 + 5u^2v^2(u^2 - v^2) + 4uv(1 - u^4v^4) = 0. \tag{14.3}$$

One is easily led, Hermite remarked, to consider the following function

$$\Phi(\omega) = \left(\varphi(5\omega) + \varphi\left(\frac{\omega}{5}\right)\right)\left(\varphi\left(\frac{\omega+16}{5}\right) - \varphi\left(\frac{\omega+4.16}{5}\right)\right) \times$$

$$\left(\varphi\left(\frac{2\omega+16}{5}\right) - \varphi\left(\frac{3\omega+16}{5}\right)\right).$$

Then the five quantities

$$\Phi(\omega), \Phi(\omega+16), \Phi(\omega+2.16), \Phi(\omega+3.16), \Phi(\omega+4.16)$$

are the roots of an equation of degree 5 whose coefficients are rational functions of $\varphi(\omega)$, which he exhibited explicitly:

$$y^5 - \left(2^4 5^3 u^4 (1 - u^8)^2\right) y - \left(2^6 \sqrt{5^5} u^3 (1 - u^8)^2 (1 + u^8)\right) = 0,$$

which is an equation in Bring's form. Better yet, given a specific value for c in Bring's equation, the corresponding value of u turns out to be given by the equation

$$c = \frac{2}{\sqrt[4]{5^5}} \frac{1 + u^8}{u^2 (1 - u^8)^{1/2}},$$

and this equation is of degree 4 in u^4.

Therefore, and this was the reason for the excitement in 1858, just as the general cubic equation can be solved by trigonometric functions so the general quintic equation can be solved by elliptic modular functions and the solution of some auxiliary equations of degree at most 4, all of which are solvable by radicals. Hermite ended his paper with some remarks about how quickly the process of numerical approximation worked in this case.

Galois did not explain his observation about the reduction in the degree of Eq. (14.3), and what he did mean was obscure to his contemporaries. It was first explained along Galois's lines by Betti in his (1853), in a way that did not impress Hermite if he ever saw it, because he gave the explicit construction just sketched.

Betti first of all showed that when p is a prime the equation relating the moduli of elliptic functions for which transformation s of order p is of degree $p + 1$. Then he showed that the (Galois) group of this equation is what would be called today $PGL(2, p)$ of order $(p + 1)(p - 1)p$ that consists of all the transformations

$$z \mapsto \frac{\alpha z + \beta}{\gamma z + \delta},$$

where $\alpha, \beta, \gamma, \delta$ are integers modulo p and $\alpha\delta - \beta\gamma$ is nonzero. Indeed, it is possible to insist that this determinant $\alpha\delta - \beta\gamma$ be a square modulo p, which reduces the group to one of order $\frac{1}{2}(p + 1)(p - 1)p$. In the case where $p = 5$, this group is of order 60 and it acts by conjugation on five subgroups of order 12, which means that there is a quintic equation equivalent to the given equation of degree 6, which is possibly what Galois meant? But Betti showed that the existence of such subgroups only happens for primes $p \leq 11$, so the reduction is limited to those cases, as Galois had claimed.

Betti's proofs were imperfect, but the crucial point to note is that Hermite's explanation was explicit and rooted in the theory of elliptic function s and Betti's explanation was group-theoretic. Which one should be taken as closer to Galois's? Which one is more likely to lead to new discoveries?

14.3   Jordan's Alternative

Jordan knew the work of Hermite and the others very well but, as we have seen, embraced and greatly expanded on the concept of a group in a series of papers in the 1860s that culminated in his *Traité des substitutions et des équations algébriques* (1870). This was not a matter of finding different routes to the same goal, but a confrontation of different approaches. Hermite deployed the considerable resources of a major branch of contemporary mathematics—elliptic function theory—and a pleasure in richness and detail. To this Jordan opposed a novel, austere, and markedly abstract approach—permutation group theory—that had few successes to its credit.

Yet Jordan's *Traité* is rightly regarded as a pivotal work by both Kiernan and Wussing. Kiernan (1971, p. 124) summed up his account of Jordan's contributions by saying

> JORDAN'S *Traité des substitutions* marks an important milestone in the development of algebra and of Galois Theory. ... The *Traité* made JORDAN'S research available generally, in a unified presentation.

Wussing saw in the *Traité* the completion of the permutation-theoretic group concept, and indeed many abstract concepts are developed in Jordan's work.

Jordan insisted on the importance of the influence of Galois's work on his *Traité*. "The point of this work", he wrote in the preface, "is to develop Galois's methods and form them into a body of doctrine [theory, perhaps, in English] by showing with what facility they lead to the solution of all the principal problems in the theory of equations". In particular, Jordan took Galois literally: an appreciation of partially symmetric expressions is not enough, one must look at the *groups* involved and the structures they have.

Brechenmacher in his (2011) places the emphasis on all the pages in which Galois's name is not mentioned: the material on linear groups, the geometrical examples, and all but the first few pages of Livre IV. This is consistent with his stressing not "the Galois theory of *general* equations" (his italics) but rather the three applications that Galois presented, equations of prime degree, primitive equations of prime power degree, and modular equations. Brechenmacher's reading is in some ways close to Kiernan's. Kiernan wrote (1971, p. 124)

> The first two-thirds of the book is ostensibly a development of Galois theory, but often this is simply a convenient framework for developing innovative, but only incidentally related, topics in group theory.

But it would be equally fair to recognise the way Jordan took Galois's ideas and shaped them into a body of theory. One could argue that the key insight Jordan took from Galois was that given a problem one should look for the group or groups it involves. For polynomial equations the adjunction process gives rise to several groups, and the equation is solvable by radicals if and only if the group is solvable, so Jordan naturally developed what is lacking in Galois's fragmentary work: a theory of solvable groups. This is his theory of composition factors. Every finite group can be fingerprinted in this way, and it can only be solvable if the factors are all prime. Seen in this light, much more of Livre III (and the purpose of Livre IV, but none of its methods) are exactly in Galois's spirit. Equally, Galois's insight into solvable quintics and almost certainly the modular equation had to do with his appreciation of linear groups, about which Jordan wrote extensively.[1]

---

[1] It is true that Galois wrote nothing about groups that arise in geometrical settings. One could say that this was a wholly independent venture on Jordan's part, or that it is another illustration of the message that given a problem one should look for a group, whether one reads this as Galois's message or Jordan's. As Jordan pointed out, the geometrical problem is presented in terms of equations, and these are usually too difficult to deal with explicitly, so it might be propitious to pass from the equations to the groups.

On this view, Jordan's *Traité* saved Galois's theory for posterity. If one mastered the *Traité* it offered the best explanation of Galois's ideas: one acquired a theory of groups, a theoretical grasp of how to translate every term in the theory of equations into a term in group theory and the language of rationally known quantities, and one wound up with a rigorous theory in which the parts contributed by Galois stood out clearly. Much as fragments of a broken statue make good sense in a well-made reconstruction, Galois's cryptic remarks made good sense in their new theoretical setting.

The solution of equations, in whatever form it was regarded, evidently attracted the attention of a number of major mathematicians who refreshed the subject and repolished its image. In the context of this book, we must also remember that Galois's second *Mémoire* also began to attract considerable interest and to stimulate new investigations, including the question much studied by Jordan of primitive equations and their groups.

But the longer-term fate of Galois theory is reflected in the situation between 1870 and the early 1900s. Jordan's *Traité* did not displace Serret's *Cours*, even as an account of the theory of polynomial equations, and the reason would seem to be that it was regarded as being very difficult. Vogt's book of 1895, which leans heavily on the equations and only introduces the idea of the group of an equation in its final chapter, was praised by Jules Tannery for presenting an up-to-date account of the theory based on Kronecker's insights now that Serret's book had become out-of-date, but he noted that Jordan's book was more appropriate for people wishing to add to the subject than those hoping to learn it. Jules Tannery was the influential Director of Studies at the École Normale Supérieure in Paris, so his opinion is a good measure of attitudes in France, and it suggests that enthusiasm for group theory per se, and for Jordan's study of finite groups of $n \times n$ matrices in particular, seems to have been lacking for some time.[2] The first book in French to pursue some of these ideas seems to be de Séguier (1904), which Wussing (1984, p. 272) hailed as 'The earliest monograph on group theory' and a conscious attempt to provide an abstract account of the subject.[3]

The long gap from 1870 to 1904 raises the question of what happened elsewhere in between, and the bulk of the answer lies in Germany and in the work of Kronecker and Klein. But we should also recall that Poisson rejected the whole thrust of Galois's approach by saying that it did not lead to a test using the coefficients of a given polynomial equation that would permit one to determine whether the equation was solvable by radicals. This is a significant objection, and it is not too much to say that opinions on it divide Hermite and Kronecker on the one hand from Jordan and Klein on the other. Indeed, even if it were to be established that Jordan's *Traité* had almost no readers and almost no influence for decades, there was one person who appreciated the importance of the idea of a group as vividly as Jordan, and who applied it energetically to the theory of equations, and that was Klein.

---

[2] See Brechenmacher (2011, 308).

[3] One may usefully consult Brunk's short essay 'Galois's commentators', available at http://www-history.mcs.st-and.ac.uk/Projects/Brunk/Chapters/Ch3.html.

14.4 Klein

Christian Felix Klein was born on 25 April, 1849 in Düsseldorf, and went to the
University of Bonn in the autumn of 1865 (at the age of sixteen and a half) intending
to study mathematics and natural sciences and specialise in physics. In Easter 1866
he became an assistant to Julius Plücker, who had become well known for his work
on cathode rays, and helped to set up and carry out demonstrations in his lectures
on experimental physics. Plücker, however, had returned to conducting research in
geometry, which he had been involved with several years before, and Klein also
assisted Plücker with his mathematical researches (Fig. 14.2).

Klein progressed very rapidly, and took his doctorate in December 1868 with
work that extended Plücker's results. By then Plücker had died, unexpectedly, and
Rudolf Clebsch, an influential mathematician in his thirties came over from Göttingen to urge upon Klein the task of publishing Plücker's posthumous geometrical
works. Klein seized the opportunity, which drew him into a circle of algebraic
geometers inspired by Clebsch, and he moved permanently out of physics and into
mathematics.[4] Klein then travelled to Paris in 1869 to learn group theory from
Jordan, although his studies were cut short by the outbreak of the Franco-Prussian
War, and he used group theory as a guiding principle to shape his view of geometry
from the early 1870s onwards.[5]

Fig. 14.2 Christian Felix
Klein (1849–1925)

Much of Klein's work in the 1870s is a response to Jordan's, and some of it is
indeed a critical response to the treatment of the quintic by Hermite, Brioschi, and

---

[4]For an interesting account of how Clebsch's contributions fit into the story about groups and
geometry, see Lê (2017).

[5]For a summary of his Erlanger Programm of 1872 and information about the geometry of the
icosahedron, see Appendix E.

Kronecker. These interests continued with his work on modular groups in the late 1870s when Poincaré emerged and Klein shifted his interests to match those of his new rival. After the collapse of his health in late 1882, Klein played himself back into mathematics by writing up some of his old papers in an enlarged and more carefully explained context, in the form of his book *Vorlesungen über das Ikosaeder und die Auflösung der Gleichungen vom fünften Grades* (*The Icosahedron and the Solution of Equations of the Fifth Degree*, 1884).[6]

Like those papers, but even more openly, the book is Klein's claim to Galois's mantle; it is his attempt to define Galois theory for the new generation, and as such it ran into nothing but opposition (in the form of neglect) from Kronecker, as we shall see. So it is surprising that Klein is hardly mentioned in Kiernan's and Brechenmacher's accounts, the more so because, as I shall argue, his book did more than any other work of the period to define Galois's legacy for the German, and soon the English-speaking, world.

14.5   Klein in the 1870s

By 1870 Hermite, Kronecker, and Brioschi were well established mathematicians, and to most readers their solution to the quintic equation seemed exemplary. Klein disagreed. For him the connection obscured more than it explained. It might be interesting that the two fields were connected, but it could not be the case that the way to advance the theory of equations led through the rich but complicated domain of elliptic function theory. He set himself the task of showing the mathematical world that a more natural, group-theoretic and indeed Galois-theoretic route was at hand, and in doing so he also set out to marginalise the theory of invariants, which was then the best way of capturing symmetries of various kinds.

In a series of papers in the 1870s and again in his book on the icosahedron (1884) he surveyed and rewrote the entire study of the quintic equation in what he explicitly called Galois theory. He argued that the right way to think about the question is in the language of group theory, which provides both conceptual clarity and computational power (through the related topic of invariants). He repeatedly contrasted this approach with the use of partially symmetric expressions or resolvents used in work in the late 1850s and early 1860s on the quintic by Hermite, Kronecker, and Brioschi and argued for the superiority of his approach.

In the paper Klein (1871, p. 1), written shortly after his return from Paris where he had met Jordan, Klein sketched how "The general theory of algebraic equations and their resolvents is beautifully illustrated by a number of geometrical examples". These, he said, make intuitive the abstract propositions of the theory of substitutions.

---

[6]The best modern explanation of the book is the German introduction by Slodowy to the re-edition of 1993.

In his (1875) he developed these ideas into a programme for the construction of all binary forms that are invariant under a group of linear transformation s. He took as an example the form of degree 12 that represents the vertices of a regular icosahedron, writing that

> I develop it as an example of *how the entire theory of these forms* can be derived from a knowledge of the linear transformation s under which they are invariant *without any complicated calculation and only with the ideas of the theory of invariants.*[7]

And he noted a remarkable coincidence, as he put it. The forms he was presenting, setting aside their particular significance, agreed precisely with those that Kronecker, Hermite, and above all Brioschi had presented in their studies of the general equation of degree 5. He returned to this theme in the final section of the paper, where he remarked that all that was new in his paper in this respect was the interpretation given to the formulae and their associated geometrical derivation.

Klein's 60-page paper (1877) is almost wholly contained in the book on the icosahedron, and need not be described in any depth, but some passing remarks are of interest. He began by saying that it was striking how easy it had proved to derive the resolvents of degrees 5 and 6 that arise in the theory of the quintic equation and to recognise their inter-connections, but, he said, it was only after talking to Paul Gordan—the leading figure in invariant theory at the time—that it occurred to him to invert the question and derive the theory of the quintic equation from a consideration of the icosahedron. Indeed, in close collaboration with Gordan, he said, he had been led to obtain from a natural source not only all the algebraic theorems and results that Kronecker and Brioschi had published—sometimes without proof—but also to introduce essential new ideas. (These concerned the precise use of elliptic functions, and will not be discussed here.)

The programmatic side of his approach came out again in Klein (1879b, p. 252)

> Neither Kronecker nor Brioschi gave any general reason *why* the Jacobian equations of the sixth degree appears as the simplest rational resolvent of equations of the fifth degree; as I regard these Jacobian equations as the bearers of a *system of 60 linear ternary substitutions that is isomorphic with the group of 60 linear permutations of five things* I have held from the beginning the principle that in every case one can characterise a priori the normal equations into which the given equations can be transformed by the creation of rational resolvents.

To understand the subsequent clash between Klein and Kronecker, we may begin by observing that Kronecker's work is often technical, and like many a gifted mathematician when young he did not stoop to making his insights clear and easy to follow, and would claim to know more than he could prove. This is very clear in his responses to Galois, which are among his earliest papers. As we have already seen in Sect. 11.4, Kronecker took a hard line on what counts as a solution. More precisely, he set some store by his distinction between, on the one hand, adjoining quantities that belong to the field that is obtained by adjoining the roots of a given polynomial to the field determined by the coefficients and on the other hand adjoining algebraic

---

[7] Here and below the italics are Klein's.

quantities that do not belong. In fact, he argued that one should properly only consider the first type of adjunctions. He claimed that this was in the spirit of Abel's work, and even called it Abel's postulate.

Klein agreed that this was a valid distinction—he called the first case that of natural irrationalities and the second that of auxiliary irrationalities—but he felt that both should be explored. This view gained plausibility when Hölder proved in his (1889) that in order to solve by radicals a cubic equation with rational coefficients and three real roots it is necessary to adjoin a complex number to the field of rational numbers, so strictly speaking the solution of the cubic must proceed in defiance of Abel's postulate.[8]

14.6   Klein's *Icosahedron*

Klein's book is not about the fact that the general quintic equation is not solvable by radicals. By the 1880s it was well known and understood that this was so and that the reason was the simplicity of the group A_5. All of this was carefully explained in Jordan's *Traité*. Rather, Klein's book is about every aspect of the theory of the quintic equation, but most obviously and importantly its solution by previous authors that Klein felt left a lot to be explained. In passing, Klein showed how to rederive their results by methods he had introduced in the 1870s. But his main aim was to show that the geometry of the icosahedron, its symmetry group and its subgroups, and the corresponding theory of forms provided a better explanation of what was involved by being more natural and direct. He also added some further geometrical considerations that illuminated other aspects of the rival methods.

Klein began by explaining the idea of a group, and then turned to the example of the regular solids

> I.1 §1 As soon as we enter upon the task of studying the rotations, etc. in question, by which the configurations which we have mentioned are transformed into themselves, we are compelled to take into account the important and comprehensive theory which has been principally established by the pioneering works of Galois, and which we term the *group-theory*. Originally sprung from the theory of equations, and having a corresponding relation with the *permutations* of any kind of elements, this theory includes, as has long been recognised, every question with which we are concerned in the case of a closed manifoldness of any kind of operations.

In particular, this was Klein's way of treating the resolvents of an equation.

> I.4 §1: A first and important portion of this theory, which distinguishes the nature of the resolvents coming generally under consideration, is formed by those reflections which, in accordance with the fundamental ideas of Galois, are usually denoted by his name, and which amount *to characterising the individual equation, or system of equations, by a certain group of interchanges of the corresponding solutions.*

---

[8]This is the irreducible case discussed in Sect. 9.4.

But

it is not sufficient in any given algebraical problem to know the *nature* of the resolvents; we require, further, to actually calculate these resolvents, and this in the simplest manner. The second part of the present chapter is concerned with this, with strict limitation to the questions immediately surrounding our fundamental problems. I show, first of all (§7), how we can actually construct the auxiliary resolvents by means of which the solution of the dihedral, tetrahedral, and octahedral equation is to be achieved. I concern myself, then, in detail with the resolvents of the fifth and sixth degrees of the icosahedral equation (§§8–15).[9]

The whole subject of resolvents was, for Klein, equivalent to the study of subgroups of a given group. In I.4 §3 Klein offered what he called some general remarks concerning a group G that is the group of an equation. He supposed that G is of order N and that there is a rational function R_0 which is unaltered by a group g_0 of ν of the permutations in G and so takes only $n' = N/\nu$ values, say

$$R_0, \dots, R_{n'}.$$

Then

$$(R - R_0) \dots (R - R_{n'})$$

is invariant under every permutation, and is a resolvent for the group. Then Klein observed that the conjugates of g_0 in G give rise to other resolvents, and that "There are as many different kinds of resolvents of the proposed equation $f(x) = 0$ as there exist different systems of associate [conjugate] subgroups within the group G". The group of the resolvent Klein called Γ, and he showed that there is map from G onto Γ, and that this map is an isomorphism when there are permutations in G that fix every R_j, and they form a self-conjugate (i.e. normal) subgroup γ of G. If no such subgroup γ exists the original equation is a resolvent of the resolvent, and nothing is gained by replacing the equation $f(x) = 0$ by its resolvent other than a restatement of the problem. But in the other case the new group—which we would call the quotient Γ/γ—is smaller and the problem of finding the roots of $f(x) = 0$ is broken into two smaller problems. "Clearly", Klein concluded, "the resolvents of the second kind are the more important".

Klein then analysed the Galois resolvent from this point of view, until he could conclude at the end of §5 that

If, then, for such an equation, a rational function of the roots is constructed which remains unaltered for the permutation S_k of a certain sub-group contained in the Galois group, and thus can be introduced as a root of a corresponding resolvent, it is sufficient to establish a rational function of the single root R_0, which will, for the corresponding ψ_ks, be

---

[9] As noted above, invariant theory contains standard mechanisms for generating one invariant from another, such as the Hessian of a form and the functional determinant of two forms. Applied to the form that specifies the 12 vertices of the icosahedron, the Hessian locates the 20 mid-face points and the functional determinant of the first form and its Hessian locate the 30 mid-edge points.

transformed into itself; for the sub-group of the ψ_ks contains at the same time all the (ψ_k^{-1})s, and, therefore, corresponds to the sub-group of the S_ks in the isomorphic co-ordination.

In I.4 §5 he went on

> I have framed the foregoing paragraph in such detail in order to be able to now marshal directly our fundamental equations in the scheme of the Galois theory, to wit, the binomial equations and the equations of the dihedron, tetrahedron, octahedron, and, icosahedron. Let us first agree that our equations are irreducible. From the considerations in the last chapter, based on the function theory, it follows, that the N-function branches, which are defined by the individual equations, on regarding in each case the right-side Z as independent variable, are all connected with one another. Therefore the hypotheses are exactly fulfilled, to which the concluding theorem of the preceding paragraph relates.

With this in place Klein proceeded to trace the solution of the quartic equation through a chain of subgroups each normal in the one before, and to show that a quintic equation is not solvable by radicals when its group is that of the icosahedron.

Looking ahead to new problems, he wrote

> I.5 §4 If we are concerned with the solution of an equation of the nth degree $f(x) = 0$, we can regard it as being the same as if a form-problem for the n variables $x_0, x_1, \ldots, x_{n-1}$ (i.e., the roots of the equation) were proposed to us. The group of the corresponding linear substitutions will be simply formed by those permutations of the xs which make up the "Galois group" of the equation; the forms F coincide with the complete system of those integral functions of the xs which, in the sense of the Galois theory, figure as "rationally known". With these remarks, nothing, of course, is primarily altered in the substance of the theory of equations. But the theorems to be developed in it acquire a new arrangement.

Nobody would write a whole book to establish that the general quintic equation is not solvable by radicals. So what was Klein doing? The answer, as he made very clear, is showing in complete detail how conceptual methods can almost completely replace the previous heavy reliance on calculations. For example, that talk of an expression taking only five values can be replaced by talk of a form connected to a subgroup of the icosahedral group. As to the detail necessary, Klein had to show that precisely the equations introduced in the other methods arise in his approach, but more naturally. To do this, he showed that the subgroups of the icosahedral group, and the invariant forms associated with them, lead in readily comprehensible ways to exactly those equations, and therefore that there is a systematic approach which directs the calculations and can even eliminate them entirely. Furthermore, Klein hoped to show that this approach is the very essence of Galois's: involving equations as the outward sign of groups and rationally known quantities, and taking explanation away from calculation and towards conceptual arguments.

14.7   Exercises

1. The group of the modular equation (14.3) is $PSL(2, 5)$, which consists of 2 by 2 matrices of determinant 1 and entries that are integers modulo 5, quotiented out by the centre. Show that there are 60 of these by regarding them as projective transformations of the projective plane over the field of five elements.

2. The elements of this group can be regarded as Möbius transformations $k \to \frac{ak+b}{ck+d}$ in the obvious way. Find transformations of orders 2, 3, and 5.
3. Find the five subgroups of order 12 identified by Galois. (You may also consult Appendix E.)
4. A periodic function of a real variable, say with period 2π, can be defined on any interval of that length and then defined everywhere else by repetition. In the same way, a doubly periodic function $f(z)$ of a complex variable is defined on a parallelogram Ω in the plane of complex numbers. If we place one vertex of Ω at the origin, and the neighbouring vertices at the complex numbers ω and τ, then the function f is defined everywhere once it is defined in Ω by the rule

$$f(z + m\omega + n\tau) = f(z), \quad m, n \in \mathbb{Z}.$$

If f is suitably defined it becomes a complex analytic function of z—an elliptic function. If you know analytic function theory, use Cauchy's theorem to show that the function must have at least two poles in Ω. Show also that two parallelograms Ω and Ω' (defined by ω' and τ') define essentially the same elliptic function if

$$\frac{\omega}{\tau} = \frac{\omega'}{\tau'}.$$

Questions

1. If you were a well-trained mathematician of the 1860s, and therefore familiar with elliptic function theory, would your sympathies be with Galois and Betti over the right way to explain the reduction of the modular equation, or with Hermite, Kronecker, and Brioschi? Why?
2. Is it reasonable to see the attitudes of Jordan and Klein to Galois theory as a generational shift in mathematics? What is involved in making such a claim? What other such shifts might have occurred in the nineteenth century?

Chapter 15
What Is 'Galois Theory'?

15.1 Introduction

At this point, one could legitimately wonder if the story of Galois theory was not essentially over. The general question of which polynomial equations are solvable by radicals has been turned into a question about which groups are solvable, which is a perfectly good theoretical resolution of the problem even if it leaves it open in any particular case—but then, how important has it ever been to solve an equation by radicals? Has not the original topic become just one of a number of topics in the emerging topic of finite group theory?

The process of successive adjunctions of auxiliary irrationals has also been spelled out fully. From a certain, later, perspective this is surely a question about finite extensions of the field of rational numbers, and if we look ahead to the work of Dedekind that is exactly what we see being developed. Once again, the original topic has seemingly been subsumed in a much larger and more diverse one, in this case with connections to cyclotomy and other emerging branches of algebraic number theory.

Something must be wrong, or no one would speak of 'Galois theory' today, whatever they might mean by it. People would speak of group theory, with an application to the theory of fields; or of field theory, with an application to the theory of groups; or of the solution of equations from a group-theoretic perspective. An explanation of the high status of Galois theory requires us to understand how it came to be regarded as an exemplary union of both group theory and field theory. If we set the reputation of its titular founder aside, we can begin by looking to see who promoted this unified point of view, and how well they got on.

© Springer Nature Switzerland AG 2018   179
J. Gray, *A History of Abstract Algebra*, Springer Undergraduate Mathematics Series,
https://doi.org/10.1007/978-3-319-94773-0_15

15.2   Klein's Influence

We have already seen that a significant place in the theory of equations was occupied by the solution of the quintic equation by means of elliptic functions. This drew on a major topic of the day, and was seen as an advance on what Galois had done; better to have solutions of a certain kind than simply proclaim that there are no solutions of a certain other kind. In this spirit, Jordan had even shown that there was a related class of functions capable of expressing the solution to polynomial equations of any degree. It could be objected that the calculation of tables of values for these functions presented considerable difficulties, but then the whole business was rather abstract: the market for solutions of specific quintic equations was very small.

The alternative promoted obscurely by Galois and explicitly by Jordan and Klein made essential use of a new and abstract concept, that of the permutation group. In this context, a remark by Paul Bachmann in *Mathematische Annalen* (1881) is worth exhibiting, because it shows how radical Klein's position was.[1]

> In spite of the fact that new editors, chief among them Herr C. Jordan [Bachmann has in mind Jordan's second Galois commentary [...] (Wussing)], have brought Galois's investigations of the algebraic solvability of equations closer to the common understanding, many may be deterred from the study of this principal part of the subject of equations by their extensive use of the very abstract substitution theory. Such persons may well welcome the present exposition of the subject which avoids the said theory as much as the nature of things allows, and bases itself essentially on the two fundamental concepts of field and irreducibility. We believe that this treatment, due, incidentally, to Herr Dedekind, gives a far more concrete view of the phases of the solution process of equations than the usual mode of presentation.

We shall look in more detail below at the split between the views of Kronecker and his former student Netto on the one hand and those of Dedekind, Weber and their followers on the other, but there was a strong preference for emphasising the successive adjunction side of the story, which Dedekind and Weber wrote up in the language of fields that they were developing.

But the quotation makes it clear that for its young author, and perhaps also for the editors of the fairly new journal (founded in 1869), substitution theory is very abstract, perhaps too abstract. There were isolated papers in German on the subject, but no book or other full-length account from which the subject could be learned and its priorities and importance assessed. That was to change with the publication of Netto's book in 1882, as we shall see in Chap. 27. But throughout the 1870s Klein's espousal of group theory was more radical, and perhaps more risky, than is often suggested.

Klein, unlike Jordan, wanted to push the role of elliptic function theory to one side. As we have seen, he argued strongly that the natural approach was to

---

[1] It is quoted in Wussing (1984, p. 274).

work with the icosahedron and the group of permutations of five objects, and not with an equation which happens to turn up in another domain of mathematics. He never remarked that the route from trigonometric functions to elliptic function s was a natural one, and he managed instead to suggest that Hermite, Brioschi, and Kronecker had not even used elliptic function theory as elegantly as they should have done. In many a less scientific context such rhetoric would have been taken as an ad hominem attack, as perhaps it was. There is no doubt that Kronecker held a very poor opinion of Klein. Petri and Schappacher (2002, p. 245) remind us that in a letter from the 1880s Kronecker dismissed Klein as a "Faiseur"—a maker of superficial phrases.[2] Equally, it is clear that Klein found Kronecker's treatment of his work unfair—as Petri and Schappacher show that it surely was—and attributed it to what he saw as a growing dogmatism in Kronecker (Klein was to make this something of an orthodoxy later in Göttingen). The failure of Kronecker to publish proofs of many of his results can only have annoyed Klein still more.

Kronecker's opposition to Klein was based on his marked dislike of general theorems that lacked concrete detail. The ideal, for him, was a theorem that would determine from the coefficients of a quintic equation alone if the equation was solvable by radicals. The transition to a discussion of groups, however interesting on other grounds, did nothing if it was not possible to determine the group of the equation. That being a task left unaddressed by Jordan and Klein, Kronecker saw their contributions as failing to achieve the required goal.

Klein's view was that calculations should proceed from a transparent base in a conceptually explicit way, and arguably that one might prefer conceptual arguments to computational ones. Today we talk about this difference in approaches as a battle between frogs and birds: Kronecker is a frog, never happier than when in the middle of things; Klein is a bird, offering the grand overview. In such questions there is usually right on both sides, and which way a mathematician goes seems to be a matter of temperament. We shall return below to the question of why Kronecker's strictures were often ignored. Here, we shall note that Klein's influence on the spread of Galois theory was considerable, and can be found in the work of Weber, Bolza, and Wiman.

If we look at Weber's *Lehrbuch der Algebra* (1896) Volume 2, Book 3, we see a heavily Kleinian presentation from a mathematician who was often close to Kronecker. The earlier books presented general group theory, which was growing into a rich subject, and an account of linear groups, much of which prepares the way for Book 3. That book opens with an account of resolvents associated to a composition series, metacyclic equations, and equations of degrees 6, 7, and 8. Then come the geometric examples that Jordan had discussed in his *Traité*, a Kleinian treatment of the quintic, and finally two chapters on the group $PSL(2, 7)$ that Klein had written about at length in his (1879).

Oscar Bolza had been driven by a downturn in the German economy to become one of the first Professors of mathematics at the new University of Chicago. He

---

[2]They are quoting from Biermann (1973/1988).

had been largely educated in Berlin in physics before turning to mathematics and studying in Strasburg and then spending a difficult year with Klein. Then, as Parshall and Rowe say in their (1994, 200) when Bolza embarked on a career in the new University of Chicago "[a]rmed with nothing more than a letter of introduction from Klein", he "became an effective and versatile propagator of Klein's mathematics". E.T. Bell reported that Bolza's lectures "did much to popularise the subject and to acquaint young Americans with the outlines of the Galois theory of equations".[3] His account, which appeared in two papers in the *American Journal of Mathematics* in 1891, is the first account in English. It is based, as he said, on the books by Serret, Jordan, and Netto, papers by Kronecker and others, and particularly on lectures that Klein had given in Göttingen in 1886.

Bolza led off with the theory of groups, and explained about groups, subgroups, conjugate and self-conjugate subgroups. He explained the effect of adjoining an irrational in terms of the effect on the Galois group, and this is used to explain how the solution of cubic and quartic equations proceeds, and in particular why the quartic equation is solved by solving a cubic equation but not a pure cubic equation. This in turn is used to explain how the self-conjugate subgroups of prime index p correspond to the solution of pure binomial equations of the form $x^p - \alpha = 0$. Bolza was then in a position to examine the decomposition of the symmetric groups of low degree, and to appeal to their composition series to decide if the corresponding equation is solvable by radicals. This section of the paper ends with a hint of the rich content of Klein's recent book (1884) on the icosahedron.

In the second part of his paper Bolza gave a general treatment of Galois theory, by which he meant the case where a specific equation and expressions in its roots (considered as numbers) is at issue, as opposed to the case where the roots are indeterminates—passing from Lagrange to Galois, so to speak. In particular, Bolza looked at the study of equations whose groups are cyclic, abelian, or metacyclic, and showed that they are always solvable by radicals.[4]

After writing this article, Bolza reviewed the English translation of Netto (1882), and found fault with the account of Galois theory. He felt that Netto had not understood the subtleties in Galois's distinction between treating the roots of an equation as symbols and treating them as numbers. He gave the example of the equation

$$x^4 + 1 = 0,$$

with roots

$$x_1 = \varepsilon, x_2 = i\varepsilon, x_3 = -\varepsilon, x_4 = -i\varepsilon, \text{ where } \varepsilon = e^{i\pi/4}.$$

---

[3]Quoted in Parshall and Rowe (1994, 201).

[4]Recall that the metacyclic case is the one in which all the roots are rational functions of any two of them; the best case is binomial equations.

The expressions x_1^2 and $x_2 x_4$ are distinct but numerically equal (to i). The expression x_1^2 is formally invariant under the group H isomorphic to S_3 that permutes 2, 3, 4; the expression $x_2 x_4$ is invariant under a different group. Moreover, the permutations of the symbols x_1, x_2, x_3, x_4 that leave $x_2^2 = i$ invariant are twelve in number and do not form a group.[5]

The group of the equation over the rational numbers is in fact

$$G = \{1, (12)(34), (13)(24), (14)(23)\}.$$

The expression $x_1 x_2^2 + x_3 x_4^2 = 0$ so it is rationally known, but it is not formally unchanged by G. The resolution of this seeming paradox, said Bolza, is that every expression that is numerically invariant under the group can be rewritten so that it is formally invariant as well. He then gave the simpler resolution, which is more obviously in line with Galois's original idea, that concentrates on the permutations of the actual roots, and forgets about the formal expressions altogether. It quickly becomes clear in the present case that looking at what can happen to ε shows that the group is the one just given. In this case, the substitutions in G that leave $x_1^2 = i$ unchanged are the identity and $(13)(24)$ and they form a subgroup of G.

Bolza, following Klein, gave a good explanation of the adjunction of natural and of accessory irrationals. A natural irrational is a numerical value of a rational expression in the roots. Adjoining one, as Galois said, reduces the group of the equation to the group H that leaves the expression invariant; in the case of G above i is a natural irrational, and adjoining it, as we saw, reduces the group of the equation to $\{1, (13)(24)\}$. If moreover the group H is self-conjugate (we would say normal) then the group of the equation that the natural irrational satisfies is the quotient group, G/H.[6]

The effect of adjoining an auxiliary irrational is covered by Jordan's theorem on the simultaneous adjunction of all the roots of $f(x) = 0$ to the equation $g(x) = 0$ and of all the roots of $g(x) = 0$ to the equation $f(x) = 0$. This says that if the group of the equation $f(x) = 0$ is reduced in this way, then the roots of $g(x) = 0$ are rational functions of the roots of $f(x) = 0$. In other words, the adjoined roots are natural after all.

Klein reinforced his influence on the development of Galois theory in the United States when he went there in 1893, and we can note that Klein's icosahedron book was translated into English in 1888, Netto's in 1892. Both were published in the United States first, not Britain. Brechenmacher's description of the Dickson network makes clear numerous other ways in which Klein's influence, although not his alone, was spreading.

---

[5]They are the elements in $H \cup (13)H$.

[6]In modern terms, a natural irrational lies in the splitting field of the original polynomial.

Hölder's two-part article 'Galois theory with applications' (Hölder 1899) for the *Encyklopädie der mathematischen Wissenschaften* and Wiman's article 'Finite groups of linear substitutions' in the *Encyklopädie der mathematischen Wissenschaften* (Wiman 1900) (see Chap. 15 below) are obviously much influenced by Klein. It will be enough to look at Hölder's. Part 1 opens with the definition of the group of an equation and some general remarks about such groups, including an indication of how to construct them explicitly by finding a primitive element. The concepts of transitivity and primitivity are explained, followed by an account of the adjunction of natural irrationals, cyclic equations (the corresponding Galois group is cyclic), pure equations (those of the form $x^n - a = 0$), factorisation through the construction of resolvents, with an indication of the role of a normal subgroup, adjunction of accessory irrationals, and of radicals. The idea of a solution by radicals is then explained, with criteria for solvability, and finally some remarks are made about equations that are not solvable. Overall, the treatment largely follows Jordan's *Traité* and Hölder (1889), in which he proved the Jordan–Hölder theorem and applied it to Galois theory.

Part 2 covers the applications. It opens with a definition of the general equation, then looks in detail at equations of the first four degrees, the unsolvability of the general equation of higher degree (the argument rests on the simplicity of $A_n, n > 4$), equations with regular groups (those for which all the roots are rational in any one), equations with commutative groups, abelian equations, cyclotomic equations. Then it looks at a topic begun by Abel, the division and transformation by elliptic functions, then at reduction of modular equations to normal form, and equations of prime degree. After that came an indication of how Sylow's theorems can be used to determine whether a group is solvable, the irreducible case of the cubic, line and circle constructions, and finally geometric equations connected to curves and surfaces possessing symmetries.

Only six books are cited: by Serret, Jordan, Petersen, Netto, Vogt, and Weber's *Lehrbuch der Algebra* (which we discuss in Chap. 22). Petersen's book (1877) in Danish was translated into German the next year and was closely modelled on Serret's *Cours* with, as the author said, information on substitution groups taken from Jordan's *Traité*.[7]

For whatever reason, Hölder did not mention the book (Borel and Drach 1895) on the theory of numbers and higher algebra. It carries a warm-hearted introduction by Jules Tannery, in which he speaks of having given lectures to third-year students at the École Normale in 1891–1892 in the hope of engaging their interest in various problems, and says that now indeed two students "who are certainly among the most distinguished of those that I have had the honour and pleasure to meet" are the authors of this book.

Number theory is presented first as a theory of congruences modulo a prime, and the theory of 'Galois imaginaries' is introduced, without the term finite field

---

[7]Petersen was a very energetic producer of textbooks for students in Copenhagen.

being used or indeed implied.[8] Quadratic congruences lead up to the theorem of quadratic reciprocity; Kronecker's proof of it is given. The treatment concludes with the introduction of the Gaussian integers and some brief remarks about quadratic forms.

Higher algebra was taken to mean polynomial algebra, algebraic numbers and algebraic integers, all of which led to a chapter on substitution groups, another on the algebraic solution of equations and solvable groups, and finally one on normal equations and abelian equations.[9]

Galois theory is therefore presented as a theory of the solution of polynomial equations, in which older material about rational expressions is repackaged as being about subgroups of the symmetric group, and the subject is pointed towards equations having particularly tractable groups. The examples of equations of low degree is not given, but nor is there much influence of the abstract approach of Dedekind. Rather, the account can be read as a simplified presentation of some of Jordan's ideas illuminated with some of Kronecker's explicit methods.

Finally, we should not neglect the question of finding the 'Galois group' of a given equation. To this day there is no algorithm that will take the coefficients of a polynomial and determine the Galois group of the corresponding equation. Instead, there is a method that for equations of low degree will quickly give enough information for the group to be found. It works very well for equations up to degree 8, and thereafter requires a little more patience as alternative possibilities can arise that can be difficult to distinguish. It has now been improved and applied to determine all the Galois groups of equations of degree less than 20.[10] The limitations have as much to do with the limitations on computing power as anything else.

The method is described in many places, for example (Cox 2004, ed. 2012) and K. Conrad,[11] but the original source is seldom given. It was discovered by Dedekind, and can be found in his *Mathematische Werke* 2, 414–415, as part of a letter to Frobenius of 8 June 1882. The problem of attribution seems to be that Frobenius himself forgot that he had read it, and raised it as a conjecture in a letter to Dedekind. When Dedekind reminded him that the 'conjecture' was already a theorem, Frobenius published the information, along with Dedekind's proof, in his paper of 1896. The story was picked up by I. Schur in his paper (1920), and this seems to be the source of the information that most recent authors have used; the relevant part of the Dedekind–Frobenius correspondence was only published in 1931.

---

[8]See Chap. 23 below, where it is explained that these are roots of polynomial equations with coefficients taken modulo a prime; in modern terms algebraic extensions of a finite field.

[9]An equation was said to be normal if each root of it is expressible as a polynomial function with integer coefficients of any one of them.

[10]See the paper Hulpke (1999) and the pdf of a talk by Hulpke available at
http://www.math.colostate.edu/~hulpke/talks/galoistalk.pdf.

[11]See his 'Galois groups as permutation groups' and his 'Recognizing Galois group s S_n and A_n', and other good sources on the web. Dedekind's original proof has been tightened in many places; one due to Tate is on the web. http://www.math.mcgill.ca/labute/courses/371.98/tate.pdf.

Dedekind's "very lovely" theorem, as Schur and many after him have called it, says that if an irreducible polynomial $f(x)$ of degree n is interpreted as a polynomial with coefficients modulo a prime p and if as such it factorises as a product of distinct polynomials of degrees $d_1, d_2, \ldots, d_k$, and none of these polynomial factors are multiple, then the Galois group of the original polynomial has an element of cycle type $d_1, d_2, \ldots, d_k$. Enough information of this kind for small primes can swiftly reduce the candidates for the Galois group to a unique possibility. In particular, it can make it possible to say whether the group is the full permutation group S_n, which is one of the tasks Schur had in mind. More generally, the theorem has implications for the so-called inverse Galois problem: find which groups can be exhibited as the Galois group of an extension of the rational numbers.

15.3   Concluding Remarks

In their rich rewriting of the consequences of Gauss's *Disquisitiones Arithmeticae* Goldstein and Schappacher argue for an endless process of selecting, re-presenting, and rearranging of the different parts of Gauss's book. The content, they write, has at times defined the subject, at times been cut and reshaped. They see the history of the *Disquisitiones Arithmeticae* in the 19th century in a new way, as an endlessly shifting collection of readings, of textual transmissions, of the transfer of knowledge between and especially across generations—and they reject the idea that a period of awe was followed by the steady accumulation of disciples until modern number theory was created around 1900.

Just so, every powerful mathematician involved in the responses to Galois's work between 1846 and 1900 made a selection of it, emphasised aspects and marginalised others, and then made contributions that changed what 'Galois theory' could be said to be. Kiernan was right to trace an arc from Lagrange to Artin at a time when that story was not well known, and right to bring to it figures whose contributions were no longer well known. In that story, Jordan's *Traité* is important, and Artin's re-writing of the subject provides a natural terminus. Brechenmacher is right to challenge that arc, to try to fracture it by attending more closely to Galois's specific contributions, and by looking more carefully at who read what; in short, to make 'Galois theory' many things that form not a set package that grew down the decades but a shifting family of ideas that evolved in many different ways.

In the 1850s, Hermite's response to Galois was rooted in the theory of elliptic function s that he loved, and in the deep attention to detail that he believed was essential to mathematics as a research activity. In this, as so often, he stood close to Kronecker, and to Brioschi. Kronecker saw his contributions as being much more inspired by Abel than Galois. In the 1890s, Moore, Dickson, and others saw their work as responding to a different part of Galois's; in their case to his paper (Galois 1830) and the theory they wished to present of finite fields. Galois had seen his paper as an extension of Gauss's work on cyclotomy to a new family of imaginaries.

How different should we see these approaches as being? Cyclotomic equations of prime degree remained throughout the 19th century as one of the canonical examples of equations that are solvable by radicals and Gauss's proof that they are, with its enumerations of cosets of subgroups of the appropriate cyclic group, looks like abelian Galois theory. Or perhaps one should say that after Galois, it was clear to Jordan that cyclotomy was an elementary part of Galois theory.

Kiernan's arc prevents us from seeing many things; events are tidied up, people are made to face forward. But we should also admit that room was made for other points of view, such as Kronecker's, and distinctions are made between, for example, what Galois did and what Jordan did, and between group-focussed accounts (typically French) and field-centred accounts (typically German). If we interpret Galois's work as advocating the merits of thinking about groups (whatever they might be) as a way to make conceptual progress in the face of mounting computational difficulties, then we must surely see Jordan and Klein as promoting that view, and in Klein's case succeeding to a considerable degree.

Chapter 16
Algebraic Number Theory: Cyclotomy

16.1 Introduction

We return to one of Gauss's favourite themes, cyclotomic integers, and look at how they were used by Kummer, one of the leaders of the next generation of German number theorists. French and German mathematicians did not keep up-to-date with each other's work, and for a brief, exciting moment in Paris in 1847 it looked as if the cyclotomic integers offered a chance to prove Fermat's last theorem, only for Kummer to report, via Liouville, that problems with the concept of a *prime* cyclotomic integer wrecked that hope. Primality, however, turned out to be a much more interesting concept, and one of the roots of the concept of an ideal.

16.2 Kummer's Cyclotomic Integers

Although cyclotomic integers were to be used in the study of Fermat's last theorem, Kummer, who knew more about them than anybody else, apparently took them up with a view to the study of higher reciprocity laws. He began by looking for the factors of cyclotomic integers of the simplest form, namely binomials, cyclotomic integers of the form $x + \alpha^j y$, where he has fixed a prime p and α is a primitive pth root of unity: the lowest positive power n such that $\alpha^n = 1$ is p. We can follow what he did if we are prepared to take some correct mathematics on trust (if you are not, consult Chapter 4 of Edwards 1977).

Kummer said that one cyclotomic integer $h(\alpha)$ divides another, $f(\alpha)$, if there is a third, $g(\alpha)$, such that $f(\alpha) = h(\alpha)g(\alpha)$. A cyclotomic integer is prime if every time it divides a product it divides one of the factors. Because the norm of a product is the product of the norms, a factor of a cyclotomic integer has a norm that divides the norm of the given cyclotomic integer. Cyclotomic integers of norm 1 are called

© Springer Nature Switzerland AG 2018

J. Gray, *A History of Abstract Algebra*, Springer Undergraduate Mathematics Series,
https://doi.org/10.1007/978-3-319-94773-0_16

units and they divide every cyclotomic integer, so they are ignored when division is of interest.

Unfortunately, cyclotomic integers are unwieldy things. Kummer took a prime p and a primitive pth root of unity, α, and defined a cyclotomic integer to be an expression of the form

$$a_0 + a_1\alpha + \ldots + a_{p-1}\alpha^{p-1},$$

where the a_j are ordinary integers. Minor irritation is caused by the fact that the identity $1 + \alpha + \ldots + \alpha^{p-1} = 0$ implies that

$$a_0 + a_1\alpha + \ldots + a_{p-1}\alpha^{p-1} = (a_0 + c) + (a_1 + c)\alpha + \ldots + (a_{p-1} + c)\alpha^{p-1},$$

for any integer c. To minimise the irritation this causes one can always insist that the term in α^{p-1} has coefficient zero.

Kummer defined the norm of the cyclotomic integer $a_0 + a_1\alpha + \ldots + a_{p-1}\alpha^{p-1} = f(\alpha)$ to be

$$Nf(\alpha) = f(\alpha)f(\alpha^2)\ldots f(\alpha^{p-1}),$$

and said he took the term from Dirichlet's work. The cyclotomic integers $f(\alpha^j)$ he called the conjugates of $f(\alpha)$. This precisely generalises what we do when finding the norm of a Gaussian integer—we multiply it by its conjugate—but at a rising cost as p increases. Even to compute the norm of a cyclotomic integer when $p = 7$ is tiresome if not done on a computer, so the examples are toy examples.

Exercise. Let $p = 3$. Compute the norm of the cyclotomic integer $f(\alpha) = 3 - 7\alpha + 2\alpha^2 = 1 - 9\alpha$ as follows:

- write $f(\alpha) = 1 - 9\alpha$;
- $f(\alpha^2) = 1 - 9\alpha^2 = 10 + 9\alpha$;
- compute the norm, which is the product

$$f(\alpha)f(\alpha^2) = (1 - 9\alpha)(10 + 9\alpha) = 10 - 81\alpha - 81\alpha^2 = 91.$$

Now, $91 = 7 \cdot 13$, and because, for any prime p, $N(f(\alpha)g(\alpha)) = N(f(\alpha))N(g(\alpha))$, it follows that either $f(\alpha) = 3 - 7\alpha + 2\alpha^2$ is irreducible or its factors have norms 7 and 13.

Given $f(\alpha) = x + \alpha^j y$, with x and y integers, how does one find its divisors? Kummer argued as follows. Let $h = h(\alpha)$ be a divisor of $f = f(\alpha)$, then the norm of h divides the norm of f, which is an integer. This imposes a severe constraint on what h can be.

Next, Kummer let h be a prime cyclotomic integer. Now it follows that h itself divides $N(f)$, the norm of f. Therefore h divides at least one of the prime factors of $N(f)$. Kummer was able to show that in fact all the integers that h divides are multiples of p. Continuing in this fashion, he was able to show that $N(h)$ is itself a prime integer congruent to 0 or 1 mod p, and in fact that, conversely, if h is a cyclotomic integer whose norm is a prime integer then $h = h(\alpha)$ is itself prime and it divides a binomial $x + \alpha^j y$.

Here are two examples, taken from not Kummer's work but Lamé's, that we shall need later (they are given in Edwards 1977, 86). The prime is $p = 5$.

1. $N(\alpha + 2) = 11$. This says $\alpha + 2$ is prime; that a cyclotomic integer $g(\alpha)$ is divisible by $\alpha + 2$ if and only if $g(-2) \equiv 0 \bmod 11$; and that 11 is a product of four factors ($\alpha + 2$ and its three conjugates).
2. $N(\alpha + 4) = 5 \cdot 41$. This says that $\alpha + 4$ is a product of a factor of norm 5 and one of norm 41. It turns out that the only factors of norm 5 are units times $\alpha - 1$, because here $p = 5$, and $\alpha + 4 = (\alpha - 1)\alpha^2(3\alpha^2 + 2\alpha + 1)$. Since α^2 is a unit it must be that $N(3\alpha^2 + 2\alpha + 1) = 41$ and $3\alpha^2 + 2\alpha + 1$ is prime.

Here's the first one in detail (I leave the second one as an instructive exercise).

$$N(\alpha + 2) = (\alpha + 2)(\alpha^2 + 2)(\alpha^3 + 2)(\alpha^4 + 2) =$$

$$(\alpha + 2)(\alpha^4 + 2)(\alpha^2 + 2)(\alpha^3 + 2) = (\alpha^5 + 2\alpha^2 + 2\alpha + 4)(\alpha^5 + 2\alpha^3 + 2\alpha^2 + 4) =$$

$$(2\alpha^4 + 2\alpha + 5)(2\alpha^3 + 2\alpha^2 + 5),$$

because $\alpha^5 = 1$. It is now helpful to proceed by collecting like powers of α:

α^4	α^3	α^2	α^1	$\alpha^0 = 1$
10		4	4	
4	4		10	
	10	10		25

The product is therefore $14\alpha^4 + 14\alpha^3 + 14\alpha^2 + 14\alpha + 25$, but this is the same cyclotomic integer as the one obtained by reducing every coefficient by 14, and that is the ordinary integer 11, so the norm of $\alpha + 2$ is $N(\alpha + 2) = 11$.

Kummer worked extensively in this fashion, weeks of tedious but not actually difficult work looking for prime cyclotomic integers, for successive values of the prime p. Eventually in 1844 he published a table of prime factorisations of all primes $q \equiv 1 \bmod p$ in the range $p \leq 19$, $q < 1000$. (The table is reproduced in Edwards 1977, pp. 102–103.) Up to this stage he had found prime factors corresponding to every value of the norm that turned up, and he was surely beginning to relax. Time, perhaps, to look for a theorem: all cyclotomic integers can be factorised (uniquely up to order) into prime cyclotomic integers?

Then came the case $p = 23$, the next prime after 19. The prime number $47 = 2 \cdot 23 + 1$ is a possible norm, but the techniques Kummer had available for finding a prime cyclotomic integer with norm 47 failed. More precisely, he showed that when $p = 23$ there is no cyclotomic integer with norm 47 at all! So the uniqueness of prime factorisation fails for 47, not because uniqueness fails but because there are no primes at all that are factors of 47. It is not itself prime, but it has no prime factors—a very unexpected turn of events.[1]

Kummer then observed that the test for divisibility could be made to work even when there are provably no factors for which it tests. He pursued this idea through the unstudied cases where $q \not\equiv 0, 1 \mod p$, which took a further 2 years, 1844–1846. It led him to the idea that there are tests for congruence modulo a cyclotomic integer that reduce to congruence questions modulo a rational prime, and these tests always give answers. They can therefore be said to determine a congruence relation on cyclotomic integers with this property: a product is congruent to 0 if and only if one of its factors is congruent to 0. The tests amount to asking 'If p divides a product, does it divide one of the factors of that product?'.

He interpreted this as meaning that the rational prime p remains prime or factors in a given family of cyclotomic integers if and only if the congruence tests say that p is congruent to zero. When this happens he said that p remains prime. But if not, he said that p was no longer prime—even when he can also show that it has no cyclotomic integers as factors! He considered that p then had *ideal* factors. He could now claim that a cyclotomic integer is determined by its (possibly ideal) prime divisors because questions about the multiplication and division of cyclotomic integers have solid answers via the tests.

Kummer published his theory in 1847. Curiously, the paper has two gaps, which he only filled in a note of 1857, and that note rather over-estimated how much Kummer had done in the first paper while still under-estimating what had to be done with the second gap. Since these gaps concern the inner workings of the theory, albeit in a crucial way, they will not be discussed. The fact that they can be filled means that their existence did not significantly affect the development of the theory.

Kummer's student Kronecker regarded cyclotomic integers as quotients of the polynomial ring $\mathbb{Z}(\alpha)$ by (here we must add some words—the ideal generated by) $\alpha^p - 1$. This makes it clear that the algebra of cyclotomic integers is very like the algebra of polynomials with integer coefficients, but it does nothing to illuminate the number-theoretical questions with which we are occupied.

Kummer's theory accomplished several significant things. It shattered any naive confidence in the idea that algebraic integers would behave like rational integers when it came to multiplication and division. It showed that a rigorous theory of algebraic integers (admittedly, only the cyclotomic integers) could be created. It introduced into mathematics the idea that ideal objects could be created to rescue a theory.

---

[1] In later terminology, this means that the factors of 47 are all ideal.

16.3 Fermat's Last Theorem in Paris

In 1847, before Kummer's discoveries became widely known, they caused a wild flurry of interest in Paris, when a number of French mathematicians hoped that they were the right general setting for a final assault on Fermat's Last Theorem. The hope was that it could be shown that Fermat's Last Theorem for exponents n greater than 2 did not have a solution in cyclotomic integers, and therefore no solution in ordinary integers. Gabriel Lamé believed he had discovered such a proof and informed the Academy of Sciences in Paris that he had done so, but it emerged that he had assumed that the new integers obeyed all the rules that genuine integers obey. In particular, he had assumed that a cyclotomic integer factors into prime cyclotomic integers in an essentially unique way.

Joseph Liouville, the editor of a mathematical journal, had made it his business to keep in touch with developments in Germany, and he advised caution, but the ever-energetic Cayley joined in on Lamé's side, and even claimed to have a similar proof of his own, which Lamé may not have found very encouraging. Liouville took the opportunity to write to Kummer, and in this way learned that Kummer had already discovered that certain cyclotomic integers have no prime factorisations. This destroyed the arguments of Lamé and Cayley, although Lamé kept working for a while. Cayley merely switched to elasticity theory.

In fact, and this is the theme of Edwards's book *Fermat's Last Theorem*, cyclotomic integers were to play a role in the subsequent history of research into Fermat's provocative claim. We cannot follow it here, but at about this time Sophie Germain had divided primes into two kinds when it came to research into the theorem, and Kummer's cyclotomic integers turned out to fit very well for what became called the regular primes. It was in this way that the record for the value of a prime p such that Fermat's last theorem was known to be true for all prime exponents q with $2 < q < p$ was made to creep up until it reached around 100,000. This work is perhaps more interesting now for the influence it had on the use of computers in mathematics (see Corry 2008) than for the mathematics itself, which languishes today in the shade of the successful and very different resolution of Fermat's last theorem in Wiles (1995) and Taylor and Wiles (1995), but it was a substantial achievement in its day.[2]

---

[2]Our understanding of what Germain achieved has been deepened in recent years by the papers of Laubenbacher and Pengelley (2010), Del Centina (2008) and Del Centina and Fiocca (2012).

Chapter 17
Dedekind's First Theory of Ideals

17.1 Introduction

This chapter picks up from the previous one and looks at how Dedekind analysed the concept of primality in an algebraic number field. This was to mark the start of a sharp difference of opinion with Kronecker.

17.2 Divisibility and Primality

Algebraic number theory in the 1870s and 1880s is often viewed by historians of mathematics as being concerned with a rather evident disagreement between Kronecker and Dedekind about the nature of the new types of number being introduced into the subject. Dedekind is taken as having a deepening commitment to a truly structural point of view (see Reed (1994), Corry (1996, 2nd. ed. 2004), and Avigad 2006) while Kronecker is regarded as the advocate of formal, algorithmic, calculational, or explicit methods—the adjective chosen tends to indicate whether one disapproves or approves of his approach. Edwards has repeatedly argued that Kronecker's theory, a revision of Kummer's, is easier to work with (see Edwards (1990), for example).

The issue is both computational and conceptual. Conceptually it lies at the heart of modern structural mathematics. Not only because the terms 'ring' and 'ideal' occur here for the first time, but because the whole idea that mathematics is about sets with extra structures, and number theory is best done on the basis of—and after a training in—a large amount of abstract algebra starts here. Computationally

© Springer Nature Switzerland AG 2018
J. Gray, *A History of Abstract Algebra*, Springer Undergraduate Mathematics Series,
https://doi.org/10.1007/978-3-319-94773-0_17

because even structuralists want honest answers: they disagree about how to get them. A case in point, that belongs to the next generation (Hilbert's) is the non-constructive existence proof. It is asked if a such-and-such exists. The answer is given: if it did not, there would be a logical contradiction, therefore it exists. No effort is wasted on looking for the such-and-such, which may be very hard to find, but on the other hand nothing is known about it other than the bare fact that it exists. Conceptually the argument is very attractive, but computationally it is likely to be useless.

The issue here is what to make of Kummer's ideal numbers. Recall that they arise as the result of a test for divisibility, the problem being to restore unique factorisation for a new class of (cyclotomic) integers. The test for divisibility can be carried out, the result is either a 'yes' or a 'no', but it does not exhibit the divisor if the number is found to be divisible. Indeed, there is no account given of what the divisor is. Conceptually we can look at this result in two ways. Either the meaning of the adjective 'divisible' has been stretched, or there is a lack of an object referred to by the noun 'divisor'.

Kummer, although he admitted that the result was "peculiar", decided to think of it in the adjectival way. It worked in the setting he created it for, cyclotomic integers. But when Kronecker and Dedekind independently came to extend these ideas to general algebraic integers (the solutions of polynomial equations with integer coefficients and leading coefficient 1) they found problems. In particular, Dedekind agreed with Kummer that ideal numbers were necessary, but he found Kummer's approach unacceptable, and that for a mixture of reasons. Their multiplication was especially difficult, and he remarked[1]:

> Because of these difficulties, it seems desirable to replace Kummer's ideal numbers, which are never defined in themselves, but only as the divisors of existing numbers ω in a domain σ, by a really existing substantive ....

This is, interestingly, both a computational and a conceptual objection. But it was not just computational ease that Dedekind sought. He wanted an object to which the noun 'divisor' referred. He was to go on to define an ideal number as a certain set of complex integers. He then had to show how his newly defined objects did what was expected of them—they divided other numbers according to well-accepted rules for the concept of division. He also had to show they were useful, and one of his first achievements was to show how his ideal numbers enabled one to transfer the theory of quadratic forms to the study of complex integers. It was in this way that the concept of an ideal was launched on its way to becoming one of the central ideas of modern algebra.

Dedekind's really existing numbers were defined as infinite sets of complex numbers. For example, in Dedekind's theory, all the complex numbers of the form $p + q\sqrt{-5}$ which can be written in the form $m \cdot 3 + n\left(1 + \sqrt{-5}\right)$ for some ordinary integers m and n, form an ideal number. The great example is Gauss's introduction

---

[1] Dedekind (1877, §10).

of the Gaussian integers in 1831, which Dedekind recalled for his readers. As we noted in Sect. 7.4, these are integers of the form $m + ni$ where $i^2 = -1$ and m, n are ordinary integers, and Gauss had shown that there is a perfectly good notion of prime Gaussian integer. He then used these new integers to solve problems in the theory of biquadratic reciprocity. This was Dedekind's exemplar.

Dedekind's first presentation of his theory of ideal numbers came out in 1871, in Supplement X to the second edition of Dirichlet's *Zahlentheorie*. By presenting his theory of ideals as an appendix to the standard introduction to Gaussian number theory, Dedekind hoped to reach the widest possible audience. But he became disappointed with the result, and convinced that almost no-one had read it. He re-wrote it for a French audience in 1876, and again for later editions of the *Zahlentheorie*.

In 1882 Kummer's former student Leopold Kronecker presented his rival version, a divisor theory in the spirit of Kummer, in his monumental *Grundzüge* (Kronecker 1882).[2] This theory sticks with Kummer's emphasis on tests for divisibility and so is ontologically parsimonious—no new nouns!—and it has always attracted a minority of influential adherents, but even Edwards writes (1987, 19) that 'it seems that no-one could read [it]'.

As one might expect, the Dedekind and Kronecker approaches have a lot in common. Both men were concerned with algebraic integers. Kronecker preferred to work directly with the polynomial of which the algebraic integers were the roots. Dedekind preferred to work with the ideals he had defined and their abstract properties, such as being prime or dividing another ideal. This clash of styles continues to draw strong feelings from some number theorists, renewed in recent years with the advent of powerful computers. Kronecker had a deep preference for procedures that exhibited specific answers, an attitude that was probably a deeper objection for him to Dedekind's approach than the latter's use of infinite sets, despite Kronecker's steady turn towards constructivism.

A good way to think of the conceptual clarity that Dedekind valued is to follow his own criticisms of his approach. He began in 1871 where the research frontier was, with problems of divisibility. Whence the need to define a prime ideal, because a good theory of division divides 'integers' into their 'prime factors'. But division is not the most basic aspect of our ordinary dealings with integers. After addition, their most fundamental property, and possibly subtraction, comes multiplication. Division comes after multiplication, and Dedekind came to feel that his theory should start with a definition of how to multiply ideals, from which a division theory would follow. Consistent with this, he deepened his objection to another aspect of Kronecker's theory: the way integers are written.

It happens that algebraic integers can be written as polynomials, just as it happens that integers can be written in decimal notation. And it happens that the decimal notation is good for certain problems (it is very good at highlighting even numbers, and those divisible by 5, for example) just as it is no use for others. But by and

---

[2]See also Edwards (1990).

large mathematicians are not interested in properties of numbers that are peculiar to their decimal representation (numbers that are a string of 1s, for example). Dedekind came to feel that algebraic number theory should not base itself on how algebraic numbers are written—and that is an attitude antithetical to Kronecker's explicitly computational approach.

17.3 Rings, Ideals, and Algebraic Integers

Opinions differ about what to make of $\mathbb{Z}$. Is it the great example of a *group*, because it is closed under addition and one can subtract, or is it the great example of a *ring*, because it is also closed under multiplication? In this book it is helpful to see it as an example of a commutative ring with a unity, in which it makes sense to talk of prime elements and irreducible elements, and in this case those two concepts coincide.

In the nineteenth century there was no general concept of a ring. One of the topics in this book is, indeed, how that general concept emerged and was taken to be important. But it will be helpful to define a ring here. A *ring* is a set of elements with two binary operations, addition and multiplication, such that it is an abelian group under addition, and multiplication satisfies the associative law—that is, $a \cdot (b \cdot c) = (a \cdot b) \cdot c$—and the distributive law—that is, $a \cdot (b + c) = a \cdot b + a \cdot c$.

If, moreover, multiplication is commutative the ring is said to be commutative. From now on unless otherwise stated the term 'ring' will be taken to mean a commutative ring. If there is an element 1 such that for every element r of the ring $r \cdot 1 = 1 \cdot r = r$ the ring is said to have a multiplicative unit (sometimes called a unity). Even here the literature is not uniform, and most people these days insist that a ring has a multiplicative unit.

Two substantial motivational examples now stretch before us. On one side are the polynomial rings. Starting from $\mathbb{Z}$ or $\mathbb{Q}$ (or any commutative ring or field) one adjoins unknowns and obtains, by the operations of addition, subtraction, and multiplication, polynomials in those unknowns. For example, $x^3 y + y^3 x - 3xyz$ is a polynomial in three variables, and an element of the polynomial rings $\mathbb{Z}[x, y, z]$ and $\mathbb{Q}[x, y, z]$. The most important single fact about polynomial rings is that they admit unique factorisation into irreducible elements. These rings interest geometers because their elements have an obvious geometric interpretation.

On the other side are the rings of algebraic integers that interest number theorists. An element α in the ring $\mathbb{Z}[x]$ is an algebraic integer if it satisfies a monic polynomial with coefficients in that ring. In other words, for some integers a_j, $j = 1, \ldots, n$ we have

$$\alpha^n + a_1 \alpha^{n-1} + \ldots + a_{n-1} \alpha + a_n = 0.$$

What makes α an algebraic *integer* is that the leading coefficient of the polynomial it satisfies is 1 (which is what it means for the polynomial to be monic). Elements that satisfy non-monic polynomials

$$a_0\alpha^n + a_1\alpha^{n-1} + \ldots + a_{n-1}\alpha + a_n = 0$$

are called algebraic numbers. The analogy is with ordinary integers and rational numbers. When the need arises to distinguish an ordinary integer from an algebraic integer an ordinary integer (a member, that is, of $\mathbb{Z}$) will be called a rational integer.

Many numbers are obviously algebraic integers. For example $\sqrt{2}$, which satisfies the equation $x^2 - 2 = 0$. It's not quite so obvious that sums and products of algebraic integers are algebraic integers. For example, $\sqrt{3} + \sqrt{5} + \sqrt{7}$ is surely an algebraic integer, but finding the polynomial of degree 8 that is satisfies is time-consuming, and the search for the appropriate monic polynomial for something like $\sqrt[3]{3} + \sqrt[8]{5} + \sqrt[5]{7} + \sqrt[10]{1221}$ would surely only be undertaken if the answer was expected to be unusually interesting.

We shall see Dedekind's proof that sums and products of algebraic integers are algebraic integers later. You will find it easier to follow if you do the next two exercises.

Exercise. Let $\alpha = \sqrt{3} + \sqrt{5}$.

- Calculate α^2, and notice that a term involving $\sqrt{3 \cdot 5}$ occurs.
- Calculate α^3, and notice that no new square roots occur.
- Show that no new square roots occur when you calculate α^n, $n > 3$, and deduce that each power of α is a linear combination of just these square roots: $\sqrt{3}$, $\sqrt{5}$ and $\sqrt{3 \cdot 5}$.
- Explain why this implies that α satisfies a polynomial of degree 4 with integer coefficients, and find it explicitly.

Exercise. Let $\alpha = \sqrt{3} + \sqrt{5} + \sqrt{7}$.

- Calculate α^2, and note that terms involving $\sqrt{3 \cdot 5}$, $\sqrt{5 \cdot 7}$, and $\sqrt{7 \cdot 3}$ occur.
- Calculate α^3 and notice that no new square roots occur.
- Show that no new square roots occur when you calculate α^n, $n > 3$, and deduce that each power of α is a linear combination of just these square roots: $\sqrt{3}$, $\sqrt{5}$, $\sqrt{7}$, $\sqrt{3 \cdot 5}$, $\sqrt{5 \cdot 7}$, $\sqrt{7 \cdot 3}$.
- Explain why this implies that α satisfies a polynomial of degree 8 with integer coefficients. (For true inner satisfaction, explain to yourself why it does not satisfy a polynomial with integer coefficients of degree 7 or even 6.)

On the other hand, it was a matter of some pride in the nineteenth century to exhibit any real numbers that were provably not algebraic integers (they are called transcendental numbers). The first to do so was Liouville, who showed in 1844 that the number

$$\sum_{n=1}^{\infty} 10^{-n!} = 0.1100010000\ldots,$$

is not algebraic. Even this is not obvious and involves knowing about the continued fraction expansion of algebraic integers. It is much harder to establish that any well-known number is transcendental. Hermite started the story by proving that e is transcendental in 1873, and he was followed almost 10 years later by Lindemann, who proved in 1882 that π is transcendental, thus finally proving that we cannot square the circle by straight edge and circle.

Ring theory and group theory differ in many ways and agree in many ways. An important distinction that arises early in the theory is this. In group theory subgroups of a group are interesting and normal subgroups are very interesting. In ring theory the place of a normal subgroup is taken by an ideal. An ideal A in a ring R is a subset closed under addition and subtraction and under multiplication by arbitrary elements of the ring ($a \in A$, $r \in R \Rightarrow ar \in A$). Interest in ring theory attaches much more to the study of ideals than subrings.

Notice also that if we insist that a ring has a multiplicative unit then the set of even integers $2\mathbb{Z}$ is not a ring! It is therefore not a subring of the ring of integers. But it is an ideal, and that saves the day.

17.4   Dedekind's Theory in 1871

In his first presentation Dedekind was concerned to show how his ideas improved Gauss's difficult theory of quadratic forms. His treatment of that part of the theory was clumsy and computational, and will not be discussed, but he followed it with the first general presentation of his theory of ideals, and that is worth considering.

He defined algebraic number and algebraic integer in the usual way and then gave a proof that the sum, difference, and product of two algebraic integers is again an algebraic integer. He supposed that α and β are algebraic integers defined by polynomial equations of degrees a and b respectively. Consider the set of all $\omega_{ij} = \alpha^i \beta^j, 0 \le i \le a - 1, 0 \le j \le b - 1$. Let ω be respectively the sum, difference, and product of α and β. Then each of the ab products $\omega\omega_{ij}$ can be written as a sum $\sum r_{ij}\omega_{ij}$ with coefficients r_{ij} that are rational integers. Eliminating the ω_{ij} from these equations yields a monic equation of degree ab for ω, as required.

Dedekind went on to show that the roots of a polynomial with coefficients that are algebraic integers are again algebraic integers (with a proof that reads like showing that an algebraic extension of an algebraic extension is finite and therefore algebraic), and conversely that an irreducible equation satisfied by an algebraic integer in a field of algebraic numbers has coefficients that are algebraic integers.

He then showed that any algebraic number can be multiplied by a rational integer so that the product is an algebraic integer, and all these rational integers are multiples of the least such. He defined a unit as an integer that divides every other integer, said that two integers are essentially distinct if one is not a multiple of the other by a unit, and defined divisible (α divides β if there is an integer γ such that $\beta = \alpha\gamma$).

When Dedekind came to the concept of prime he observed that the concept would make no sense if it was interpreted as meaning divisible only by itself and units, because the algebraic integer α is divisible by the algebraic integers $\alpha^{1/n}$ for all n. But the concept of relatively prime could be made to work: two algebraic integers α and β are relatively prime if every number that is divisible by α and β is divisible by $\alpha\beta$.

With these definitions in place, Dedekind turned to build up his general theory. He called a module a system (i.e., in later language a set) **a** of numbers closed under addition and subtraction. (It is indeed a $\mathbb{Z}$ module in modern terminology.) He said that if the difference of two numbers belonged to **a** then they were congruent modulo **a**, written $\omega \equiv \omega' (\bmod\, \mathbf{a})$.

He then said that if all the members of a module **a** are members of a module **b** then **a** is a multiple of **b** and that **b** divides **a**. To remember this, note that 6 is a multiple of 3 and the set of multiples of 6 is a subset of the multiples of 3. The set of elements common to **a** and **b** he called the least common multiple of the two modules. He called the set of all sums of the form $\alpha + \beta$, where α is in **a** and β is in **b**, the greatest common divisor of **a** and **b**.

He then introduced the concept of an ideal in a number field. This is a set of algebraic numbers in a given number field that is closed under addition, subtraction, and multiplication by the algebraic integers in the number field. Some of these ideals are generated by a single element; he called these principal ideals. He did not say at this point, but it is true that every ideal in the ring of rational integers is principal, but this is not true in general. The algebra of modules extends to ideals.

The crucial definition is that of a prime ideal in the collection A of algebraic integers in a given number field.[3] A *prime* ideal is an ideal divisible only by itself and A. He promptly showed that if the least common multiple of a collection of ideals is divisible by a prime ideal then so is at least one of those ideals. After a few pages he finally deduced that every ideal is the least common multiple of all the powers of the prime ideals that divide it. Therefore an ideal **a** is divisible by an ideal **b** if and only all the powers of the prime ideals that divide the ideal **b** also divide the ideal **a**. Two ideals are relatively prime when their greatest common divisor is A. Finally he showed that if **a** and **b** are two ideals then there is always an ideal **m** relatively prime to **b** such that the product **am** is a principal ideal.

---

[3]I have replaced Dedekind's symbol $\mathfrak{O}$ with A throughout for ease of use.

Chapter 18
Dedekind's Later Theory of Ideals

18.1 Introduction

Here we look at how Dedekind refined his own theory of ideals in the later 1870s, and then at the contrast with Kronecker.

The crucial advance was to define both ideals and operations on ideals that mimic properties of ordinary arithmetic. Let $\mathbf{a}$ and $\mathbf{b}$ be two ideals in a commutative ring R with a unit. In particular, the *sum* of $\mathbf{a}$ and $\mathbf{b}$, written $\mathbf{a} + \mathbf{b}$, is $\{a + b, a \in \mathbf{a}, b \in \mathbf{b}\}$, and the *product* of $\mathbf{a}$ and $\mathbf{b}$, written $\mathbf{ab}$, is $\{\sum_j a_j b_j, a_j \in \mathbf{a}, b_j \in \mathbf{b}\}$ (the sums are finite sums of any length). It's a good idea to check that these sets are indeed ideals of R.

18.2 The Multiplicative Theory

In 1877 Dedekind had a second go at defining ideals and doing number theory with them. With an eye on Gauss's theory of quadratic forms he began with an account of Gaussian integers. A unit is defined as a number that divides every number, and there are four units, $1, i, -1$ and $-i$. A prime Gaussian integer p is one with the property that if p divides a product mn then p divides either m or n. The prime Gaussian integers are the ordinary (rational) primes of the form $4n + 3$, the number $1 + i$, which divides the rational prime 2 (which is therefore no longer prime in this ring), and the numbers $m \pm ni$ that divide rational primes of the form $4n + 1$. Dedekind observed that this theory immediately gave the result that rational primes of the form $4n + 1$ are sums of two squares, a result first discovered by Fermat.

© Springer Nature Switzerland AG 2018
J. Gray, *A History of Abstract Algebra*, Springer Undergraduate Mathematics Series,
https://doi.org/10.1007/978-3-319-94773-0_18

Finally, a theory of congruences modulo a Gaussian integer can be constructed just as one does with the rational integers.

Fig. 18.1 Richard Dedekind (1831–1916). Photo courtesy of the Archives of the Mathematisches Forschungsinstitut Oberwolfach

Dedekind then went on to consider analogous domains. For example, the five domains $\{m + n\theta : m, n \in \mathbb{Z}\}$, where θ satisfies one of these equations

$$\theta^2 + \theta + 1 = 0, \ \theta^2 + \theta + 2 = 0, \ \theta^2 + 2 = 0, \ \theta^2 - 2 = 0, \ \theta^2 - 3 = 0,$$

allow one to determine a greatest common divisor of two numbers, and so have a familiar theory of divisibility. But the same is not true of numbers in the domain

$$\mathbb{O} = \{m + n\theta , \ m, n \in \mathbb{Z}, \theta^2 = -5\}.$$

Here, he said, one needs Kummer's theory of ideal numbers.

Certainly, said Dedekind, one can add, subtract, multiply and sometimes even divide such integers. One can define the norm of $m + n\theta$ to be $m^2 + 5n^2$, and the norm of a product will be the product of the norms of the factors. One can define a number to be decomposable if it is the product of two factors neither of which is a unit. But factorisation ceases to be unique. Dedekind invited his readers to consider these 15 numbers:

$$a = 2, \ b = 3, \ c = 7,$$

$$b_1 = -2 + \theta, \ b_2 = -2 - \theta, \ c_1 = 2 + 3\theta, \ c_2 = 2 - 3\theta, \ d_1 = 1 + \theta, \ d_2 = 1 - \theta,$$

$$e_1 = 3 + \theta, \ e_2 = 3 - \theta, \ f_1 = -1 + 2\theta, \ f_2 = -1 - 2\theta, \ g_1 = 4 + \theta, \ g_2 = 4 - \theta.$$

These numbers are all indecomposable, as is seen by looking at their norms, but they admit products that pair up:

$$ab = d_1 d_2, \ b^2 = b_1 b_2, \ ab_1 = d_1^2, \ \ldots, \ ag_1 = d_1 e_2.$$

Dedekind gave ten such equations, showing that a number can be factorised in two or three different ways. However, if one pretends that

$$a = \alpha^2, \ b = \beta_1 \beta_2, \ c = \gamma_1 \gamma_2,$$

then all the other twelve numbers factorise in terms of $\alpha, \beta_1, \beta_2, \gamma_1$ and γ_2 as if unique factorisation were present. Dedekind therefore set himself the task of explaining how this could be and unique factorisation genuinely restored, allowing, of course, for the presence of units.

First he showed that the number 2 behaved in the domain A as if it was square of a prime number α, in that it could only be factored as a square. To do this, he considered the congruence

$$\omega^2 \equiv N(\omega)(\mathrm{mod} \ 2).$$

Certainly there is no such number in A, but one may follow Kummer, said Dedekind, and introduce an ideal number with this property; call it α. It will be the case that $\omega = m + n\theta \in \mathbb{O}$ is divisible by α if and only if $N(\omega)$ is even, in which case $m \equiv n \ (\mathrm{mod} \ 2)$. Contrariwise, ω is not divisible by α whenever $N(\omega)$ is odd. By looking at the norms Dedekind in this way built up a specification of the properties of an ideal factor whose square was 2. This did not, of course, say that there was such an ideal number.

A similar argument produced specifications for the ideal factors of $b = 3$ and $c = 7$. In fact, said Dedekind, one finds that;

1. rational primes congruent mod 20 to one of 11, 13, 17, 19, remain prime in A;
2. θ behaves like a prime number, and 2 like the square of an ideal prime number;
3. rational primes congruent mod 20 to one of 1 or 9, split in A into a product of two distinct prime factors (not ideal ones);
4. rational primes congruent mod 20 to one of 3 or 7, split in A into a product of two distinct ideal prime factors;

5. the factorisation of every number other than 0 and ± 1 is either one of the primes
   just listed or, for the purposes of divisibility, behaves like a product of such real
   or ideal primes.

The question then, said Dedekind, was to meet all these specifications. One
could, he said, do what he believed Kronecker advocated, and enlarge the domain
A by adjoining, in the manner of Galois, $\beta_1 = \sqrt{-2 + \theta}$ and $\beta_2 = \sqrt{-2 - \theta}$. The
new domain $\mathbb{O}' = \{m + n\theta + m_1\beta_1 + m_2\beta_2\}$ contains A and all the factorisations
one might want to restore unique factorisation. However, he said, this method was
not as simple as one might like because the new domain was more complicated than
A and moreover was somewhat arbitrary.

Dedekind therefore suggested regarding the ideal number α as the infinite set
generated by 2 and $1 + \theta$, denoted $\{2, 1 + \theta\}$. Similarly, he defined these ideal
numbers:

$$\mathbf{b}_1 = \{3, 1 + \theta\}, \ \mathbf{b}_2 = \{3, 1 - \theta\}, \ \mathbf{c}_1 = \{7, 3 + \theta\}, \ \mathbf{c}_2 = \{7, 3 - \theta\}.$$

These sets are what he again called ideals: they are closed under addition and
subtraction, and under multiplication by elements of A.

Next Dedekind defined products and quotients of ideals. The product of two
ideals $\mathbf{a}$ and $\mathbf{b}$ he defined as consisting of arbitrary sums of products of the form
$ab, a \in \mathbf{a}, b \in \mathbf{b}$. So the generator of the product of two principal ideals is the
product of the generators of the individual ideals. One ideal, $\mathbf{a}$, divides another, $\mathbf{b}$
if every member of $\mathbf{b}$ is a member of $\mathbf{a}$. In particular, one principal ideal divides
another when the generator of the first divides the generator of the second.

Now he could show that with his definition of ideal factors, the 15 numbers he
introduced could be written in terms of the ideals they correspond to or factor into,
and unique factorisation was apparently preserved. Then it only remained to show
that the concept of a prime ideal could be defined and the factors proved to be prime
factors. Dedekind repeated his earlier definition: an ideal other than A is prime if its
only factors are itself and A. It is then clear, indeed from the earlier consideration
of norms, that the ideal factors are, after all, prime ideals, and unique factorisation
has been restored to the ring A.

18.3   Dedekind and 'Modern Mathematics'

Dedekind revised his theory four times over the course of his working life, and
each time he set more and more value by the generality and uniformity of his
theory. He developed it to apply to all sorts of rings of algebraic integers, and
relied on the fact that the introduction of prime ideals allowed him to establish a

prime factorisation theorem for ideals.[1] This is when the concepts of an 'integer' being 'prime' and 'irreducible' diverged, and it is interesting to see that when this happened the stronger and more traditional word went with the deeper structural property.

Avigad and Corry have emphasised how firmly Dedekind's successive revisions of his theory make it clear he identified with a structuralist approach.[2] His dislike of relying on choices of elements not only denied algorithmic considerations a fundamental role, it drove his critique of the concepts he himself employed. As we have just seen, he re-wrote his theory to accord with the logical priority of multiplication over division, but this moved his work further away from the original problems.

I have argued in *Plato's Ghost* that Dedekind's approach is recognisably modernist. It expresses a need for internal definitions of objects that are tailored to meet internally set goals of the theory. The concepts of integer and prime are stretched; and indeed Dedekind's very definition of an integer is both novel and set-theoretic. He first defined the set of natural numbers, and then defined a natural number as a member of this set. Here, as with his definition of the real numbers, Dedekind's key tool was his confidence in the use of infinite sets of more familiar objects. Like his famous cuts in the theory of real numbers, his ideals in algebraic number theory are infinite sets of more basic elements.

That said, Corry made a very useful distinction between Dedekind's modernist tendencies and those of later, more structuralist, mathematicians when he wrote (1996, 131): 'Dedekind's modules and ideals are not "algebraic structures" similar to yet another "structure": fields. They are not "almost-fields" failing to satisfy one of the postulates that define the latter.' Rather, numbers remained the focus of Dedekind's enquiries, and modules and ideals were only tools. For all his abstract conceptual way of thinking, he was a long way from the groups, rings, and fields formulation of modern algebra that is due almost entirely to Emmy Noether and her school, and which will be discussed in Chap. 28.

The subsequent reception of the approaches of Dedekind and Kronecker cannot be explained by their contrasting professional situations. Precisely because Kronecker was well established in Berlin, and although he attracted only a few students they were generally very good, whereas Dedekind was in self-imposed isolation in Braunschweig, one might have expected Kronecker's methods to carry the day. Both men in different ways tied themselves to the tradition established by Gauss and Dirichlet. But Dedekind was a lucid writer, and Kronecker a difficult one. Dedekind's friend Heinrich Weber also wrote his influential three-volume textbook of algebra largely in Dedekind's manner, and Hilbert learned number theory from Weber. Hilbert's account of the subject in his hugely influential *Zahlbericht* (Hilbert 1897) set the terms of research for over 50 years, and since Hilbert preferred Dedekind's ideals to Kronecker's divisors, their survival was assured.

---

[1] See Avigad (2006).
[2] See Corry (1996).

Both Kronecker's and Dedekind's approaches carried great weight as examples of how existence questions in mathematics could be treated. They arose from genuine research questions in a high point of contemporary German mathematics, the algebraic theory of numbers, and so mathematicians could not dismiss them as mere philosophy. When Hilbert made Dedekind's approach the mainstream one it not only represented a victory for naïve set theory, he may also, as Goldstein, Schappacher, and Schwermer have pointed out, diverted attention from the number theory involved with elliptic and modular functions that was preferred by Kronecker (and Hermite in France). It was in this spirit that van der Waerden wrote: "It was Evariste Galois and Richard Dedekind who gave modern algebra its structure; that is where its weight-bearing skeleton comes from."[3]

18.4 Exercises

1. Prove Dedekind's claim that the greatest common divisor of two integers exists in each of the following five domains $\{m + n\theta \ : m, n \in \mathbb{Z}\}$, where θ satisfies one of these equations:

$$\theta^2 + \theta + 1 = 0 \,, \ \theta^2 + \theta + 2 = 0 \,, \ \theta^2 + 2 = 0 \,, \ \theta^2 - 2 = 0 \,, \ \theta^2 - 3 = 0.$$

2. Verify the equations

$$ab = d_1 d_2, \ b^2 = b_1 b_2, \ ab_1 = d_1^2, \ \ldots, \ ag_1 = d_1 e_2$$

found by Dedekind (see above) and discover and verify some more.
3. Check that the same equations hold with Dedekind's interpretation in terms of ideals:

$$\alpha = \{2, 1 + \theta\}, \ \mathbf{b}_1 = \{3, 1 + \theta\}, \ \mathbf{b}_2 = \{3, 1 - \theta\},$$

$$\mathbf{c}_1 = \{7, 3 + \theta\}, \ \mathbf{c}_2 = \{7, 3 - \theta\}.$$

4. Verify by example some of the five claims Dedekind made about the factorisation of numbers in A.

Question

1. What topics in number theory seem to you to belong more to 'analytic' than 'algebraic' number theory in the second half of the nineteenth Century?

---

[3]See the foreword to Dedekind (1964).

Chapter 19
Quadratic Forms and Ideals

19.1 Introduction

One of the successes of Dedekind's theory was the way it allowed Gauss's very complicated theory of the composition of quadratic forms to be re-written much more simply in terms of modules and ideals in a quadratic number field, which in turn explained the connection between forms and algebraic numbers. Here we look at how this was done.

19.2 Dedekind's 11th Supplement, 1871–1894

As we have said already, Dedekind produced four editions of Dirichlet's *Lectures*. Here we have space only to look at his treatment of composition of forms.

This topic is not discussed at all in the first edition. In the second edition (1871) the new tenth Supplement, entitled 'On the composition of binary quadratic forms', begins by explaining how forms are composed and how genera are composed, then introduces the concepts of algebraic integers, modules, and ideals in a ring (as we would say, but not Dedekind) of algebraic integers. The Supplement closes with an account of modules in quadratic fields and the composition of quadratic forms.

Dedekind was unhappy with the poor reception this Supplement received, and produced an improved account in a paper in French (Dedekind 1877) and then rewrote the material for the third edition of the *Lectures*. The account of composition of forms in the second edition was considerably rewritten, and although we do not know why, it may well have been because the closing sections that compare the Gaussian approach with one involving modules come across as an unilluminating exercise in elementary algebra. The effect is to show that one could do it either

© Springer Nature Switzerland AG 2018   209
J. Gray, *A History of Abstract Algebra*, Springer Undergraduate Mathematics Series,
https://doi.org/10.1007/978-3-319-94773-0_19

way, and either way there is a lot of work to check that everything proceeds correctly. What may have been intended as a conceptual rewrite looks more like book keeping.

The third edition (1879) is dedicated to Dedekind's friend Heinrich Weber. In the foreword to this edition Dedekind says that his earlier presentation of the theory of ideals "was presented in such a condensed form that the wish for a more detailed exposition was expressed to me from many directions". This, he said, he was now happy to provide, and to thank Weber for his help with this and the (1877) account. The (1871) account of 110 pages had now grown to a revised Supplement X on the composition of forms of 49 pages and a Supplement XI on algebraic integers of 178 pages.

The fourth and final edition (1894) keeps this division, but Supplement XI now grew to 224 pages. Most of the enlarged and improved material has to do with the general theory of algebraic numbers. The changes to the material we are focussing on here are mostly cosmetic: some ideas are introduced in earlier sections rather than being introduced when needed. The significant improvement over the second edition is clear in both the third and fourth editions: there is a step by step account of modules in a quadratic field, how they correspond to binary quadratic forms, how they can be multiplied, and how this corresponds to the composition of the corresponding forms in terms of the norms of the modules. The conceptual structure stands out, the necessary elementary algebra is clearly motivated.[1]

Dedekind's huge 11th supplement to Dirichlet's *Lectures*, (1894), entitled *Über die Theorie der algebraischen Zahlen* ('On the theory of algebraic numbers'), presents the theory in considerable generality. Dedekind gave an account of finite-dimensional extensions of the rational field, which we looked at in Sect. 17.3 above and will look at again in Sect. 23.3 below. He then defined integers in such fields, explained the theory of modules and ideals (prime ideals, norm of an ideal) and ideal classes. At the end he turned to quadratic fields, and that is the only case I shall consider here. This is not too perverse a selection because, as Dedekind himself said "It stands in the closest connection with the theory of binary quadratic forms, which is the principal object of this work". A translation of this part of the supplement will be found in Appendix F.

The supplement has an ambition that matches its scope. It opens with the remark that the concept of the integer has been greatly extended in the nineteenth century, starting with Gauss's introduction in 1831 of the Gaussian integers (not Dedekind's term). Dedekind carefully reviewed their theory, established the concept of a norm of a Gaussian integer, defined primes in this setting and proved the unique factorisation theorem for them. He repeated the proof that rational integers of the form $4n + 1$ cease to be prime as Gaussian integers, but those of the form $4n + 3$

---

[1] I have decided to concentrate on the fourth edition (1879) rather than the third (1879) because the fourth edition was reprinted in Dedekind's *Mathematische Werke*.

remain prime. This led on to a discussion of integers of the form $m + n\theta$, where m and n are rational integers and $\theta^2 + 5 = 0$.

All of this was preparatory to the theme of the supplement, which was to pick up from Kummer's introduction of the cyclotomic integers and in particular the idea of cyclotomic integers and to provide a general theory of algebraic integers of any kind.

As part of that theory, Dedekind described (in §168) what he called a *module*: a system of numbers (a subset of a field) closed under addition and subtraction, and therefore under multiplication by the rational integers $\mathbb{Z}$. In short, what we would today call a $\mathbb{Z}$ module. It is what (in §177) he called an *ideal* if it is closed under multiplication by the algebraic integers in the field. The *norm* of an ideal **a** in a ring of algebraic integers A is the number of elements in $A/\mathbf{a}$ – thus the norm of the prime ideal (p) in $\mathbb{Z}$ is the number of elements in $\mathbb{Z}/p\mathbb{Z}$, which is p.

To study quadratic fields, starting in §186, Dedekind introduced the 'system', $\Omega = \mathbb{Q}(\sqrt{d})$, where d is a square-free integer. This is a field with discriminant D, where $D = d$ if $d \equiv 1 \mod 4$ and $D = 4d$ if $d \equiv 2, 3 \mod 4$. The algebraic integers in these fields are:

1. when $d \equiv 1 \mod 4$: of the form $\frac{1}{2}(m + n\sqrt{d})$, $m, n \in \mathbb{Z}$ and both of the same parity (both odd or both even);
2. when $d \equiv 2, 3 \mod 4$: of the form $(m + n\sqrt{d})$, $m, n \in \mathbb{Z}$.

The algebraic integers form a module, and Dedekind observed that if we define $\Theta = \frac{D+\sqrt{D}}{2}$, then $A = \mathbb{Z}[\Theta]$, which he wrote as $[1, \Theta]$.

Dedekind was interested in the factorisation of ideals in A, in particular when does an ideal prime in $\mathbb{Z}$ cease to be prime as an ideal in A? Note that it becomes the ideal $A(p)$ consisting of all the elements in A that are multiples of p. First we do a little calculation. If p factorises in A then it must factorise as

$$p = (a + b\sqrt{d})(a - b\sqrt{d}) = a^2 - b^2 d.$$

Modulo p this says that

$$a^2 \equiv b^2 d \mod p,$$

so d must be a square mod p. When it is, say

$$c^2 \equiv d \mod p,$$

with $-\frac{1}{2}(p - 1) \le c \le \frac{1}{2}(p - 1)$ and consider

$$(c + \sqrt{d})(c - \sqrt{d}) = c^2 - d.$$

This will be a multiple of p, in fact $\pm p$. This suggests looking at these ideals $[p, c + \sqrt{d}]$ and $[p, c - \sqrt{d}]$ as potential factors of $A(p)$.

Dedekind introduced the notion of a conjugate of an ideal. The conjugate of an ideal **a** has, as its elements, the conjugates of the elements of **a**. An ideal may or may not equal its conjugate.

Here is what he found: an ideal **p** is prime if its norm divides p^2, where p is a rational prime. Assume $p \neq 2$. Then there are three cases (in what follows, **p**$'$ denotes the conjugate ideal of **p**):

1. $N(\mathbf{p}) = p$: in this case $A p = \mathbf{p}\mathbf{p}'$, $\mathbf{p} \neq \mathbf{p}'$ and D is a square mod $4p$;
2. $N(\mathbf{p}) = p^2$: in this case $A p = \mathbf{p}$, and D is not a square mod $4p$;
3. p divides D, in which case $A p = \mathbf{p}^2$.

In the first case the rational prime p factors into two distinct prime ideals in A, in the second case it remains prime, and in the third case it factors as a square of a prime ideal.

Next, he dealt with the rational prime 2. He found that:

1. if 2 does not divide D, and $d \equiv 1 \bmod 8$ then $A2 = \mathbf{p}\mathbf{p}'$, $\mathbf{p} \neq \mathbf{p}'$;
2. if 2 does not divide D, and $d \equiv 5 \bmod 8$ then $A2 = \mathbf{p}$;
3. if 2 divides D then $A2 = \mathbf{p}^2$.

Next, Dedekind gave a formula for the number of ideal classes. Presently I omit this.

Dedekind concluded his very long paper with a leisurely account of how his theory re-writes the classical theory of binary quadratic forms. For this he worked with modules in a quadratic field Ω. The module $[\alpha, \beta]$, where α and β are two algebraic numbers (not necessarily integers), is all algebraic numbers of the form $x\alpha + y\beta$, where $x, y \in \mathbb{Z}$. Simple algebra shows that there is always an algebraic integer ω and a rational number m, not necessarily an integer, such that $[\alpha, \beta] = m[1, \omega]$. This ω satisfies an equation

$$a\omega^2 - b\omega + c = 0, \quad \text{where } a, b, c \in \mathbb{Z}, a > 0.$$

The sign of b is chosen for convenience, and the greatest common divisor of a, b, c is 1. This implies that there are rational integers h and k such that

$$a\omega = h + k\theta, \text{ and } d = b^2 - 4ac = Dk^2.$$

He had earlier drawn attention to the situation where all the elements of a module **m** are contained in another module **n**. In this case he said that **m** was *divisible* by **n**, or **n** *goes into* **m**. But then he wrote this as

$$\mathbf{m} > \mathbf{n} \text{ or } \mathbf{n} < \mathbf{m}.$$

So we must be careful: his $>$ is our $\subseteq$, his $<$ our $\supseteq$!

He now associated to a module **m** the set of all algebraic numbers v such that $v\mathbf{m}$ is divisible by **m**. It is a module, which he denoted $\mathbf{m}^o$. In particular he was interested in **n**, the set of all v such that $[v, v\omega]$ is divisible by $[1, \omega]$. After some algebra he deduced that

$$\mathbf{n} = [1, a\omega] = [1, k\theta] = Ak + [1],$$

where $A = \mathbf{o}$ is the ring of quadratic integers and a is the integer we met before.

Now, an algebraic number μ is in $\mathbf{m}$ if there are rational integers x and y and a rational number (not necessarily an integer) m such that

$$\mu = m(x + y\omega).$$

Take the norms and we get

$$N(\mu) = \mu\bar{\mu} = m^2(x + y\omega)(x + y\bar{\omega}).$$

This multiplies out to give

$$N(\mu) = N(\mathbf{m})(ax^2 + bxy + cy^2).$$

This shows that to the module $\mathbf{m}$ there corresponds the quadratic form $(a, \frac{1}{2}b, c)$. The correspondence proceeds via a number of choices to get the basis for the module $\mathbf{m}$, and so other quadratic forms can be obtained, but although they differ from this one they are equivalent to it.

Now, modules can be multiplied. Dedekind showed that the product of a module and its conjugate is $\mathbf{m}^o N(\mathbf{m})$. Then, after a long calculation, he showed that: given two modules and the quadratic forms they correspond to, the product of elements from the modules corresponds to the quadratic form obtained by Gaussian composition of the two associated quadratic forms. Dedekind concluded his paper by indicating how one can define equivalence relations among modules and so study the orders and genera that make up the class group. This was to be widely accepted as the way, henceforth, that Gauss's theory of composition of forms should be treated.

19.3   An Example of Equivalent Ideals

It is true that every ideal in the ring of integers $\mathbb{Z}[\Theta]$ has a basis of the form $[a, b + g\Theta]$ where a and b are rational integers

$$0 \leq b < a, \ 0 < a, \ \text{and} \ a|b^2 - g^2\Theta^2.$$

Moreover, such a basis is unique. I shall assume these results without proof.

We say that two such ideals J_1 and J_2 are equivalent if there are integers α_1 and α_2 in $\mathbb{Q}[\Theta]$—not necessarily rational integers—such that

$$J_1\alpha_1 = J_2\alpha_2.$$

In words, two ideals are equivalent if they become equal on multiplication by principal ideals.

Notice that if we are interested in ideals that are equivalent to $[a, b + g\Theta]$ we have

$$[a, b + g\Theta] = g[(a/g), (b/g) + \Theta],$$

where a/g and b/g are integers, so the ideals $[a, b + g\Theta]$ and $[(a/g), (b/g) + \Theta]$ are equivalent. Therefore we can assume that it is enough to work with ideals of the form $[a, b + \Theta]$ with $g = 1$.

For simplicity, I now work in the cases where $D \equiv 2, 3 \bmod 4$, and write $\Theta = \theta = \sqrt{-d}$. You should write out the corresponding results in the case $d \equiv 1 \bmod 4$.

Exercises

1. Calculate $[a, b + \theta](b - \theta)$ and deduce that the ideal $[a, b + \theta]$ and the ideal $[c, -b + \theta]$ are equivalent (note that $\theta^2 = b^2 - ac$).
2. Calculate $[a, b + \theta](a - b + \theta)$ and deduce that the ideal $[a, b + \theta]$ and the ideal $[a - 2b + c, b - c + \theta]$ are equivalent.
3. Show that the ideals $[3, 2 + \theta]$ and $[7, 3 + \theta]$ are equivalent.

To the ideal $[a, b + \theta]$ we associate the quadratic form

$$\frac{(ax + (b + \theta)y)(ax^2 + (b - \theta)y)}{a} = ax^2 + 2bxy + (b^2 - \theta^2)y^2.$$

If we can take $\theta = \sqrt{-d}$, the quadratic form is

$$ax^2 + 2bxy + \frac{b^2 - \theta^2}{a}y^2 = ax^2 + 2bxy + (b^2 + d)y^2 = ax^2 + 2bxy + cy^2,$$

where $\theta^2 = b^2 - ac$.

You should now check that the equivalences just established correspond to the transformation s of quadratic forms S_2 and S_1 respectively that were defined in Sect. 3.2. This gives us a quick way of identifying the equivalence class of an ideal.

Some ad hoc notation may help. I shall write the ideal $[a, b + \theta]$ as $[a, b; c]$, where $c = \frac{b^2 + d}{a}$. Then we have the equivalences

$$[a, b; c] \sim [c, -b; a], \quad [a, b; c] \sim [a - 2b + c, b - c; c],$$

$$[a, b; c] \sim [a + 2b + c, b + c, a],$$

which we can call s_2, s_1, and s_1^{-1} respectively.

If we can now show that if two ideals are equivalent then so are the corresponding quadratic forms, and vice versa, then it remains to check that composition of forms corresponds to multiplication of ideals, and a weight is lifted from the theory of binary quadratic forms. In fact, those claims are valid, but it takes just a little too long to prove them here. And you may recall that we did not *prove* the reduction theorem for quadratic forms in Chap. 3. For this, you may read the Dedekind extract in Appendix F, where, however, he worked with modules.

In the case we keep returning to, where $d = -5$, we find that there are two equivalence classes of ideals. In one class go the principal ideal and all the ideals equivalent to them, and in the other the ideals equivalent to $[2, 1 - \theta]$. In this case, a first step in finding ideals equivalent to $[a, b + \theta]$ is to consider $[a, b + \theta](b - \theta) = [ab - a\theta, b^2 + 5]$ and to reduce that ideal to canonical form. For example:

$$[7, 3 + \theta](3 - \theta) = [21 - 7\theta, 14] = (7)[2, 3 - \theta] = (7)[2, 1 + \theta],$$

which is as far as that reduction can be taken.

Chapter 20
Kronecker's Algebraic Number Theory

20.1 Introduction

Here we look at the Kroneckerian alternative to Dedekind's approach to 'ring theory' set out in his *Grundzüge* and later extended by the Hungarian mathematician Gyula (Julius) König. This leads us to the emergence of the concept of an abstract field.

20.2 Kronecker's Vision of Mathematics

Kronecker is one of those mathematicians who seldom wrote anything simple, and usually addressed only important questions in the fields he cared about (elliptic functions, number theory, and algebra). Through Dirichlet he was introduced to the circle around the banker Alexander Mendelssohn, where he came to know the composer Felix Mendelssohn Bartholdy and Alexander von Humboldt (Wilhelm von Humboldt, who founded Berlin University in 1810, was Alexander's older brother).

In later life Kronecker inclined more and more to a philosophy of mathematics that few shared. This contributed to a painful split in the Mathematics Department at Berlin where, with Kummer and Weierstrass, he was one of the three most senior professors until 1890, when Kummer retired; Kronecker died in 1891. He advocated speaking of mathematical objects only when a construction had been given for them, which could include finite formal expressions, and this is often encapsulated in his

© Springer Nature Switzerland AG 2018
J. Gray, *A History of Abstract Algebra*, Springer Undergraduate Mathematics Series,
https://doi.org/10.1007/978-3-319-94773-0_20

dictum that "The good Lord made the natural numbers, every thing else is the work of man".[1]

Kronecker's *Grundzüge* has a justified reputation for difficulty. Even a work like Edwards (1990), closely as it sticks to the spirit of Kronecker's endeavour, does not set out to do justice to the size and scope of that enterprise. The problem is compounded by Kronecker's notoriously difficult style; it is easy to get lost in the march of detail. Edwards comments (Edwards, 1990, p. 355) that

> Kronecker's theory … did not win wide acceptance. The presentation is difficult to follow, and the development leaves gaps that even a reader as knowledgeable as Dedekind found hard to fill.

Even Hermann Weyl, who in this matter is a supporter of Kronecker's, had to admit that "Kronecker's approach … has recently been completely neglected" (Weyl 1940, p. iii). Finally, there is the conscious philosophy that comes with the school and cannot fairly be cut from it. The paradigm for all who worked in this tradition is not algebra but arithmetic, and it will be worth attending carefully to what they meant by that. There is real excitement for the historian here, and we can get a sense of it by attending to the sheer ambition of his project, and why it held arithmetic in such high regard.[2]

It might be best to try and set aside what one thinks one knows about Kronecker's philosophy of mathematics. This is usually expressed in negative terms: his factorisation theory eschewed Dedekind-style naive set theory and could therefore happily announce that some number was divisible without having an object that represented its divisors (Dedekind was appalled by this); Kronecker was a strict finitist with no place for transcendental numbers, even, on some views, algebraic numbers. Thus Klein said of Kronecker that[3]:

> he worked principally with arithmetic and algebra, which he raised in later years to a definite intellectual norm for all mathematical work. With Kronecker, who for philosophical reasons recognised the existence of only the integers or at most the rational numbers, and wished to banish the irrational numbers entirely, a new direction in mathematics arose that found the foundations of Weierstrassian function theory unsatisfactory.

He then alluded briefly to what has become one of the best-known feuds in mathematics, the last years of Kronecker and Weierstrass at Berlin, offered his own wisdom as an old man on these matters, and observed that although Kronecker's philosophy has always attracted adherents it never did displace the Weierstrassian point of view. Finally he quoted with approval Poincaré's judgement that Kronecker's greatest influence lies in number theory and algebra but his philosophical teaching have temporarily been forgotten.

The matter is, as so often, better put positively. The first thing is the enormous range of the project. This was emphasised by his former student Eugen Netto when he surveyed Kronecker's work for an American audience, on the occasion of the

---

[1] This remark is to be found in Weber's obituary of Kronecker (Weber 1893).

[2] For an overview of Kronecker's work and its influence one can consult Neumann (2006).

[3] See Klein (1926–1927, pp. 281, 284).

World's Columbian Exposition in Chicago in 1893, 2 years after Kronecker's death (Netto 1896). Netto quoted Kronecker as having said that he had thought more in his life about philosophy than mathematics, and that the expression of his philosophical views was to be found in his ideas about arithmetic.

So far as possible, Kronecker wanted a common method for dealing with all the problems of mathematics that come down to properties of polynomials in any finite number of variables over some field, usually the rational numbers. In the strict sense in which Kronecker intended to be understood, the ground field is at most an algebraic extension of a pure transcendental extension of finite transcendence degree of the rationals. We shall see that there were those, like Molk, who accepted this starting point, and others, like König, who preferred to start with the complex numbers.

So Kronecker's subject matter included all of algebraic number theory, and, when interpreted geometrically, the theory of algebraic curves and, insofar as it existed, the theory of algebraic varieties of any dimension. This is why he occupies what might otherwise seem an unexpected place in the history of early modern algebraic geometry. The fact that the ground field is not the complex numbers, nor even algebraically closed, need not be an insuperable problem: a great deal of algebraic geometry can still be done by passing, if need be, to successive algebraic extensions. What Kronecker could not do, according to his lights, is pass to the full algebraic closure of the rational field. While this would not necessarily be a significant mathematical problem for any one drawn to this approach, the historian cannot escape so easily, however, as we shall see when we discuss the limited references Kronecker actually made to geometry.

The analogy between algebra and arithmetic, which will be discussed a little below, is a real one, and by refining the question they share of finding common factors Kronecker sought to exploit to the benefit of all the various aspects. It is the analogy with algebraic number theory that drove him to call his theory arithmetic, rather than merely algebraic. The basic building blocks were two things: the usual integers and the rational numbers, on the one hand, and variables on the other. These were combined according to the usual four laws of arithmetic; root extraction was to be avoided in favour of equations (for example, the variable x and the equation $x^2 - 2 = 0$, rather than $\sqrt{2}$).

Kronecker himself set out the thinking that led him to his general programme in a fascinating preface to a paper of 1881, *Über die Discriminante algebraischer Functionen einer variabeln* ('On the discriminants of algebraic functions of a single variable'). The preface is a lengthy historical account indicating how much he had already proposed in lectures at the University of Berlin (and who his audience had included) and at a session of the Berlin Academy in 1862. The guiding aim, which he traced back to 1857, was to treat integral algebraic numbers (for which the modern term is algebraic integers). These he defined as roots of polynomial equations with leading term 1 and integer coefficients. He encountered certain difficulties, which is where the discriminants come in. The resolution of these problems came with the insight that it was a useless, even harmful restriction to consider the rational functions of a quantity x that satisfies an algebraic equation of degree n only in the form of polynomials in x (i.e. as linear homogeneous functions of $1, x, \ldots, x^{n-1}$). It would be better, he realised, to treat them as linear homogeneous forms in any n linearly independent functions of x. This made it possible to represent complex numbers by forms, in which every algebraic integer appeared as an integer while circumventing the difficulties. The insight may be put another way: an irreducible polynomial of degree n with distinct roots defined n quantities at once, and it can be shown that it cannot share a subset of these roots with any other irreducible polynomial. By picking on one root, problems arise that can be avoided by treating all the roots simultaneously. (If you like, and to over-simplify, study $\pm\sqrt{2}$, but not just $\sqrt{2}$.)

He discussed these results with Weierstrass, who was his friend at the time, and Weierstrass urged him to apply the same principles to algebraic functions of a single variable and if possible to the study of integrals of algebraic functions, taking account of all possible singularities. This set him on the road to a purely algebraic treatment, shunning geometric or analytic methods. He sent the first fruits to Weierstrass in October 1858, but Weierstrass's own results rendered his superfluous in his own eyes and so he refrained from further publication. He was brought back to the topic by discovering how much his thoughts coincided with those of Dedekind and Weber (an agreement which did not, he noted, extend to the basic definition and explanation of the concept of a divisor). Therefore he presented his old ideas, abandoned in 1862, for publication in 1881.

Kronecker's *Grundzüge* is a lengthy work, and an unrelenting one (you can consult Edwards (1990) for a thorough mathematical commentary, where you will learn amongst other things of the unproven claims Kronecker made). This mixture of great claims for the rigour and immediacy of the theory and the absence of proofs at crucial points surely contributed to the work's poor reception. Dedekind's rival version, attacked by Kronecker for its abstraction, was not so embarrassed. Kronecker does not seem to have had the knack of conveying in print what was known, what can be done, and what might be discovered in a way that would drive future research.

Kronecker's *Grundzüge* is about objects which are so common in mathematics that it is hard to know what to call them. They are polynomials in several variables and their quotients by other such polynomials. A collection of rational functions in some (finite set of) indeterminates R_i closed under the four operations of arithmetic Kronecker called a *Rationalitätsbereich*, which I translate as domain of rationality. A *ganz* or integral function in a domain is a polynomial in the Rs. The old name of 'rational functions' (which Kronecker used) survives, although they are not thought of as functions but expressions. The coefficients may be integers, rational numbers, or elements of some 'field' (another term we shall have to return to). The crucial thing about these objects is that they can be added, subtracted, multiplied, and divided (of course, one cannot divide by zero). Kronecker also insisted that there was no question of order here, such expressions are not greater or less than others (which rules out use of anything like the Euclidean algorithm). He also explicitly wished to avoid geometrical language.

Kronecker's *Grundzüge* places three obstacles in the way of comprehension. One is the number of unproven claims that are made. Another is the style, which mixes up what is proved with unproven claims about what is true. The third is the delicate, and often heavily computational nature of the material. Each of these difficulties calls for comment. The existence of unproven claims is not necessarily a barrier to the acceptance of a work; it may function as a challenge to later workers. However, the failure to meet these challenges, coupled with the opportunity of switching to a rival theory that did not have these disadvantages, was to prove crucial in the demise of Kronecker's approach. As for the style, unexpected though it may be in a mathematician with a strong axe to grind about what is the right way to do mathematics, it is typical of the period. Long papers and books were designed to be read; they were seldom presented in the style of definition, theorem, proof that came in later, with Landau. It is tempting to imagine that it was the obstacles presented by such works as Kronecker's *Grundzüge* that pushed Landau to accentuate the division between mathematics and literature. It is also noticeable that Hilbert's writings were much more lucid and carefully structured so that they could be easily understood. The final obstacle would be a virtue if the computational machine gave acceptable answers. The problem seems to have been that it did not.

Edwards's analysis of Kronecker's theory praises it for being based on the idea of divisors, and in particular the idea of greatest common divisors (when they exist). The greatest common divisor of two elements of a field is, Edwards points out (1990, v), independent of the field: if two objects have a third as their greatest common divisor this third object remains their greatest common divisor even if the field is extended. This is not the case for prime elements: an element may be prime in one field but factorise in an extension of that field. Kronecker claimed to have a method for factorising a given divisor as a product of prime divisors. His successor Hensel gave a proof that this method works in the context of algebraic number fields, and echoed Kronecker's claim that the method worked in general. But he never gave such a proof, and Edwards wrote in 1990 that he did not know of one.

Edwards makes light, however, of what seemed to every one at the time to be an immediate problem with divisor theory: there simply may not be a greatest common divisor of two elements. Rings and fields with greatest common divisors include all the so-called 'natural domains'; these are the domains which are either the rational numbers or pure transcendental extensions of the rational numbers. They are to be contrasted with algebraic number fields, which typically do not have greatest common divisors. All writers (Kronecker, Molk, König) give the same example, due originally to Dedekind, because it is the simplest: algebraic integers of the form $m + n\sqrt{-5}$. It is easy to show that $2 - \sqrt{-5}$ is irreducible, but it is not prime. Indeed, $(2 - \sqrt{-5})(2 + \sqrt{-5}) = 9 = 3 \cdot 3$, but $2 - \sqrt{-5}$ does not divide 3 (the solutions, x and y, of the equation $(2 - \sqrt{-5})(2 + \sqrt{-5}) = 3$ are not integers). So the algebraic integers 9 and $3(2 - \sqrt{-5})$ have no greatest common divisor: their common divisors are 1, 3 and $2 - \sqrt{-5}$, but neither of 3 and $2 - \sqrt{-5}$ divides the other. It was exactly this problem that caused Dedekind to formulate his theory of ideals, which invokes ideals that are not principal (generated by a single element) precisely to the get round the problem. In König's terminology, rings and fields having greatest common divisors are called *complete*, those like $\mathbb{Q}(\sqrt{-5})$, are *incomplete*. In the Kroneckerian approach the whole analysis grinds to a halt with incomplete domains. Netto (1896, p. 252) observed that finding irreducible factors was an open question once the existence of greatest common divisors failed.

Kronecker's most powerful influence was exerted on the handful of students around him at any time in Berlin. For that reason it is worthwhile looking at the

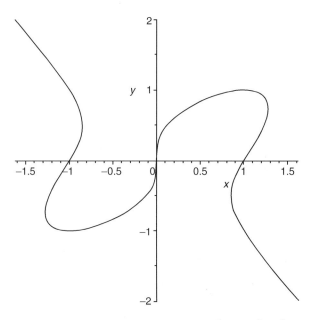

Fig. 20.1 The three real inflection points of the curve $x(1 - x^2) = y(y^2 - x^2)$ lie on the x-axis

posthumous volume of his lectures (Kronecker 1901), edited by Hensel. This was largely based on his lectures in the 1880s, but the lectures have the advantage over the *Grundzüge* of being both more elementary, clearer about what has been proved, and more geometric. It may be the case that the lectures were more influential than the *Grundzüge*, or at least that the combination of lectures and conversation was more potent (Kronecker was a sociable mathematician, see Biermann 1973/1988). Best known among his followers are Hensel (who knew him personally, succeeded him at Berlin, and edited this volume of Kronecker's *Vorlesungen über Zahlentheorie* in 1901) and Landsberg. Another was Netto, whose two-volume *Algebra* (1896–1900) is his account of ideas in the *Grundzüge*. Volume 1 covers the algebra of polynomial equations in a single variable (but no group theory), the approximation of the roots, explicit methods for equations of degrees less than five, and the cyclotomic equations. Volume 2 first deals with algebraic functions in several variables and equations in several unknowns, the theory of elimination in various forms, and culminated in an account of Kronecker's theory of characteristics and Hermite's theory of quadratic forms. Then it returns to algebraic functions of a single variable, presents Hilbert's irreducibility theorem, the theory of cyclic and abelian equations, the concept of algebraic numbers, the algebraic solution of algebraic equations, the theory of inflection points on cubic curves, and concludes with a treatment of the solvable and the general quintic equations (Fig. 20.1).[4]

---

[4]Compare the account in Chap. 15.

20.3 Kronecker's Lectures

In his *Vorlesungen über Zahlentheorie* (1901, Lecture 13) Kronecker defined a domain of rationality determined by an indeterminate R as the totality of all products and quotients of polynomials in R (division by 0 being excluded). He denoted it (R)—we would write it as $\mathbb{Q}(R)$. He showed that it consisted precisely of all rational functions of R with integer coefficients. If division was not allowed, a subdomain of (R) was constructed which he called a domain of integrity (*Integritätsbereich*) and denoted $[R]$—we would write it as $\mathbb{Z}[R]$. Kronecker observed that if in particular the indeterminate R is set equal to 1, then $(R) = (1)$ is the usual rational numbers, and $[R]$ is the integers. The same construction can also be carried out with finitely many indeterminates.

Divisibility in any domain of integrity had the natural meaning that m divides a if and only if there is an integer c in the domain such that $a = cm$. Kronecker chose to write this in the formalism of congruences to a modulus, deliberately echoing Gauss, and therefore spoke of modular systems (*Modulsysteme* in German). When $a = cm$ he said that a was contained in the modular system (m), and more generally he said that a form a was contained in a modular system $(m_1, m_2, \ldots, m_r)$ or was divisible by the modular system if $a = c_1 m_1 + c_2 m_2 + \cdots + c_r m_r$, for some integers $c_1, c_2, \ldots, c_r$. He said that two modular systems were *equivalent* if each was contained in the other.

He then turned to divisors. The greatest common divisor of the modular systems $(m_1, m_2, \ldots, m_r)$ and $(n_1, n_2, \ldots, n_s)$ was, he showed, the modular system

$$(m_1, m_2, \ldots, m_r, n_1, n_2, \ldots, n_s),$$

and the composition of two modular systems $(m_1, m_2, \ldots, m_r)$ and $(n_1, n_2, \ldots, n_s)$ was, he also showed, the modular system $(m_i n_j ; \ 1 \leq i \leq r, 1 \leq j \leq s)$. This was later called the product of the two modular systems.

In Lecture 14, Kronecker restricted his attention to polynomials in a single variable and integer coefficients. In this domain a generalisation of the Euclidean algorithm permitted him to find the greatest common divisor of two polynomials, as follows. Given $f_1(x)$ and $f_2(x)$, with $f_1(x)$ of higher degree than $f_2(x)$, one can write $f_1(x) = q(x) f_2(x) + r(x)$, where the degree of $r(x)$ is less than that of $f_2(x)$, but it is not certain that the coefficients of $q(x)$ and $r(x)$ will be integers. By following this complication through, Kronecker deduced that: If $f_1(x)$ and $f_2(x)$ are polynomials with integer coefficients, then by successive division one can find a polynomial $f_n(x)$ with integer coefficients and two integers s_1 and s_2 such that $f_n(x) \equiv 0 \pmod{f_1(x), f_2(x)}$ and $s_1 f_1(x) \equiv s_2 f_2(x) \equiv 0 \pmod{f_n(x)}$. The polynomial is the greatest common divisor of $f_1(x)$ and $f_2(x)$ if and only if $s_1 = s_2 = 1$.

Divisor or modular systems $(f_1(x), \ldots, f_n(x))$ that are equivalent to a system with just one element $(f(x))$ he called modular systems of the first level or rank, all others of the second level. An example of the first kind was $(3x - 3, x^2 - 1, x^2 +$

$x - 2$); of the second kind $(m, x - n)$, where $m > 1$. A modular system was said to be *pure* if the defining terms have no non-trivial common factor, otherwise *mixed*. So $(3, x - 1)$ is pure, and $(3(x^2 + 1), (x - 1)(x^2 + 1))$ is mixed. To use Dedekind's language harmlessly here, an element of a domain of integrity belonged to a modular system if it was in the ideal the modular system generated.

Pure divisor systems of the first level lead to the study of polynomials with integer coefficients and their irreducible factors. In Lecture 15, Kronecker proved the unique decomposition theorem in this context. Let the given polynomial be

$$F(x) = c_0 + c_1 x + \cdots + c_n x^n.$$

The Euclidean algorithm finds the greatest common divisor of the coefficients, m say. Factoring it out leads to the polynomial with integer coefficients

$$f(x) = a_0 + a_1 x + \cdots + a_n x^n.$$

The possible degrees of any factor of $f(x)$ can be found easily from the factors of n, but finding the coefficients of any factor would be harder. Kronecker offered an argument based on the Lagrange interpolation formula to show how one could find out if $f(x)$ had a divisor of any given degree μ, and to determine the coefficients of the divisor if it exists, in a finite number of steps. This insistence on exhibiting a finite process is characteristic of Kronecker.

He then defined a prime function as an integer or a polynomial that is not divisible by any other in the domain, and proved that if a product $\varphi(x)\psi(x)$ is divisible by a prime function $P(x)$ then at least one of $\varphi(x)$ and $\psi(x)$ is so divisible. Finally he showed that a polynomial with integer coefficients can be written in essentially one way as a product of prime functions.

In Lecture 20 he broached the generalisation to modular systems in more than one variable, without going into the proofs in detail. He showed that an integral quantity $F(x, y)$ can be factored uniquely into irreducible or prime functions, by treating it as a polynomial in y with coefficients that are polynomials in x. Kronecker now admitted what he called arbitrary, not merely integral coefficients (it is not clear this means complex numbers!) so every element in the domain of integrity $\{x, y\}$ therefore corresponds to an algebraic equation $F(x, y) = 0$ and so to an algebraic curve. Similarly, the elements in the domain of integrity $\{x, y, z\}$ correspond to algebraic surfaces.

Kronecker considered the equations divisible by the modular system generated by $(f_1(x, y), \ldots, f_n(x, y))$ in $\{x, y\}$, and showed that they corresponded to curves through the common points of the curves with the equations $f_1(x, y) = 0, \cdots, f_n(x, y) = 0$. Such points he called the fundamental or base points of the system. Two modular systems are equivalent if they generate the same ideal; for this to happen in $\{x, y\}$ it is necessary but not sufficient that they have the same base points; he gave the example of (x^2, y) and (x, y^2), which are not equivalent but have the same base point (the coordinate origin $(0, 0)$). Similar if vaguer remarks followed about $\{x, y, z\}$.

The defining elements of a modular system in two variables may have a common curve and also meet in isolated points. If there are no such isolated points, Kronecker called the system pure, otherwise mixed.

Rather than recapitulate Kronecker's definitions, here is one of his examples. In the domain $\{x, y, z\}$ (with arbitrary coefficients) a quantity $f(x, y, z)$ is a prime divisor of the first level if and only if it is irreducible, and so the equation $f(x, y, z) = 0$ defines an indecomposable algebraic surface F in 3-dimensional space. If $g(x, y, z)$ is another integral quantity from the same domain and G the corresponding surface, then either g is divisible by f, or (f, g) is a modular system of the second level. In the first case, the surface F is part of the surface G. In the second case the modular system (f, g) corresponds to the complete intersection of the two surfaces F and G, and therefore to a space curve C. In this latter case, if the modular system is a prime modular system then the curve is irreducible.

If a third quantity $h = h(x, y, z)$ is taken in $\{x, y, z\}$ then either h is divisible by the prime modular system (f, g) or the modular system (f, g, h) is of the third level, and the three surfaces F, G, and H have only isolated points in common. Whence the theorem: an irreducible space curve and an algebraic surface either have a finite number of common points, or else the curve lies completely in the surface. Similar considerations allowed Kronecker to give what he called a complete overview of the geometrical interpretation of the purely arithmetical idea of a prime divisor: divisors of the first, second and third kinds in the domain $\{x, y, z\}$ correspond to algebraic surfaces, algebraic curves, and points; prime divisors of these kinds correspond to indecomposable surfaces, irreducible space curves, and isolated points.

The *Grundzüge* goes over the ground of the *Vorlesungen*, in a more visionary way. As far as geometry is concerned, he noted (§21) that there was a connection with the theory of hypercomplex numbers. When three variables are taken as coordinates of space, divisors of the first level (*Stufe*) are either numbers or polynomials in x, y, z, the vanishing of which represents a surface. Modular systems of the second level represent either a number or a curve, of the third level, sets of points. A modular system of level n was defined by Kronecker to be of the principal class if it was defined by n elements. So, in the principal class of each divisor system of the second level are those curves which are the complete intersection of two surfaces. Kronecker commented that this, surprisingly, is a higher viewpoint from which the representation of integers as norms of complex numbers and the isolated representation of geometric figures are seen to be intimately related.

20.4   Gyula (Julius) König

Let us now look at the work of those who took up Kronecker's ideas. An influential, if perhaps unexpected, follower was Jules Molk, who studied in Berlin from 1882–1884, where he was drawn above all to the teaching of Kronecker. On his return to Paris he took his Doctorat ès Sciences at the Sorbonne in 1884; we may read this thesis, lightly revised, in *Acta Mathematica* (Molk 1885). It is a summary, with a few simplifications, of Kronecker's ideas, coupled with strongly worded claims for its merits; Netto (1896, p. 247) called it a very thorough and well arranged presentation. Some years later, Molk arranged for an extensive reworking of Lansdberg's article on divisor theory in the *Encyclopädie der Mathematischen Wissenschaften* for the French *Encyclopédie des Sciences Mathématiques pures et appliquées*, of which Molk was editor-in-chief. The authors of that article were Kürschák and Hadamard. In 1903 and 1904 the Hungarian mathematician Gyula König wrote the first textbook on the subject; in Molk and König we have two valuable guides to the *Grundzüge*, with the essays by Landsberg and by Hadamard and Kürschák to take us further. A paper and a book by the English mathematician F.S. Macaulay bring this journey to an end.

The work of Gyula (Julius) König, published simultaneously in German and his native Hungarian, has become almost forgotten. König spent his working life in Budapest, having studied at Vienna and Heidelberg. He did not claim any personal acquaintance with Kronecker or reliance on correspondence or letters. Rather, he seems to have set himself the task, as he turned fifty, of writing a useful book sorting out an important topic for which no guide existed. Such a work, if successful, will draw others into the field who will go on to discover better results, simpler and more general methods, and if it does not attain the status of a classic gradually the work will be covered up and forgotten. Such, at all events, was the fate of König's *Einleitung in die allgemeine Theorie der algebraischen Gröszen* (1903). In view of its importance in its day, it is worth saying a little about König himself.

The man and his work are well described in Szénássy (1992), where he rates a chapter to himself. Szénássy calls König 'a great man of the nation' (p. 333) and credits him with establishing Hungarian mathematics as a significant force. This he did as much by his own work as by his magnetic personality and the breadth of his organisational work: training teachers and engineers as well as professional mathematicians, lecturing on everything from pure analysis to economics and history of mathematics. Szénássy writes (1992, p. 241) that Hungarian "secondary school education benefited for decades from his textbook on algebra". König helped found the Hungarian Mathematical Society, worked with publishers, and was three times Rector of the Technical University. In research, it was his habit to work on one area of mathematics at a time, publish several papers and then a monograph summarising the field, and then move on. He worked on algebra, then analysis and partial differential equations, and finally on Cantorian set theory, where he is better remembered for his unsuccessful attempt on Cantor's continuum hypothesis than for several smaller but secure contributions (see Moore 1982, p. 86).

Szénássy's discussion of König's *Einleitung* is rather brief, and although he points out the debt to Kronecker and the extent of the new material, much of it by König himself, it masks the importance of the book by listing its main topics in unduly modern language. In fact, the book possesses two aspects of interest to us. One is the novel mathematical concepts it introduces; the other is the insights of a sharp critic of the period.

From the standpoint of the early history of field theory, König's book introduced some useful terminology and made some interesting distinctions. He based his account on the twin concepts of an orthoid domain and a holoid domain. An *orthoid domain* corresponds exactly to our concepts of a field (of characteristic zero) and a *holoid domain* to our (commutative) ring with a unit 1 such that no sum of the form $1 + 1 + \ldots + 1$ vanishes; Szénássy incorrectly glosses a holoid domain as an integral domain. König gave no rationale for the terms; presumably he had in mind the Greek roots 'holo' for whole or entire and 'orth' for straight or right. He advocated the terms holoid and orthoid to express general properties of domains, by analogy with the integers and rational numbers.

As König saw it, a field, a *Körper* in Dedekind's terminology, is an orthoid domain (certain vaguenesses in Dedekind's definitions, and certain methodological differences, aside). But a field or orthoid domain is not the same concept as Kronecker's domain of rationality. König argued first that Kronecker's natural domains were obtained by taking any finite set of μ elements from any holoid or orthoid domain, and forming the function field in them (so non-trivial relations may exist among these generators). This gave an orthoid domain that, he said, Kronecker called a domain of rationality. It might be that just one element was chosen, and it was equal to 1, in which case the absolute domain of rationality was obtained (i.e. the rational numbers). If the μ quantities were completely undetermined the resulting domain was the natural domain of rationality in μ indeterminates.

It follows that every domain of rationality is an algebraic extension of a natural domain of rationality. His proof was to pick x and form polynomials in x with at least one non-zero coefficient. Either none vanish or one at least does. In the first case, the domain of rationality is (x), i.e. $\mathbb{Q}(x)$. In the second case, x is an algebraic number and (x), i.e. $\mathbb{Q}(x)$, is an algebraic extension. The proof for any finite number of elements x follows by induction.

Conversely, if all the elements of an orthoid domain can be written in the form $r_1\omega_1 + \cdots r_n\omega_n$, where the r_i belong to a natural domain of rationality $D = (x_1, \ldots, x_m)$, then the orthoid domain is a domain of rationality. For this, he first showed that every ω in the orthoid domain satisfies a polynomial equation with coefficients in the natural domain of rationality D. In particular, this is true of the quantities $\omega_1, \ldots, \omega_n$, so D and the domain of rationality in which $x_1, \ldots, x_m$ and $\omega_1, \ldots, \omega_n$ are adjoined to $\mathbb{Q}$ coincide.

The smallest number n of elements ω_i is therefore the order of the domain of rationality thought of as an algebraic extension of its underlying natural domain of rationality. The elements $\omega_1, \ldots, \omega_n$ themselves form a basis for the domain of rationality. But an orthoid domain is not necessarily a domain of rationality. For example, the domain of all algebraic numbers is an orthoid domain that is not a domain of rationality. The domains of real and of complex numbers are likewise orthoid but not, König seems to suggest, domains of rationality.

König found much to criticise. The original papers were very hard to read and remained restricted to a small circle of readers. They had therefore, he said, failed in their principal purpose and so he had set himself the task of popularising the spirit of Kronecker's method. He was pleased to offer an elementary proof of Kronecker's fundamental Theorem. From this he deduced a generalisation of the concept of resultant to what he called the *Resultantform* which enabled him to deal with multiplicities in systems of equations. Geometrically, he gave a general account of Noether's fundamental Theorem in n-dimensional space, which he connected to results of Hilbert. Arithmetically, he showed how to decompose algebraic integers in terms of prime ideals.

Kronecker's fundamental theorem is the result he proved after so much effort in 1883, and which Molk had then re-proved in much the same way, claiming however that it was elementary. König introduced it for two polynomials in one variable:

$$f(v) = \sum_i a_i x^{m-i} \text{ and } \sum_i b_i x^{n-i},$$

whose product is

$$f(x)g(x) = \sum_i c_i x^{m+n-i}.$$

The Theorem claims that there are identities connecting the products $a_i b_j$ and homogeneous linear expressions in the c_k. Similarly in general, there are identities connecting the products of the coefficients of any number of polynomials and the coefficients of the product. König went on to offer a truly elementary (and for that matter simple) proof of this result. Since Edwards says that Kronecker's paper remains obscure to him, and since he then explains just why it is so significant, it is worth digressing to explain what is going on. The matter is discussed in Edwards (1990, Part 0).

It is a famous result due to Gauss that if the coefficients of the product $f(x)g(x)$ of two monic polynomials are all integers, then so are all the coefficients of the polynomials themselves, $f(x)$ and $g(x)$. This can be generalised: if the coefficients of the product $f(x)g(x)$ are all algebraic integers, then so are all the coefficients of $f(x)$ and $g(x)$. Dedekind proved that if the coefficients of a product $f(x)g(x)$ are all integers, then the product of any coefficient of $f(x)$ and any coefficient of $g(x)$ is an integer (the constituents need no longer be monic). This he then generalised: if the coefficients of the product $f(x)g(x)$ are all algebraic integers, then the product of any coefficient of $f(x)$ and any coefficient of $g(x)$ is an algebraic integer. He published it in 1892, and it became known as his Prague Theorem (because of its place of publication). Unquestionably he did not know that it was a consequence of Kronecker's Theorem published in 1881. Either he had not read that paper or we have further evidence that it was obscure.

What Kronecker, Molk, and König all proved, in their different ways, is that the modular system generated by the products $a_i b_j$ and the c_k are equivalent in an extended sense of the term due to Kronecker. Let us call this result the ABC Theorem. What Landsberg pointed out (Landsberg 1899, p. 312) is that Hilbert's *Nullstellensatz* shows that the concepts of equivalence for systems of equations and for the corresponding modular systems are not exactly the same. Moreover, this problem had already been spotted and dealt with by Kronecker, and Kronecker's discussion is exactly the theorem we are discussing. Let the module generated by the products $a_i b_j$ be denoted AB, and that generated by the c_k be denoted C. Then certainly C is divisible by AB. Conversely, every $a_i b_j$ is the root of polynomial equation $v^m + g_1 v^{m-1} + \cdots g_m = 0$ whose coefficients are divisible by successive powers of C (g_i by C^i). Kronecker called such a function v "divisible by C in an extended sense", and proclaimed the equivalence of the modular systems AB and C in this sense. In a short paper of 1895, Hurwitz deduced that if the a_i and b_j are algebraic integers and the c_k are divisible by an algebraic integer ω, then so is every product $a_i b_j$. In modern terms this is a theorem about the integral dependence of ideals.

Chapter 21
Revision and Second Assignment

This is the second opportunity to revise the topics so far and now to discuss the second assignment, and as before it is better to raise some questions and ask you to answer them yourselves, to your own satisfaction.

Chapter 13 is a difficult one, to be navigated with care. Jordan was the first person to fully understand Galois's Mémoire. I argue that if you understand what he did you can make sense of what Galois did, by seeing Galois's work as fragments that fit into Jordan's more detailed picture.

Key technical terms are 'transitive group' and 'composition series'. Be sure you know what they mean, and then you can ask: What do they do? Where did Jordan use them, and to what purpose?

Normal subgroups and normal extensions were also introduced, and are vital. Can you say why? It's not so easy to spot quotient groups, because Jordan did not introduce the concept explicitly, but they are there if you look carefully at collections of cosets.

Chapters 14 and 15 are more methodological and reflective. The first one argues for the influence of Klein in shaping the reception of Galois theory, a fact that has not been noticed in the literature on the history of the subject. The second one then attempts to answer the question: Why do we speak of Galois theory at all? Would it not be enough to speak of field theory, or group theory, or the theory of polynomial equations? The answer I provide amplifies the previous remarks about Klein by taking the story, albeit briefly, to the start of the twentieth century.

Chapter 16 returns us to the topic of cyclotomy. More evidence is presented that it was appreciated as a topic, now for the types of new 'numbers' it introduced. What remarkable conclusion did Kummer come to?

Chapter 17 describes Dedekind's first theory of ideals, which he later abandoned, and Chap. 18 his second theory, which he much preferred. Can you state what was good about the first theory and what was not? On what grounds did he prefer his second theory? Can you see the merit in the abstract point of view that Dedekind took?

© Springer Nature Switzerland AG 2018
J. Gray, *A History of Abstract Algebra*, Springer Undergraduate Mathematics Series,
https://doi.org/10.1007/978-3-319-94773-0_21

Kronecker usually gets a bad press, although he has always had some vocal advocates (you might want to consult Edwards (1983), for example). What is happening to algebra at the hands of these two masters of the subject?

Chapter 19 is the last really hard chapter, in which number theory and the structural theory of rings come together. This is a part of a larger story that we shall not discuss, in which the precursor of modern structural algebra can be clearly seen for the first time, but enough is visible here, and it is a good time to look back over the topics so far. When did school algebra start to disappear behind something like modern algebra? Notice that not everyone agreed about this.

In Chap. 20 we note that although Kronecker probably did himself no favours with his confrontational attitudes and his difficult style of writing, he had his followers, and he is the best example of how things could have been different. Good questions to ask are: in what ways was his approach to algebra different from Dedekind's, and why did some people prefer it? A good figure to consider would be Jules Molk.

One aim of the course was to get students to think about the algebra courses they may have studied from a historical perspective. Opinions and arguments on such matters are more valuable the more mathematically insightful they are,[1] and the second assessment gives the students a chance to grapple with some celebrated pieces of mathematics in their original presentations, and then to write about them (in no more than four pages). I also expected a bibliography of sources, a sensible use of footnotes, and an intelligent use of the web.

Question 2 Choose one of the following two questions, and write an essay answering it. In your essay you should demonstrate some understanding of the mathematics, and be able to situate the people and their ideas in a historical context. EITHER Describe Galois's theory of the solution of polynomial equations by radicals as it was presented by Jordan. What made it a successful explanation of Galois's ideas?

OR Given an account of the number theory of binary quadratic forms as it was presented by Gauss, Dirichlet, and Dedekind. Why was this theory important?

I also offered the following advice.

For the Jordan essay, base your answer largely on the extracts from Jordan's *Traité*. He took several pages to describe what you must reduce to four. To overcome this difficulty, please organise your essay clearly, perhaps in this way:

- Set aside up to one page to explain why Jordan's treatment was successful; most easily, either page 1 or page 4. You may take it for granted that Jordan's account was successful; do not investigate the reception of his ideas.
- In evaluating Jordan's account you should compare his work with that of Galois himself, and perhaps with what little you know about Serret; your answer should draw on information presented elsewhere in your essay.

---

[1] Which is not to say that mathematical accuracy is all that matters!

- A page of your essay should be an account of which quintic equations are solvable by radicals. It should draw on ideas set out in the previous pages, making clear what mathematical facts you have had to assume.

For the Dedekind essay, this is your chance to explain a significant advance in the theory of binary quadratic forms.

- Although the details of what Dedekind did can be intimidating, the task he set himself, of translating a structured theory of quadratic form s into a structured theory of a particular class of modules, conforms to a particular operation familiar to us: try to bring out the key steps. Make an assessment of what is difficult and what, although perhaps long, is more routine, and let these judgements guide your presentation.
- Set aside one page for your investigation of the historical context, either page 1 or page 3. Make clear what was the evolving contemporary attitude to the subject: what was important? what was difficult? where was progress being made?
- The fourth and final page of your essay should be an account of why the theory of binary quadratic forms is so interesting: give some substantial mathematical example(s). It should draw on ideas set out in the previous pages, making clear what mathematical facts you have had to assume. You need not restrict your attention to the theory of binary quadratic forms, but do not bring in irrelevant information.

More generally:

- You may assume the truth of any mathematical facts you state provided you state them clearly and provided they are widely known outside the present context (e.g. the fundamental theorem of algebra, the group A_5 is simple, the fundamental theorem of algebra, the prime factorisation theorem for rational integers).
- Mathematical facts that belong more intimately to Galois theory or number theory (for example, which you learned for the first time in this course) should be treated as an opportunity to display your understanding of the theory. There is one exception: you need not prove the theorem of quadratic reciprocity.
- As a general rule do not go back before 1800. It is enough to say, for example, '(known earlier to Lagrange)'. But do not exclude results just because they were known earlier, especially if you think they are among the highlights of the theory.
- Biographical information should be kept to a minimum.

Chapter 22
Algebra at the End of the Nineteenth Century

22.1 Introduction

Here we begin to look at the major change that happened to algebra in the nineteenth century: the transformation from polynomial algebra to modern algebra. A vivid impression of the subject is given by the book that described the state of the art around 1900, Weber's *Lehrbuch der Algebra*, much of which described Galois theory and number theory as it then stood.

22.2 Heinrich Weber and His *Textbook of Algebra*

Heinrich Weber's name is known to mathematicians these days, if at all, in connection with class field theory and the Kronecker–Weber theorem, which says that every abelian extension of the rationals is a subfield of a cyclotomic extension. But in his day he was regarded as a leading algebraist, the finest exponent of the Riemannian tradition in complex analysis, and as the co-author, with his friend Dedekind, of one of the most important papers on the subject of algebraic curves and function fields. In particular, his three-volume *Lehrbuch der Algebra*, published between 1895 and 1896 (second edition 1904–1908), may be taken as the definitive presentation of the subject for its time. (Vol. 1 appeared in French in 1898).[1]

Weber was born in 1842, graduated in mathematics from the University of Heidelberg and took his PhD from there in 1863. He taught in a number of places, including the Polytechnic in Zurich (later known as the ETH, where he met and became friends with Dedekind), Königsberg, and Göttingen, before becoming a Professor at the University of Strasburg in 1895. Among his pupils at Königsberg

---

[1] For a careful analysis of the book, see Corry (1996, pp. 40–45).

© Springer Nature Switzerland AG 2018
J. Gray, *A History of Abstract Algebra*, Springer Undergraduate Mathematics Series,
https://doi.org/10.1007/978-3-319-94773-0_22

were Hilbert and Minkowski. Strasburg, it should be said, was a contentious position. The town, as part of Alsace and Lorraine, had been taken from the French by the Germans in the Franco-Prussian war of 1870–1871, and was to be reclaimed by the French after the First World War (when it became Strasbourg again).

Fig. 22.1 Heinrich Martin Weber (1842–1913). Photo courtesy of the Archives of the Mathematisches Forschungsinstitut Oberwolfach

Volume I of the *Lehrbuch der Algebra* opens with chapters on 'The Foundations': rational functions, determinants, roots of algebraic equations, symmetric functions, linear transformations and invariants, and the Tschirnhaus transformation. Book two (of volume I) discusses the reality of the roots of a polynomial equation and how they can be found or approximated—we shall not be concerned with it but notice that it was regarded as belonging to algebra at the time. Book three, the last in volume I, was on algebraic quantities: Galois theory, the application of permutation groups to equations, cyclic equations, cyclotomy, the algebraic solution of equations, and, last of all, the roots of metacyclic equations.

Volume II began with a book on groups: general group theory, then abelian groups, the groups of the cyclotomic fields, cubic and quartic abelian fields, the constitution (i.e., structure) of general groups. Book two covered linear groups including the polyhedral groups and congruence groups. Book three looked at applications of group theory: metacyclic equations, the geometry of cubic and quartic curves, the general theory of curves of degree 5, and finally two chapters on the group G_{168}. Book four was on algebraic number theory: numbers and functions in an algebraic field, general theory, fields and subfields, lattices, the class number, cyclotomic fields, cyclotomic fields, abelian fields and cyclotomy, and transcendental numbers (e and π had recently been shown to be transcendental).

Volume III ('dedicated in heartfelt friendship' to Dedekind, Hilbert, and Minkowski) was much more analytic. Book 1: elliptic integrals, theta functions, elliptic functions, modular functions, transformation theory, and the quintic equation. Book 2: quadratic fields, algebraic numbers, ideals, composition of forms and ideals, the genus of quadratic forms, the class number of ideals in quadratic fields and the application to genera. Book 3: elliptic functions and quadratic forms, the Galois theory of the class equation, calculation of class invariants, complex multiplication. Book 4: class field theory. Book 5: algebraic functions, functionals, values of algebraic functions, algebraic and abelian differentials, and the Riemann–Roch theorem.

All this in roughly $700 + 850 + 725 = 2275$ pages.

What are we to make of this? Obviously there are some unfamiliar words: metacyclic, for example, although you may recall it was introduced by Jordan in his *Traité*.[2] There are some buzzwords that may not be known to you: class field theory, the Riemann–Roch theorem. Some unexpected topics: the geometry of algebraic curves, and, even more unexpected, all that analysis. You would be correct in presuming that they are included because of some deep link between these topics and more mainstream algebra. Even that's worth noticing: they are there because they link to algebra, not because they are part of algebra.

Can we say that any of this is familiar or expected? The topics of polynomial equations and Galois theory are reassuringly present, perhaps at intimidating length, as is something about groups in general, and another large tranche of material on algebraic number theory.

And, not to look ahead too much because historians should not do that, there are missing words—ring, for example—and field theory seems tied to algebraic number theory and Galois theory rather than being an independent subject.

This suggests that we should look into the material on Galois theory and algebraic number theory to see how it connects to what we know has been going on beforehand and in what way it reformulates or advances the subject. Then we should try and see if the unfamiliar material paints an interesting picture of what algebra was taken to be just over 100 years ago.

We should also note that in 1893 Weber published a more advanced, research, paper on Galois theory which takes a more abstract approach to the subject. In the preface to that paper he wrote (1893, p. 521):

> In the following, an attempt is made to present the Galois theory of algebraic equations in a way which will include equally well all cases in which this theory might be used. Thus we present it here as a direct consequence of the group concept illumined by the field concept, as a formal structure completely without reference to any numerical interpretation of the elements used. Thus this development which follows is even independent of the fundamental theorem of algebra on the existence of roots. The theory under this interpretation appears simply as a direct formalization which first gains meaning and

---

[2]Weber explained that an equation was metacyclic when it was completely solvable by a chain of cyclic equations. It follows that the roots of such an equation are expressible in terms of nested radicals.

existence by the substitution of some one element with a numerical interpretation. In this way, this method of procedure is applicable to every conceivable case in which the given hypotheses hold, running from Function Theory on the one hand to Number Theory on the other.[3]

We can also observe that even a mathematician as insightful as Georg Frobenius could write to Weber in 1893 to say[4]

> Your announcement of a work on algebra makes me very happy .... Hopefully you will follow Dedekind's way, yet avoid the highly abstract approach that he so eagerly pursues now. His newest edition (of the *Vorlesungen*) contains so many beautiful ideas, ... but his permutations are too flimsy, and it is indeed unnecessary to push the abstraction so far. I am therefore satisfied, that you write the *Algebra* and not our venerable friend and master, who had also once considered that plan.

Frobenius, as the man who would emerge as the leader of his generation of mathematicians in Berlin, was a master of technical as well as conceptual mathematics, notably in the areas of algebra and abelian function theory, so his reluctance to go all the way with Dedekind is understandable, but this letter still testifies to the disturbing novelty of Dedekind's approach.

22.3 Galois Theory

Weber began Volume 1, Book 3, with a quick introduction to fields in the manner of his friend Dedekind. These are settings where addition, subtraction, multiplication and division (other than by zero) are permitted, whether of numbers or functions. The field consisting of zero alone is explicitly excluded from further consideration. This means that a multiplicative unit was not necessarily part of a field by definition, or so it would seem; nor can I see an opinion either way on the commutativity of multiplication—it is present in all the examples Weber gave. It also seems that all the fields Weber will consider are infinite and of characteristic zero (the concept of characteristic does not appear).

On to algebraic functions, irreducibility, adjunction of elements to a field, primitive and imprimitive fields, normal fields and Galois resolvents. By now it looks like we're into deep water. Then comes a section on permutation groups, Galois groups, transitive and intransitive groups, primitive and imprimitive groups.

So far, it looks as if this is a story about polynomial equations and their roots, which are taken to generate extension fields of the rational numbers, and in which permutation groups will play a supporting role. There are some terms we may think we know the meaning of, and some that are new, such as normal fields.

The chapter on permutation groups runs through the basics (not entirely in the modern way, as we shall see) and then introduces the idea of subgroups (divisors,

---

[3] As quoted in Kiernan (1971, p. 136).
[4] Quoted in Corry (1996, p. 128) from Dugac (1976, p. 269). On Frobenius, see Hawkins (2013).

they are called), the reduction of a Galois resolvent and the concept of a normal subgroup of a group, and finishes with imprimitive groups. This is likely to be reassuring because familiar, but not easy.

Then comes a chapter on the classical examples: the cubic equation, the quartic equation, what Weber called abelian equations and their reduction to cyclic equations, Lagrange resolvents, the solution of cyclic equations, and the division of angles into equal parts.

I'm going to skip over most of the chapter on cyclotomy, and note only that Weber discussed the Gaussian integers. Here he showed that irreducible Gaussian integers are prime, and that this was true because there was an algorithm for finding highest common factors. To make the point more secure, he showed that the same result was true for the algebraic integers in the field $\mathbb{Q}$ with a cube root of unity adjoined, say $\rho = \frac{1}{2}(-1 + \sqrt{-3})$. These integers can be written in the form $\frac{1}{2}(m + n\sqrt{-3})$. He concluded by listing (up to multiplication by units) all the prime algebraic integers of this form with a norm less than 200.

Next we come to the chapter on the algebraic solution of equations. Here we find such topics as the simplicity of the alternating group and the question of when the radicals are real, alongside investigations into metacyclic equations that conclude the book and indeed Volume I. Here he showed, for example, that Galois's group G_{20} of order 20 is a metacyclic group, and he discussed ideas that Kronecker had introduced about the specific form of the roots of a metacyclic equation.

Let us now take a more detailed look at the treatment of Galois theory. The first comment Weber made that is worth noting is this: if an irreducible function $f(x)$ has a common factor with a function $F(x)$ whose coefficients lie in the same field, then $f(x)$ divides $F(x)$. It follows that if $F(x)$ vanishes for a single root of $f(x)$ then it vanishes for all of them, and if $F(x)$ is of lower degree than $f(X)$ then it must be identically zero. Finally, the decomposition of a function into irreducible factors is essentially unique.

Weber moved quickly to consider an algebraic field $\Omega(\alpha)$, where α satisfies a (possibly reducible) polynomial equation $F(x)$ without repeated roots, whose coefficients are in the field Ω. If the polynomial is reducible then α satisfies a unique irreducible polynomial among its factors, say $f(x)$ of degree n, and this raises the possibility of adjoining not α but any of the other roots: $\alpha_1, \ldots, \alpha_{n-1}$, obtaining the fields $\Omega(\alpha_i)$. These fields are called conjugate fields and they may or may not be isomorphic—a question Weber would return to.

Weber let t stand for any of the αs and observed that every element of $\Omega(t)$ can be written as a polynomial χ of degree at most $n - 1$ in t. In this way one obtains the n quantities $\Theta_i = \chi(\alpha_i)$, one in each conjugate field; these are called conjugate quantities. Their sum is called the trace (German: *Spur*) and their product the norm (German: *Norm*). Conjugate numbers have the same trace and norm. A consideration of the polynomial

$$(t - \Theta)(t - \Theta_1) \ldots (t - \Theta_{n-1}) = \Phi(t)$$

shows that every element in $\Omega(\alpha)$ is a root of an equation of degree n whose other roots are the corresponding conjugate quantities.

Finally Weber deduced that one of two situations occurs:

1. The function $\Phi(t)$ is irreducible, and the conjugates of a number in $\Omega(\alpha)$ are all distinct, or
2. The function $\Phi(t)$ is reducible, when it is an irreducible polynomial of degree n_1 raised to the power n_2, and the conjugates of a number in $\Omega(\alpha)$ form n_2 systems of n_1 identical numbers.

The led him to the important definition of a primitive element of quantity in the field $\Omega(\alpha)$: it is an element different from all of its conjugates and therefore satisfies an irreducible equation of degree n. By an earlier result, there are infinitely many primitive elements. Whence the theorem:

Theorem (Theorem of the Primitive Element). *Every element of a field $\Omega(\alpha)$ can be written rationally in terms of an arbitrary primitive element of the field.*

Weber next observed that the fields $\Omega(\alpha_i)$ are all of the same degree, so if one is contained in the other those two are identical. He defined a normal field to be one identical with all its conjugates. He remarked that normal fields are much easier to study than arbitrary fields, and that it was the great achievement of Galois to show how an arbitrary field can be reduced to a normal field, which is why normal fields are sometimes called Galois fields. It would be a good historical exercise to say what Weber meant by this and how much confidence can be attached to it as a judgement.

Likewise, Weber called an equation a normal or Galois equation if it is irreducible and all its roots can be expressed rationally in terms of any one of them. So a primitive element of a normal field of degree μ is a root of a normal equation of degree μ. He then defined a Galois resolvent to be a normal equation.

Since all the conjugates of a normal field are related rationally, it follows that a permutation of the roots is known completely once it is known on one root, so the group of permutations has order μ—and Weber showed that these permutations do indeed form a group. As a result he was led to review the basic results about permutation groups.

Weber then turned to define the Galois group of an equation. He took an equation $F(x)$ with m roots $\alpha, \alpha_1, \ldots, \alpha_{m-1}$ that are elements of a normal field $\Omega(\rho)$ of degree μ. Then any permutation of the roots of the normal equation that defines the field $\Omega(\rho)$ is also a permutation of the roots of $F(x)$, and as such can be thought of as a permutation of the symbols $\{0, 1, \cdots, m - 1\}$. This permutation group Weber called the Galois group of the equation $F(x) = 0$, adding that if $F(x)$ is irreducible then the group is also called the Galois group of each of the fields $\Omega(\alpha_i)$.

He then showed that the Galois group had properties that distinguished between reducible and irreducible equations $F(x)$. Specifically, the Galois group is transitive if and only if the equation is irreducible. Weber concluded the chapter by establishing that primitive fields have primitive groups and imprimitive fields have imprimitive groups.

Weber now looked further into the theory of permutation groups. He called a subgroup Q of a permutation group P a divisor of P, which gave him this handy result: the degree of a divisor of a group is a divisor of the degree of P. This result (which we would call Lagrange's theorem) he attributed rather more accurately to Cayley. He denoted a normal field $\Omega(\rho) = \Omega(\alpha_1, \ldots, \alpha_m)$ by N, and proved that to every divisor (subgroup) Q of P there were what he called corresponding elements of N, and that corresponding elements in N correspond to a divisor (subgroup) Q of P. By corresponding in this context he meant elements of N that are left fixed by every element or permutation in the subgroup Q. He then showed that if $Q\pi$ is a coset of Q in P—something he called a near group (German: *Nebengruppe*)—then the elements of the coset are precisely those that send a function ψ corresponding to Q to the same function ψ_1. Weber also noted that $\pi^{-1}Q\pi$ is a permutation group, which he said was derived from or conjugate to Q.

Weber now turned to the reduction of a Galois resolvent. He supposed that P was a Galois group of degree p, Q a subgroup of degree q and index j, so $qj = p$, and ψ a function corresponding to Q. He denoted the quantities conjugate to ψ by $\psi_1, \ldots, \psi_{j-1}$, and his first important result in the chapter was that these quantities are the roots of an irreducible equation of degree j. From this he deduced what he called Lagrange's theorem: Every element of the field N that is invariant under the group Q is contained in the field $\Omega(\psi)$ when ψ is a function corresponding to Q.

He then deduced that when ψ is adjoined to Ω then the group of N is reduced to Q. So N is a field of degree p over Ω and of degree q over $\Omega(\psi)$. The reduction of the Galois group corresponds to a factorisation of the Galois resolvent. He observed, without offering an example, that if Q is normal in P and R is normal in Q then R is a divisor (subgroup) of P but not necessarily normal. However, if R is a divisor (subgroup) of Q that is also normal in P then R is normal in Q.

Of particular interest, he said, was the case when all the conjugate groups $\pi^{-1}Q\pi$ are the same, in which case the field $\Omega(\psi)$ is a normal field over Ω. Such a subgroup is therefore called, he said, a normal divisor (German: *normal Theiler*), adding in a footnote that Galois had spoken of a proper decomposition, whence the expression 'proper divisor' that was also in use, while more recent authors called a normal divisor a 'distinguished or invariant subgroup'.

A sharper flavour of Weber's version of the theory may be given by observing that his account of the quartic equation was driven by a search for transitive subgroups of the permutation group S_4. These subgroups were then tied to the resolvents that enter the explicit solution of the fourth degree equation. The burden of explanation was evidently shifted from expressions to subgroups. In line with what we discussed in Chap. 15, Weber's presentation plainly owes much more to Klein than Kronecker.

22.4   Number Theory

Weber discussed the general theory of algebraic number fields in Volume II, Book 4, of his *Lehrbuch*. He reviewed the basic definitions of algebraic numbers, algebraic integers, and prime factorisation, and discussed field extensions. He then moved on to define what he called Dedekind ideals in the collection of all algebraic integers in an algebraic number field. I shall denote a given field of algebraic numbers by K and the algebraic integers in that field by A.

Given a fixed K and A, a (Dedekind) ideal is a family of algebraic integers closed under addition and subtraction and under multiplication by any algebraic integer in A. So A, the family of all algebraic integers in the field, is an ideal, as is the family of all multiples of an algebraic integer v, written vA. Such an ideal is called a principal ideal.

Weber then observed that if **a** and **b** are two ideals in A, then so is their product, written **ab** and defined to be all finite sums of the form $\sum_j a_j b_j$ with $a_j \in$ **a** and $b_j \in$ **b**. Moreover, he said, this ideal deserves to be called the greatest common divisor of A, because it is the largest ideal that is a sub-ideal (or *divisor*, German: *Theiler*) of both **a** and **b**.

Weber then followed Dedekind in defining the norm of an ideal and the equivalence of ideals (see below), and he then proved that the number of ideal classes in a number field is finite. Then he moved on to the factorisation of prime algebraic integers in an extension field, and a great deal more. Some 200 pages after he began he reached the particular cases of abelian and cyclotomic fields.

In Volume III, Book 2, Weber returned to the topic and looked at quadratic fields. I shall concentrate on this part of his *Lehrbuch*, but treat it in a very selective way. He defined the discriminant of such a field (following Kronecker) as any non-zero (rational) integer congruent to 0 or 1 mod 4, and so capable of being a value of an expression of the form $b^2 - 4ac$. He fixed a square-free integer d, and considered the field $\mathbb{Q}(\sqrt{d})$. The algebraic integers in this field are of the form $m + n\sqrt{d}$, where are m and n are rational integers if $d \equiv 2$ or 3 (mod 4), and of the form $\frac{1}{2}(m + n\sqrt{d})$, where are m and n are rational integers and both even or both odd if $d \equiv 1$ (mod 4). This distinction between $d \equiv 2, 3$ and $d \equiv 1$ mod 4 runs right through the theory from now on. Weber now took the discriminant of the field, Δ, to be $4d$ in the 2, 3 case and d in the 1 case.

He now showed how to pass between ideals in a quadratic field (better, the associated ring of quadratic integers) and the quadratic forms of the appropriate discriminant. He took an ideal **a** with a basis (α_1, α_2) and defined $\lambda = \alpha_1 t_1 + \alpha_2 t_2$. The norm of the ideal is $N(\mathbf{a})$, and the norm of λ is $N(\lambda)$, where

$$N(\lambda) = N(\mathbf{a})T,$$

where $T = at_1^2 + bt_1 t_2 + ct_2^2$, and T is a primitive binary quadratic form with discriminant $b^2 - 4ac = \Delta$.

Conversely, Weber showed how to start from a primitive quadratic form of discriminant Δ and arrive at an ideal with a basis (α_1, α_2). More precisely he showed (I omit the proof) that from the form $at_1^2 + bt_1t_2 + ct_2^2$ one defines the number $\omega = -\frac{b+\sqrt{\Delta}}{2a}$ and the required ideal is $(a, a\omega)$.

Weber, following Dedekind, next recalled that two ideals $\mathbf{a}$ and $\mathbf{b}$ are equivalent if there is a number (not necessarily an integer) η in the field such that $\eta\mathbf{a} = \mathbf{b}$. If it is stipulated that η be an integer, then a stricter definition of equivalence is defined, but the distinction evaporates if all norms are positive or if there is a unit of norm -1. He now disentangled the algebra and showed that two equivalent ideals give rise to two equivalent quadratic forms. A similar argument but in reverse shows that two equivalent forms give rise to two equivalent ideals. It then followed from the general theory in Volume II that the form corresponding to the product of two ideals was the composition of the forms corresponding separately to each ideal. In (our symbols) if the form corresponding to the ideal $\mathbf{a}$ is (a, b, c) and the form corresponding to the ideal $\mathbf{a}'$ is (a', b', c') then the form corresponding to the ideal $\mathbf{a}\mathbf{a}'$ is the composite of the two forms (a, b, c) and (a', b', c'). In other words: the complicated operation of the Gaussian combination of forms is reduced entirely to the multiplication of the corresponding ideals.

Weber did much more, but I shall not pursue him here. He defined an *order* (German: *Ordnung*), and noted that while Dedekind had chosen the word 'order' because he was interested in Gauss's 'orders' in the theory of quadratic forms, Hilbert had recently preferred the new term 'ring' for the same object. The collection of all algebraic integers in $\mathbb{Q}(\sqrt{d})$ is an order. The Gaussian composition extends to orders and genera, and it is possible to re-write it in the language of ideals. Weber considered the theory of prime factorisation, quadratic reciprocity in this new setting, and concluded with a discussion of the class number of a quadratic field.

Optional Note on the Kronecker–Weber Theorem

The theorem states that a field extension K of the rationals, $\mathbb{Q}$, that has an abelian Galois group is contained in a cyclic extension $\mathbb{Q}(e^{2\pi i/n})$ for some integer n. A proof would go too far afield, but here is an interesting example.

Suppose that $K = \mathbb{Q}(\sqrt{5})$ and consider the Gauss sum (see the exercises to Chap. 6)

$$\tau = \sum_{j=1}^{5} \left(\frac{j}{5}\right) e^{2\pi i j/5}.$$

On the one hand, $\tau \in \mathbb{Q}(e^{2\pi i/5})$. On the other hand, $\tau^2 = 5$, so $\tau = \pm\sqrt{5}$. So $K < \mathbb{Q}(e^{2\pi i/5})$.

Chapter 23
The Concept of an Abstract Field

23.1 Introduction

Here we look at three aspects of the theory of abstract fields: the discovery
that all finite fields are 'Galois' fields; Dedekind's presentation of the 'Galois
correspondence' between groups and field extensions; and the emergence of the
concept of an abstract field.

23.2 Moore, Dickson, and Galois Fields

In 1830, Galois had published a short paper on arithmetic modulo a prime p.[1] In it
he considered expressions of the form

$$a_0 + a_1 i + a_2 i^2 + \cdots a_{v-1} i^{v-1} \qquad (A)$$

in which the a_js are integers modulo p, and i is a quantity that Galois called
imaginary (whence the later name 'Galois imaginaries' for the roots of equations
with coefficients in a finite field) and said was a root of an irreducible polynomial
mod p of degree v, and so, in what he noted was Gauss's notation, a solution of the
equation

$$F(x) \equiv 0 \mod p.$$

---

[1] See Galois (1830, pp. 428–435). This paper was a part of Galois's study of what he called
primitive equations, the topic of his second Mémoire. Accordingly it cannot be pursued further
here.

© Springer Nature Switzerland AG 2018
J. Gray, *A History of Abstract Algebra*, Springer Undergraduate Mathematics Series,
https://doi.org/10.1007/978-3-319-94773-0_23

He remarked that (A) has p^ν possible values that enjoy the same properties as natural numbers, and he then claimed that values of the a_j can be chosen so that every one of these expressions other than zero is a power of $\alpha = a_0 + a_1 i + a_2 i^2 + \cdots a_{\nu-1} i^{\nu-1}$. His unstated argument may well have been that every expression of this form satisfies a congruence of the form $\alpha^k = 1$ for some n dividing $p^\nu - 1$, and the roots of these equations for $n < p^\nu - 1$ do not exhaust the collection of elements of the form (A)—"as in the theory of numbers", as he put it.

He went on[2]

> One sees now the remarkable consequence that all algebraic quantities which can appear in the theory are roots of equations of the form
>
> $$x^{p^\nu} = x.$$

In other words, every expression of the form in (A) satisfies the equation in the quoted remark. This result has become known as the existence of a primitive element for a finite field, or in other words, the theorem that the multiplicative group of a finite field is cyclic. The reason is that one has

$$(F(x))^{p^n} = F(x^{p^n}),$$

so if x is one of the roots of F the others are the successive powers of x.

He then claimed the converse, that the roots of the congruence

$$x^{p^\nu} \equiv x \quad \mathrm{mod}\ p$$

all depend on a single congruence of degree ν. This means that this congruence has exactly p^ν distinct roots.[3] Finally he made some remarks about how to find a suitable irreducible congruence of degree ν, which would lead to an explicit representation of a primitive element.

His programmatic remarks are striking. He wrote[4]:

> The principal benefit of the new theory that we have just expounded is to carry over to congruences the property (so useful in ordinary equations) of admitting precisely as many roots as there are units in the order of their degree

I made tacit use of this fact as a shortcut at several points in the early chapters. Galois went on[5]:

> It is above all in the theory of permutations, where one forever needs to vary the form of [the] indices, that consideration of the imaginary roots of congruences appears to be

---

[2] I take the quote below from Neumann (2011, p. 65).

[3] In modern language, if $\mathbb{F}_p$ denotes the field of p elements, then expressions of the form in (A) are elements of the set $\mathbb{F}_p(x)$ modulo the polynomial $F(x)$, and that when $F(x)$ is irreducible over $\mathbb{F}_p$ this set is a field of p^ν elements, as Galois claimed.

[4] Neumann (2011, p. 73).

[5] Neumann (2011, p. 73).

indispensable. It gives a simple and easy means of recognising the cases in which a primitive equation is soluble by radicals, of which I shall try to give an idea in two words.

We shall not follow him here but proceed to the final page of this short paper, where Galois talked about primitive equations of degree p^v that are solvable by radicals.[6] He had already stated that only equations of prime power degree can be both primitive and solvable by radicals, and at this point he considered the case $v = 1$. He noted that the substitutions of the group are of the form (x_k, x_{ak+b}) – which we might write as $k \mapsto ak + b$ – as k runs through the whole numbers (strictly speaking, modulo p). Galois noted that there will be $p(p - 1)$ of these substitutions.

It is a melancholy reflection that these ideas of Galois, published in 1830 and again by Liouville in 1846, did not find their ideal reader until Jordan in the early 1860s.

By the end of the century it was generally agreed that Galois had established that his process does produce a field of the given size. The converse (that every finite field with p^n elements is generated by a single element of order p^n for some prime p and some integer n) was taken to be an open conjecture until it was established by the American mathematician E.H. Moore in his (1896). He had presented the result at the Chicago Congress of Mathematics in 1893, but the *Mathematical Papers* were only published three years later. It follows that up to isomorphism there is only one field of each possible order.

Moore argued as follows in his (1896, pp. 211–216). Let μ be a non-zero element of a finite field F, then there is an integer c such that the sum of c copies of μ vanishes, but no smaller sum vanishes (c was later called the characteristic of the field). This number c is the same for all non-zero μ (it is the value of c for the element 1 of the field), and it is a prime integer, say q. The elements $0, 1, 2, \ldots, (q - 1)$ form a field of q elements, F_q. Now suppose that there are h elements of the field, say $v_1, \ldots, v_h$, but not $h + 1$, for which the equation

$$n_1 v_1 + n_2 v_2 + \cdots + n_h v_h = 0,$$

where the n_j are elements of F_q, forces $n_1 = \cdots = n_h = 0$. (The elements v_j are said to be linearly independent.) Then there are q^h expressions of the form $n_1 v_1 + n_2 v_2 + \cdots + n_h v_h$, and they form a field. Furthermore, every element of the field must be of this form (else h would not be maximal). Finally, every element of the field satisfies a polynomial of degree at most h, else for some element μ the elements $1, \mu, \ldots, \mu^h$ would be linearly independent. Therefore the finite field is a Galois field.

---

[6]For a careful discussion of the meaning of the term 'primitive' in Galois's work as it applies to equations and to groups, see Neumann (2006).

23.3   Dedekind's 11th Supplement, 1894

After the long introduction to the supplement, Dedekind began in §160 with the definition of a field (German: *Körper*) of real or complex numbers.[7] He defined this as a collection of numbers closed under addition, subtraction, multiplication, and division (division by zero being excluded). If a field A is contained in another field B Dedekind called A a divisor of B and B a multiple of A, natural choices of names that might also reveal his ultimate focus on number theory. He observed that the intersection of a family of fields is again a field, which he called the greatest common divisor (gcd) of the fields, and he defined the concept of the least common multiple of a family of fields as the gcd of all the fields containing the given fields—recall that by definition they are all subfields of $\mathbb{C}$.

He then discussed what he called a permutation of a number field, and we would call a homomorphism, and explained that it can be thought of as a mapping (*Abbildung*, in German) of one field to another. He defined it as obeying these rules:

$$(u + v)' = u' + v', \ (uv)' = u'v',$$

with analogous rules for subtraction and division. The image of a field under a permutation is a field, and a permutation is a one-to-one map, so a field is isomorphic (as we would say) with its image. This allowed Dedekind to consider permutations of a field to itself, and to look at the collection of all permutations of a field to itself, which indeed form a group.

In the case where a field B is an extension (or, as he said, a multiple) of a field A Dedekind also looked at the permutations of B that leave the elements of A fixed; he said these permutations of B were single-valued on elements of A but could be n-valued on other elements of B. Conversely, the elements of a field B for which a given collection of permutations is single-valued form a subfield of B. Indeed, this is true if B is replaced by a family of fields.

In §164 Dedekind began "the detailed investigation of the relationship between different fields", adding the striking comment "and herein lies the real subject of today's algebra" ("*der eigentliche Gegenstand der heutigen Algebra*"). He now looked at finite degree extensions of one field by another, and proved what have become the familiar theorems about them, such as that if B is an extension of A of degree m and C is an extension of B of degree n then C is an extension of A of degree mn. The realisation that an extension field can be thought of as a vector space with coefficients in the subfield is a driving force behind many of the proofs, along with basic facts about linear dependence and independence.

---

[7]See the translation of §§160–166 in Dean (2009) and Dean's commentary for a proof of the results summarised here, and a thoughtful commentary and a disagreement with Kiernan's account of the same material.

In §166 he came to the theorem that if a field B is an extension of A and G is the group of permutations of B that are single-valued on A then the order of the group is equal to the degree of B over A.

Now he considered the case in which there is a collection of fields A', A'', ... that have A as a subfield and are themselves subfields of B, and deduced that "the complete determination of all these fields A', A'', ... and the investigation of their mutual relations is completely settled by the determination of all groups G', G'', ... contained in the group G.

The statement that there is a one-to-one correspondence between the lattice of intermediate fields of two fields A and B and the corresponding groups is today known as the fundamental theorem of Galois theory. To be correct, some assumption has to be placed on the field extensions, and in modern accounts that is that the extension of A by B is normal.[8] Dedekind recognised this by insisting that all the homomorphisms of intermediate fields that are single-valued on A fix only A and no larger field.[9]

As Dean (2009) points out, this is a clear statement of the fundamental theorem, at least for subfields of $\mathbb{C}$, and it is strikingly conceptual. The crucial intermediate step between Dedekind's account and a modern one, say Artin (1942), was to be taken by Steinitz, as we shall shortly see.

23.4 Kürschák and Hadamard

József Kürschák was a Hungarian mathematician educated in Budapest. He organised seminars with König, and König thanked him for his help during the writing of his *Einleitung*. He also shared something of his mentor's breadth of interests and his influence. One sign of this is a prize competition organised for school leavers in mathematics and physics, which is named after him. In the early 1890s he came into contact with Hadamard because of his study of the relation between the simple pole of a power series and its coefficients. This stimulated Hadamard to investigate what conditions on a power series yield particular types of singularity. One supposes, in the absence of evidence, that it was Molk who encouraged them to expand upon Landsberg's article in the *Encyclopädie der Mathematischen Wissenschaften* when he came to organise the French version for his *Encyclopédie*. At all events, it is a much larger article than the German original, and its comments provide an interesting view of how all this material was regarded in 1910–1911. It is not clear, however, what impact this article had; references to it are hard to find. Let us take it, then, as a snapshot of the times.

---

[8] In the setting of subfields of $\mathbb{C}$ the condition that all extensions be separable is met automatically.

[9] Dedekind expressed matters somewhat differently, see the commentary in Dean (2009, pp. 15–16).

The first part was published on 30 August 1910. The title is interesting in itself: *Propriétés générales des Corps et des Variétés Algébriques* ('General properties of fields and algebraic varieties'). Landsberg's had been 'IB1c Algebraic varieties. IC5 Arithmetic theory of algebraic quantities', so it is clear that the editors had not been sure where to place it, because IB is devoted to algebra and IC to number theory. Generally speaking, Hadamard and Kürschák kept Kronecker's approach at a distance, setting off material specifically of that nature in square brackets [...].

They reviewed the proliferating terminology carefully. The term 'known quantities' had given way to Dedekind's 'field', which was the same thing as König's orthoid domain. Kronecker's domain of rationality was also a field, more precisely a finite or algebraic extension of the rationals (the terms 'finite' and 'algebraic' were regarded as synonyms by Hadamard and Kürschák). There were two sorts of field: number fields and function fields, but every field contained the field of rational numbers, which they denoted R. Hadamard and Kürschák regarded the simplest function field as the function field in n variables, which they denoted indifferently R. The coefficients were to be unrestricted numerically, which means complex numbers. If the coefficients were restricted in any way, say to be rational, they wrote the field R_R. They admitted this was back to front from Kronecker's approach.

What they called finite or algebraic fields in the strict sense of the word were simple algebraic extensions of any of the fields just defined, to wit, the fields that Kronecker had called derived domains of rationality, reserving the term natural for just the fields in the paragraph above. They also presented the concept of a field in purely formal terms (analogous to that of a group, they said). A commutative group is a field, they said, when it also admits an associative multiplication with a multiplicative identity, and when every element that is not a divisor of zero has a unique multiplicative inverse. Weber's definition was more restrictive, they observed: the multiplication must be commutative, and there are no zero divisors. König's orthoid domains satisfied all these conditions and were of characteristic 0. What he called a hyperorthoid domain dropped the condition about divisors of 0. For example, the domain $\{a + bx : a, b \in \mathbb{C}, x^2 = 0\}$. König called a domain pseudo-orthoid if it had no divisors of 0, but was not of characteristic 0. They arose, for example, by taking numbers modulo a prime. So pseudo-orthoid domains were also fields in Weber's sense of the term. Hadamard and Kürschák settled on the definition of field that agreed with König's orthoid domain. What is notable is that at this stage in the paper they were unable to take on board Steinitz's paper of 1910. They could only do that at the end of the paper, printed in the next fascicle and published on 15 February 1911.

A holoid domain satisfies all the defining conditions of an orthoid domain except those relating to division. From a holoid domain one can always form an orthoid domain—its field of fractions. The algebraic integers form a holoid domain. The algebraic integers in a field K likewise form a holoid domain, as do subdomains generated over the rational integers by a finite set of algebraic integers. Such domains were called 'Art' or 'Species' by Kronecker, 'Ordnung' by Dedekind, and 'rings' or 'integral domains' by Hilbert.

After this came several more advanced sections, here omitted. Then, in the second fascicle they plunged into the theory of modular systems, and in due course Hilbert's invariant theory and the ABC Theorem (see Sect. 20.4). Hadamard and Kürschák showed that Hilbert's third Theorem (later much better known under the name *Nullstellensatz*, see Sect. 25.5) implies Noether's fundamental theorem (the $AF + BG$ Theorem). The second fascicle, published on 15 February 1911, is also of interest because by now the authors had had time to take on board Steinitz's *Algebraische Theorie der Körper*, and it is clear from the clarity of the exposition and the new generality of expression just why Steinitz's paper had the foundational effect that it did.

23.5 Steinitz

Ernst Steinitz's *Algebraische Theorie der Körper* ('The algebraic theory of fields') first appeared as an article in Crelle's *Journal für Mathematik* in 1910, and was judged so important that it was reprinted as a book in 1930 with additional material by Reinhold Baer and Helmut Hasse. It is the article that mathematicians know of if they know anything of the history of field theory, and indeed Baer and Hasse speak of it as being the starting point for many far-reaching researches in the domain of algebra and arithmetic. They also called it a milestone in the development of algebraic science, and an exceptionally clear introduction.

It opens with an introduction in which Steinitz says that his definition of a field is the same as Weber's: a system in which elements can be added, subtracted, multiplied, and divided (only division by zero is excluded), the associative and commutative laws for addition and multiplication are obeyed, and so is the distributive law. However, he said, whereas in Weber's day the purpose was Galois theory, now the field concept itself is taken as central, so the aim of the paper was to survey all possible fields.

The approach is to identify the simplest possible fields and then all their extensions. The first step singles out the prime fields and the concept of the characteristic of a field. There are simple extensions, when a single element is adjoined, and these may be algebraic or transcendental. Algebraic extensions are finite-dimensional, and form the subject matter of Galois theory. Kronecker had shown how to avoid working with irrationals by working modulo an irreducible polynomial, said Steinitz, and he proposed to do so here.

He then noted that matters get more delicate with questions that go one way over the rationals or perhaps fields of zero characteristic, and another over fields of characteristic p, and he indicated that he would deal with this too. This is perhaps the most important extension of the theory of fields in Steinitz's treatment: a thorough-going acceptance of the existence of fields of finite characteristic and a re-writing of the theory to include them systematically.

If we turn now to the paper or book itself, we see that the first chapter deals with the basic concepts: that of a field, an isomorphism, subfields, extension fields, adjunction of elements, and integral domains. Steinitz defined a *prime field* as one that has no subfields, and he proved that every field has a unique prime subfield. This prime subfield is either such that there is a prime number $p > 0$ such that $px = 0$ for every x in the subfield, in which case the *characteristic* of the field is p, or the only number with this property is 0, in which case the *characteristic* of the field is 0.

I skip ahead to the discussion of the adjunction of an element. Steinitz, following Kronecker, preferred to work with the rational functions in a variable x over a field K and congruences modulo a polynomial $f(x)$ of degree n that is irreducible over the field, and he showed that indeed $K(x)/(f(x))$ is a field. It is an algebraic extension of the ground field K of degree n. In this way he adjoined all the roots of the polynomial.

At the cost of providing an inadequate account of what Steinitz did, one that omits any discussion of his analysis of separable and inseparable extensions of a field of characteristic p, I turn instead to give a brief look at the additional material by Baer and Hasse.[10] This concerned Galois theory.

They took a ground field K and a normal algebraic extension field L. A normal extension was defined by Steinitz as one with the property that an irreducible polynomial over K either remains irreducible over L or factors completely. They proved that an intermediate field is a normal extension if and only if it is identical to its conjugates, which is to say that it is mapped to itself by all the elements of the Galois group.[11]

They took the subject matter of Galois theory to be how fields intermediate between K and L relate to subgroups of the Galois group G of the extension, which is defined as the automorphisms of L that leave K pointwise fixed.

Given a subgroup H of G, they defined $L(H)$ to be the subfield of L that is fixed pointwise by the elements of H.

Given an intermediate field M, they defined the group $G(M)$ to be the subgroup of G that fixes M pointwise. The natural questions are then to compare $G(L(H))$ and H, and to compare $L(G(M))$ and M. It is straight-forward that

$$G(L(H)) \supseteq H, \text{ and } L(G(M)) \supseteq M,$$

---

[10] An extension is *separable* if the splitting field is generated by a polynomial with distinct roots. To give an example of an *inseparable* extension, consider the polynomial $y^3 - x$ over the field $\mathbb{F}_3(x)$, where $\mathbb{F}_3$ is the field of three elements, and the extension generated by α, a root of this polynomial. We have $y^3 - \alpha^3 \equiv (y - \alpha)^3$ mod 3, so α is the only root.

[11] A remark about names: it seems that Klein and his students in the 1880s were the first to use the term Galois group, but they applied it specifically to the groups $PSL(2, p)$, $p = 5, 7, 11$ that derive from a part of Galois's letter to Chevalier that we have not studied. We shall see in the next chapter that Weber spoke of Galois groups in his *Lehrbuch der Algebra*.

so the central question becomes: when are these objects equal in pairs? The answer they give is that these pairs are equal if and only if L is a finite extension of K obtained from a polynomial with simple roots.

It will be noted that while Steinitz in 1910 had moved away from Galois theory and towards a theory of fields in their own right, in 1930 his editors moved the emphasis back to Galois theory.

Chapter 24
Ideal Theory and Algebraic Curves

24.1 Introduction

Polynomials in two variables define algebraic curves in the plane, and algebraic curves in the plane generally meet (perhaps in complicated ways) in points. What is the connection between the geometry and the algebra? More precisely: given two plane algebraic curves, we can consider the curves that pass through the common points of the two given curves and ask: Does this force the curve to have an equation of the form

$$a(x, y)f(x, y) + b(x, y)g(x, y) = 0,$$

where the two given curves have equations $f(x, y) = 0$ and $g(x, y) = 0$? Notice that this question is—or can be seen as—a question about membership in the ideal generated by $f(x, y)$ and $g(x, y)$ in the polynomial ring $\mathbb{C}[x, y]$.

In this chapter we shall see how this question was answered, not entirely successfully, in the late nineteenth century by two mathematicians: Alexander Brill and Max Noether (the father of the more illustrious Emmy). The generalisation to more variables was very difficult, and was chiefly the achievement of Emanuel Lasker, who was the World Chess champion at the time, with his theory of primary ideals. We shall give an example of his fundamental result taken from the English mathematician F.S. Macaulay's fundamental work on polynomial rings. With these results, the basic structural features of polynomial rings were all in place, and with the equally rich theory of number fields so too were all the basic features of commutative algebra.

© Springer Nature Switzerland AG 2018
J. Gray, *A History of Abstract Algebra*, Springer Undergraduate Mathematics Series,
https://doi.org/10.1007/978-3-319-94773-0_24

24.2 The Brill–Noether Theorem

In his (1874), Max Noether first criticised a number of important papers for
answering the original question (given f and g, when can F be written in the form
$F \equiv Af + Bg$?) by saying that F must vanish at the common zeros of f and g.
He rightly found this altogether too naive. Consider the case where $f(x, y) = y$
and $g(x, y) = x^2 - y$; the curves $f = 0$ and $g = 0$ have a common point at the
origin and the line $h(x, y) = y - x = 0$ passes through that point, but it cannot
be written in the required form because no expression of that form can contain the
monomial x.

Noether concentrated on the case where f has a q-fold point and g has an r-fold
point at a common point of intersection, where $q \leq r$ and what Noether meant by
a q-fold point is that, in local coordinates, all the terms of ϕ of degree less than
$(q - 1)$ vanish and at least one term of ϕ of degree q does not vanish.

His aim was to show that the coefficients of F reflect the conditions on the
coefficients on f and g. He argued that because all the terms of f of degree less
than q vanish (there are $\frac{1}{2}q(q + 1)$ such terms) there are $\frac{1}{2}q(q + 1)$ conditions of the
coefficients of F. More conditions on the coefficients of F and A arise in the range
q to $r - 1$ inclusive, and more conditions on the coefficients of F, A and B arise in
the range r to $q + r - 1$, at which stage the number of coefficients in A and B equals
the number of coefficients of F for the first time, the analysis can stop, and so the
total number of conditions is $\frac{1}{2}q(q + 1) + (r - q)q + \frac{1}{2}q(q - 1) = qr$.

Noether's argument was therefore a simple case of counting constants. But it
is easy to see that the result as stated is incorrect, as the above example given by
$f(x, y) = y$ and $g(x, y) = x^2 - y$ shows. For a polynomial in x and y to be of the
form $Af + Bg$ it is necessary and sufficient that the coefficients of the first power
of x vanishes, but in the case at hand we have $q = 1$ and $r = 1$ suggesting that it is
enough that the constant term of the polynomial F vanish.

Noether's argument requires that the intersections are simple, which means that
the curves have no common tangents. He did in fact impose this condition, but
he missed the other condition needed for his theorem to be valid, which is that
the curves are given by homogeneous expressions. To see this, define g_* to be x^2.
Clearly, the polynomial is of the form $Af + Bg$ if and only if it is of the form
$Af + Bg_*$ but Noether's conclusion is in error for g, and correct for g_*.

In the 1890s the newly-founded German Mathematical Society (the *Deutsche
Mathematiker-Vereinigung* or DMV) asked various experts to write lengthy reports
on the state of their subject. Brill and Noether were asked to survey the theory of
algebraic functions in ancient and modern times and they wrote a magisterial 457-
page account (Brill and Noether 1894).

The topic of intersections of curves was covered in pp. 347–366. They now
took the fundamental question to be: given homogeneous polynomials $f(x_1, x_2, x_3)$
and $g(x_1, x_2, x_3)$ with no common factor, what are the necessary and sufficient
conditions on a homogeneous polynomial $F(x_1, x_2, x_3)$ so that $F \equiv Af + Bg$,
where A and B are likewise homogeneous polynomials in x_1, x_2, x_3? Once again,

their answer was expressed in necessary and sufficient conditions at each of the common points of the curves defined by the equations $f = 0$ and $g = 0$: the terms of F must agree with the terms of expressions in local coordinates of the form $A'f + B'g$ up to a certain degree, where A' and B' are arbitrary homogeneous forms of any indefinitely large but finite degree.

The reason for passing to complex projective space was to be able to take account of all the points of intersection of the curves. Appendix H gives more detail on the necessity for the use of homogeneous coordinates and projective space, and an indication of the subtleties involved.

First of all, Brill and Noether reduced the problem to a binary one by eliminating x_1 from the equations $f(x_1, x_2, x_3) = 0$ and $g(x_1, x_2, x_3) = 0$, by using the resultant $R(x_2, x_3)$ of f and g. This resultant can be written as $R = Cf + Dg$, where C and D are polynomials.[1] Brill and Noether assumed that f and g have no common factor, so $R \not\equiv 0$ (and also, for the proof, but not the result, that $(x_2, x_3) = (0, 0)$ is not a common point of $f = 0$ and $g = 0$). Then F will be such that $F \equiv Af + Bg$ if and only if F is such that the identity $FD \equiv A_1 f + BR$ is satisfied, where A_1 is another homogeneous polynomial. They then said (1894, p. 352):

> Now evidently, separate conditions arise for this latter identity corresponding to the separate factors of R, and it immediately follows that the conditions for F can only be such as arise at the individual intersection points of $f = 0$ and $g = 0$.

Therefore the original identity can be satisfied if and only if the terms of F agree with the terms of a development of $A'f + B'g$ – where A' and B' are arbitrary forms of any fixed degree – up to a given, sufficiently high degree. For example, when the curves $f = 0$ and $g = 0$ have no multiple tangents, the intersection point is q-fold for f and r-fold for g, and of qr-fold multiplicity for the curves through that point (which will happen if the relevant factor of R is of degree only qr) then the comparison need proceed only as far as dimension $q + r - 2$. In particular, if F itself has a $q + r - 1$-fold point there, the comparison is established identically.

We see that the original problem was taken to be firmly geometric, and so the answer was unproblematically given in terms of an analysis of the intersection points. There was no suggestion that there might be other than purely local conditions to check. As we shall see, this assumption is a flaw in the proof that later workers had to confront.

As for the problem of curves with multiple tangents, Noether himself had addressed the issue in his (1884), where he appealed to the idea that complicated singularities can be resolved into simple ones by a particular technique. This was widely believed, but it is clear from Bliss's Presidential address to the American Mathematical Society in 1921 that there was still work to be done on the resolution of singular points.[2]

---

[1] Resultants are discussed in Appendix I.

[2] General satisfaction is usually attributed to Walker, see Bliss (1923) and Walker (1950).

Netto's Theorem

In 1885 Netto produced an interesting variant of the Brill–Noether theorem that was to prove to be a curtain-raiser to Hilbert's *Nullstellensatz* of 1893. His argument is not so difficult to follow and by following it we can see some of the difficulties that only Hilbert was able to overcome.

Netto considered two curves, which we may call C and C_1, given respectively by the polynomial equations $f(x, y) = 0$ and $f_1(x, y) = 0$ with no common factor that meet in a finite number of points, and a curve C' given by a polynomial equation $F(x, y) = 0$ that passes through the intersection points of C and C_1. He proved the theorem that there is an integer r such that

$$F(x, y)^r = f(x, y)g(x, y) + f_1(x, y)g_1(x, y),$$

or, in "Kronecker's convenient way of writing", as he put it,

$$F(x, y)^r \equiv 0 \quad (\text{mod } f(x, y), f_1(x, y)).$$

Loosely, Netto argued that there is a convenient linear change of variable such that the result of eliminating say y from the equations $f(x, y) = 0$ and $f_1(x, y) = 0$ is an equation $R_1(x) = 0$ that has multiple points precisely where the curves C and C_1 have common points and moreover the degree of multiplicity of R_1 at such a point is precisely the degree of the multiplicity of the corresponding multiple point. Netto also argued that the same is true for the equation $R_2(y) = 0$ obtained by eliminating x.

More precisely, the equations $R_1(x) = 0$ and $R_2(y) = 0$ have the property that their solutions give the x and y coordinates of the points where the curves meet, but the method can produce extraneous factors that must be eliminated by inspection. The need for the change of variable is to eliminate some of these extraneous factors. It is purely a consequence of the method for finding $R_1(x)$, and I shall assume it has been done without introducing (as Netto did) new symbols for the variables (see also the examples in Appendix I).

24.3 The Failure of the Brill–Noether Theorem to Generalise

Much of the historical importance of the ideas of Brill and Noether lies in how they were generalised. In the present context the difficult question was how to extend them to the very different situation in several variables, and this was successfully tackled by Lasker in Germany and Macaulay in London.

König was the first to prove a generalised Noether Theorem (see König 1904, Chapter VII, § 12). It applied to ideals defined by k polynomials $F_1, \ldots, F_k$ in k variables that are what Kronecker had called of the principal class, which means that the varieties corresponding to the ideals $(F_i, \ldots, F_k)$, $1 \leq i \leq k$, are of codimension $k - i + 1$. In other words, the variety corresponding to $F_1, \ldots, F_k$ is of dimension 0 (it is a finite set of points), the variety corresponding to $F_2, \ldots, F_k$ is of dimension 1, and so on.[3]

König gave necessary and sufficient conditions for a polynomial to belong to the ideal. It is plainly necessary, he observed, that the polynomial vanishes at the common zeros of $F_1, \ldots, F_k$, but this is not sufficient. In fact, as Hilbert had shown in his *Nullstellensatz*, if F vanishes at those points, then some power of it belongs to the modular system. König replaced Hilbert's proof, which he found rather complicated, with a simpler argument using the elimination theory he developed in his book (an argument that is nonetheless too complicated to include here). König's test, like Brill and Noether's original one, was therefore purely local: membership of the module was tested for by looking at each point of the corresponding variety.

It was soon observed that König's theorem fails when the number of functions defining an ideal exceeds the number of variables—I will give Macaulay's later example below. The failure of the Brill-Noether theorem to generalise is part of the story that involves Emanuel Lasker and his discovery of primary ideals. This is rather complicated and we can only look at it briefly.

24.4 Lasker's Theory of Primary Ideals

Emanuel Lasker was a student of Max Noether who also studied under Hilbert. He became the world chess champion in 1893, at the age of 25, and in 1904 he emigrated permanently to New York, where he became a regular player at the Manhattan Chess Club. He defended his title successfully for nearly 27 years, before eventually surrendering it to José Capablanca, who for many years had been an opponent of his there. His paper on the primary decomposition theorem is his major contribution to mathematics.

He submitted this paper (Lasker 1905) to *Mathematische Annalen* from New York. In it he treated the theory of polynomials in several variables, which he regarded as intimately connected to the theory of invariants. In a lengthy historical introduction, Lasker traced two approaches current in the subject, one due to Dedekind and Hurwitz and the other to Kronecker, Weber, and König. He also noted other traditions, such as the geometric one started by Cayley and Salmon, and made rigorous by Noether, but, he said, Noether's ideas had had to wait 20 years before they were developed by Hilbert. Lasker placed his own work in the line of descent from Dedekind's ideal theory.

His paper deals with homogeneous polynomials (called forms) in m variables. They have either arbitrary complex coefficients or integer coefficients, in which case

---

[3] In modern terminology, this is the condition that the F_i form a regular sequence. See Eisenbud (1995).

he called the forms integer forms (*ganzzahlig*). He defined a 'module' to be a set of polynomials closed under addition and multiplication by forms in the ring of all forms, and an 'ideal' to be a set of integer forms closed under addition and multiplication by integer forms. This clashes with modern terminology, and I shall instead use the modern term 'ideal' throughout (integer forms will not be discussed).

Lasker's paper is best remembered for the theorem that every ideal is an intersection of primary ideals in an essentially unique way. This theorem is the analogue of the unique factorisation theorem for integers. More precisely, Lasker defined an ideal P to be prime (1905, p. 50) if $p \cdot q \in P$ implies either $p \in P$ or $q \in P$. To define a primary ideal Q he first introduced the idea of the variety (which he called a *Konfiguration* or *Mannigfaltigkeit*) corresponding to it: this is the set of points at which all the forms in the ideal vanish. He then said (1905, p. 51) that if P was a prime ideal of the variety V then the ideal Q is a primary ideal corresponding to P if its corresponding variety is a subset of V and $a \cdot q \in Q$ and $a \notin P$ implies that $q \in Q$.

Lasker then stated his principal result (p. 50) as:

Theorem (Lasker's Theorem VII). *Every ideal M is representable in the form*

$$M = Q_1 \cap Q_2 \cap \ldots \cap Q_k \cap R$$

where $Q_1, Q_2, \ldots, Q_k$ are primary ideals and R is an ideal whose corresponding variety contains no points.

In the proof, Lasker made considerable use of Dedekind's idea of one ideal being residual to another: the residual ideal of M with respect to N, denoted $N : M$, consists of those forms f such that $f \cdot F \in N$ whenever $F \in M$. He supposed that the variety corresponding to M has as its irreducible components $C_1, C_2, \ldots, C_j$ and formed the ideal M_{C_1} of all forms F such that the residual ideal $M : (F)$ does not consist of forms vanishing on all of C_1. By definition, these forms are precisely those for which there is a form Φ such that F does not vanish on C_1 and $F.\Phi \in M$. Lasker's achievement was in showing that the ideal M_{C_1} is primary; it is then taken as Q_1.

Lasker next took the residual ideal of M_{C_1} with respect to M, which he denoted M'_{C_1}. The forms it contains do not vanish on all of C_1. The forms in M belong both to M_{C_1} and M'_{C_1} and this opened the way to the primary resolution. Once all the varieties $C_1, C_2, \ldots, C_j$ had been considered, Lasker was left with an ideal of the form (M, ϕ) where ϕ is a form in $(M'_{C_1}, \ldots, M'_{C_j})$ that vanishes on none of $C_1, C_2, \ldots, C_j$. By looking at the dimensions of the varieties that arise Lasker showed that the preceding analysis can be applied to (M, ϕ) repeatedly until the process is exhausted and the primary decomposition is obtained.

The components of Lasker's primary decomposition are not entirely unique. Macaulay called a primary ideal among the $Q_1, Q_2, \ldots, Q_k$ irrelevant if it contains the intersection $Q_1 \cap Q_2 \cap \ldots \cap Q_k$, and can therefore be omitted. The others, which he called relevant, divide into ones he called isolated and embedded. A Q_j

is isolated if its corresponding variety is not contained in another relevant variety of higher dimension; otherwise he called it embedded. The isolated primary ideals are unique, but the embedded ones are not.

The potential occurrence in the primary decomposition theorem of a primary ideal R whose corresponding variety is empty marks a major development in the theoretical understanding of ideals, because it shows immediately that no test for membership of an ideal defined by a set of functions can be confined a priori to local membership tests at points on the variety defined by the given functions. For these to be enough, it must be shown that the variety R does not occur in the primary decomposition. Retrospectively, this places all discussions of ideal membership, such as Noether's $AF + BG$ Theorem in doubt. Prospectively, it invites consideration of conditions which can ensure that such an ideal R does not arise.

24.5 Macaulay's Example

The simplest case of an irretrievable failure of Noether's Theorem is that of a rational quartic curve in space (see Eisenbud 1995, p. 466). Let $[x, y, z, w]$ be the coordinates of a point in $\mathbb{CP}^3$. Let the curve C be defined as follows:

$$C = \left\{ \left[s^4, s^3t, st^3, t^4 \right] : s, t \in \mathbb{C} \right\}.$$

Suppose that a hyperplane H in $\mathbb{CP}^3$, with equation $f_H = 0$, meets the curve at four points P_1, P_2, P_3, P_4. We note that the quadric Q with equation $f_Q = xw - yz = 0$ is the only quadric to contain C, although there are several cubic surfaces, for example $y^3 - x^2z$.

If the Brill–Noether theorem were true, it would say that any quadric through the four points must have an equation of the form $Q + uf_H = 0$, where $u = 0$ is a linear expression in x, y, z, w. This is a 4-dimensional family of quadrics. However, there is a 9-dimensional family of quadrics in projective 3-space, and so a 5-dimensional family through the four points where the curve meets the hyperplane. So some quadrics through the four points are not of the form required by the putative Brill and Noether theorem, and the failure of Noether's Theorem is complete.

For example, if the equation of the hyperplane is taken to be $y = z$, then the quadric with equation $y(w - x) = 0$ passes through the four points

$$[1, 0, 0, 0], [0, 0, 0, 1], [1, 1, 1, 1], [1, -1, -1, 1],$$

but its equation is not of the form $Q + uH = 0$. This does not contradict König's extension of the Brill–Noether theorem, because the module defining the variety $C = \left\{ \left[s^4, s^3t, st^3, t^4 \right] : s, t \in \mathbb{C} \right\}$ is not of the principal class.

Macaulay gave many significant examples in his *Tract* (1916), among them (see his §44) the counter-example to the Brill–Noether theorem in higher dimensions consisting of the smooth rational quartic in $\mathbb{C}P^3$. Its ideal is

$$I = (xw - yz, \ y^3 - x^2z, \ z^3 - yw^2, \ y^2w - xz^2).$$

Modulo $y - z$ this ideal is $I' = \left(xw - y^2, \ y\left(y^2 - x^2\right), \ y\left(y^2 - w^2\right), \ y^2\left(w - x\right)\right)$, which is the intersection of $(x, y), (y, w), (w - x, y - x), (w - x, y + x)$ and (for example) $I' + (x, y, z, w)^3$. The function $y(w - x)$ vanishes at the four points but it is not in I'; in fact it generates the ideal of the four points. Thus, for the embedded component one can take any homogeneous ideal containing I' and a power of (x, y, z, w) but not containing $y(w - x)$. The failure of the Brill and Noether theorem now appears as the presence of an embedded irrelevant component which is not even unique (although not entirely arbitrary).[4]

24.6   Prime and Primary Ideals

First, some definitions. Recall that an ideal J is primary if $ab \in J \Rightarrow a \in J$ or $b^n \in J$ for some positive integer n. If we define $R(J)$, the radical of J, as

$$R(J) = \{a : a^n \in J, \text{ for some positive integer } n\},$$

then it is clear that $R(J)$ is a prime ideal, and we say that $R(J)$ is the prime ideal associated with J.

It is trivial that a primary ideal is prime. The simplest example of a primary ideal that is not prime is the ideal $J = \langle x, y^2 \rangle$ in the ring $\mathbb{C}[x, y]$. Here $y^2 \in J$ but $y \notin J$. In fact, this example does more, because the radical of J is $R(J) = \langle x, y \rangle$, and the square of this ideal $R(J)^2 = \langle x^2, xy, y^2 \rangle$, and we have

$$R(J)^2 \subset J \subset R(J),$$

with strict inclusions, so J is an example of a primary ideal that is not even a prime power.

It is more work to show that there are examples of prime power ideals that are not primary, but one is $J = \langle xy - z^2 \rangle$ in $\mathbb{C}[x, y, z]$. I omit the proof.

---

[4]For a modern discussion, see Eisenbud (1995, p. 466).

Chapter 25
Invariant Theory and Polynomial Rings

25.1 Introduction

Another, related branch of algebra that mingles polynomials with geometry was called invariant theory. We shall see how this field was decisively rewritten by the young David Hilbert in work that made his name in the international mathematical community.

25.2 Hilbert

Fig. 25.1 David Hilbert (1862–1943). Photo courtesy of the Archives of the Mathematisches Forschungsinstitut Oberwolfach

David Hilbert was born in Königsberg on 23 January, 1862. Königsberg was a small town in East Prussia, best known for being the home town of the philosopher Immanuel Kant. As the only boy in the family, Hilbert was sent to school when he was eight, but he did not thrive there, because the emphasis on ancient languages and the heavy demands on the use of memory was uncongenial, and he transferred to another school in his final year. In these years there was no clear sign of his later eminence. Hilbert himself said, however plausibly it is hard to determine, that 'I did not particularly concern myself with mathematics at school because I knew that I would turn to it later' (Fig. 25.1).

Hilbert then went to the university in Königsberg in 1880. The university was small, but it had a strong tradition in mathematics that had begun when Jacobi had introduced the first mathematics seminar in a German university there. Physics was also strong: after Franz Neumann the physicist Gustav Kirchhoff had worked there. When Hilbert arrived Heinrich Weber held the chair in mathematics at Königsberg, and when he left in 1883 he was succeeded by Ferdinand Lindemann, a geometer who had just become famous for his proof that π is transcendental.

Weber was to be a significant influence on Hilbert. As we saw in Chap. 22, he had broad interests. In 1876 he and Dedekind had published the posthumous edition of Riemann's *Werke*, with its generous selection of unpublished material. This was the major source of information for the many who came to respond to the challenges Riemann had left for future generations. In 1882 Weber and Dedekind published a major paper on the algebra and geometry of algebraic curves, from a novel and abstract standpoint. Through Weber, Hilbert came into contact for the first time with the strong current in German mathematical life that led back to Gauss. He took lecture courses from Weber on elliptic functions, number theory, and a seminar on invariant theory (I shall define this topic below). His replacement, Lindemann, also encouraged Hilbert to study invariant theory. Hilbert took little advantage of the German University system that allowed students to study wherever they wished, and apart from a term in Heidelberg studying under Lazarus Fuchs he stayed at home. He never went to the more dynamic University of Berlin; presumably Königsberg was getting something right.

It was at Königsberg that Hilbert met two people who were to be lifelong influences. The first of these was Hermann Minkowski, a fellow student who was 2 years younger, but already a term ahead. Minkowski was a prodigy, and in 1883, when he was 19, he won a prestigious prize from the Paris Academy of Sciences for answering a question in the theory of numbers.[1] Unlike Hilbert, Minkowski had studied in Berlin for a time, and he introduced Hilbert to the traditions of mathematics represented by Kronecker, Kummer, and Weierstrass.

---

[1] The competition turned into a famous scandal. The English mathematician Henry Smith wrote to the organisers to point out that he had already published a complete solution to that precise question some years before. Next it was suggested, entirely falsely, that Minkowski had entered the competition corruptly, knowing of Smith's work. Then Smith died. The only way out for the French was to proclaim Minkowski and Smith joint winners, which they duly did.

In Easter 1884 Adolf Hurwitz, who was only 3 years older than Hilbert, was appointed to an extraordinary professorship at Königsberg. Hurwitz was a former doctoral student of Klein, then in Leipzig. Hurwitz inspired Hilbert with the desire to be a universal mathematician. As he later said to his own first doctoral student, Otto Blumenthal, "Minkowski and I were totally overwhelmed by his knowledge and we never thought we would ever come that far".[2] Hurwitz introduced Hilbert to the circle Klein was building around him, and from 1886 to 1892, when Minkowski was in Bonn, Hilbert went almost daily on mathematical walks with Hurwitz. With Hurwitz, he said, he rummaged through every corner of mathematics, with Hurwitz always in the lead.

In February 1885, Hilbert submitted his doctoral thesis on invariant theory, as suggested by Lindemann. That winter he set off on his student travels: to Leipzig to meet Klein, and then in March to Paris to meet Camille Jordan and Gaston Darboux. On his return he became a Privatdozent, a junior position that allowed him to study for his Habilitation (the necessary and almost sufficient condition to teach at a German University). Klein opposed Hilbert's decision to habilitate in Königsberg, because he wanted to broaden the younger man's education, but he failed, and Hilbert completed the process quickly and successfully, but with a thesis topic that was not propitious for his later work. Hilbert was now ensconced at the university of his choice (his beloved Königsberg) with Hurwitz as a colleague. Later Hilbert called these years in Königsberg as a time of slow ripening, but at the time he knew that staying there was not sufficiently stimulating, and in Easter 1888 he set of on his travels again, going this time to Göttingen (where Klein now was) and then, at Klein's suggestion, to Erlangen, to talk with Paul Gordan.[3]

Gordan was a former colleague of Klein's. He was sometimes referred to as the King of invariant theory for showing, in 1868, that the ring of invariants and covariants of any binary form is finitely generated as a ring (I will define these terms shortly). His proof was heavily computational and explicit, for Gordan was a master at manipulating long algebraic expressions, and no one had been able to make significant progress on the case of three variables. But Gordan's method allowed a mathematician to produce for each degree d a basis for the invariants of binary forms of degree d such that every invariant of that degree is a sum of products of powers of these invariants. Furthermore, it provided a similar basis for the covariants of every degree. Computationally the method became too bulky for forms of degree higher than 6, but in principle it was both general and explicit.

Hilbert and Gordan spent a good week together in Leipzig in the spring of 1888, after which Hilbert elatedly reported to Klein:

> With the stimulating help of Professor Gordan, meanwhile, an infinite series of brain-waves has occurred to me. In particular we believe I have a masterful, short, and to-the-point proof of the finiteness of complete systems for homogeneous polynomials in two variables. (Frei 1985, p. 39)

---

[2] See (Blumenthal 1935, p. 390).
[3] This account follows (McLarty 2012).

Hilbert now came alive as a mathematician, and remarkably quickly established the finiteness for any system of invariants for forms of arbitrary degree in any number of variables, but by methods that were not constructive. They were general existence theorems, and, as Hilbert was to discover, they have an air of the miraculous about them that some mathematicians find hard to accept.

Hilbert placed the question "is there a finite basis for these objects?" in an abstract context, and showed that there must be a basis by little more than using induction on the number of variables involved. The whole proof barely lasts two pages. As he commented in a letter to Klein when he sent one of his first notes describing his work for publication, "I have restricted the use of formulae as far as possible, and only presented the intellectual content in a crisp manner".

Gordan initially balked at the lack of explicit results, apparently saying: "This is not mathematics, it is theology", and Hilbert himself admitted that his method gave no means of finding the basis, no indication of its size, and no impression of what the elements of the basis might look like. Indeed, as we shall see below, Hilbert's first published proof had errors. Although Gordan later conceded that even theology has its uses, Hilbert felt the shortcomings of his conceptual approach rather keenly, and in the late 1880s he began to remedy the problem, moving the subject on dramatically, and by 1897, as he observed in his lectures "We can nevertheless not be satisfied with merely knowing the number of invariants, as it is even more important to also know about the in- and covariants themselves, and about the relations between them". In 1888 Hilbert published the proofs of his discoveries in the *Mathematische Annalen*, at the time edited by Klein. Not for the last time the fortunes of these very different men, Hilbert and Klein, rose together.

Gordan's approach was by no means that of someone defending an out-of-date viewpoint against a better idea. He refereed Hilbert's fuller version of the invariant theorem for the *Mathematische Annalen* and wrote:

> Sadly I must say I am very unsatisfied with it. The claims are indeed quite important and correct, so my criticism does not point at them. Rather it relates to the proof of the fundamental theorem which does not measure up to the most modest demands one makes of a mathematical proof. It is not enough that the author make the matter clear to himself. One demands that he build a proof following secure rules .... Hilbert disdains to lay out his thoughts by formal rules; he thinks it is enough if no one can contradict his proof, and then all is in order. He teaches no one anything that way. I can only learn what is made as clear to me as one times one is one. I told him in Leipzig that his reasoning did not tell me anything. He maintained that the importance and correctness of his theorems was enough. It may be so for the initial discovery, but not for a detailed article in the *Annalen*. (Hilbert and Klein 1985, p. 65)

Naturally, Hilbert was annoyed by this report and complained to Klein about it. Klein accepted the paper which became (Hilbert 1890) and replied:

> Gordan has spent 8 days here .... I have to tell you his thinking about your work is quite different from what might appear from the letter reported to me. His overall judgment is so entirely favorable that you could not wish for better. Granted he recommends more organized presentation with short paragraphs following one another so that each within itself brings some smaller problem to a full conclusion. (Hilbert and Klein 1985, p. 66)

25.3 Invariants and Covariants

The objects of study are *forms*: homogeneous polynomials in n variables, and how they transform under invertible linear transformations in the variables. The canonical example is the binary quadratic form $ax^2 + bxy + cy^2$, which transforms under a linear transformation of the variables into, say, $a'x^2 + b'xy + c'y^2$. In this setting we have $b'^2 - 4a'c' = (b^2 - 4ac)\delta^2$, where δ is the determinant of the linear transformation, and so $b^2 - 4ac$ is called an invariant of the binary quadratic form.

More generally, the *invariants* of a form are expressions in the coefficients of the form that are altered by a linear transformation in the variables only by a multiple of the determinant of the transformation. There are also the *covariants* of the form; these are expressions in the coefficients and the variables of the form that are altered by a linear transformation in the variables only by a multiple of the determinant of the transformation.

The invariants of a system of forms in n variables form a ring. Gordan's major result was that the ring of invariants of binary forms of any fixed degree is finitely generated. For example, the ring of invariants and covariants of a binary cubic form f is generated by the unique invariant, d, of a cubic form and its three covariants, f, H, and J, which satisfy an identity (called by Sylvester and after him by Hilbert a *syzygy*): $4H^3 = df^2 - j^2$. Unfortunately it would take us too far afield to define these – an indication of the complexity of the work. But, as Hilbert showed in his Lectures, d and H have this significance: the condition that the cubic have three distinct roots is that $d \neq 0$, and among cubics with repeated roots two are distinct if and only if $d = 0$, $H \neq 0$, and all three roots are repeated if and only if $d = 0 = H$.

25.4 From Hilbert's Paper on Invariant Theory (1890)

Theorem I. *If any non-terminating sequence of forms in the n variables $x_1, x_2, \ldots, x_n$ is given, say $F_1, F_2, F_3, \ldots$, there is a number m such that each form in any sequence can be put in the form*

$$F = A_1 F_1 + A_2 F_2 + \cdots + A_m F_m,$$

where $A_1, A_2, \ldots A_m$ are suitable forms in the given n variables.

[...]

In the simplest case $n = 1$ every form in the given sequence consists of a single term of the form cx^r, where c is a constant. Let $c_1 x^{r_1}$ be the first term in the given sequence for which the coefficient c_1 is different from zero. We now seek the first subsequent form whose order is less than r_1; let this form be $c_2 x^{r_2}$. Then once again we seek the first subsequent form whose order is less than r_2; this form shall be

$c_3 x^{r_3}$. *Proceeding in this way we obtain at the latest in r_1 steps[4] a form F_m in the given series after which no form has a lower order, and every form in the sequence is divisible by F_m, so m is a number having the property specified by our Theorem.*

[Hilbert then reviewed previous attempts on Theorem I., observed their growing complexity as the number of variables increased, and suggested a new approach by induction on n.]

Let $F_1, F_2, F_3, \ldots$ be the given sequence of forms in the n variables $x_1, x_2, \ldots, x_n$ and F_1 a form of order r that does not vanish identically. We first determine a linear substitution in the variables $x_1, x_2, \ldots, x_n$ that has a non-zero determinant and also converts the form F_1 into a form G_1 in the variables $y_1, y_2, \ldots, y_n$ so that the coefficient of y_n^r in the form G_1 takes a non-zero value. By means of this linear substitution the forms $F_1, F_2, \ldots$ are converted respectively into $G_1, G_2, \ldots$. If we now consider a relation of the form

$$G_s = B_1 G_1 + B_2 G_2 + \cdots + B_m G_m,$$

where s denotes an arbitrary index and $B_1, B_2, \ldots, B_m$ are forms in the variables $y_1, y_2, \ldots, y_n$, then under the inverse linear transformation this itself goes into a relation of the form

$$F_s = A_1 F_1 + A_2 F_2 + \cdots + A_m F_m,$$

where $A_1, A_2, \ldots A_m$ are forms in the original variables $x_1, x_2, \ldots, x_n$. Therefore our Theorem I. for the originally given sequence of forms $F_1, F_2, F_3, \ldots$ follows as soon as the proof for the sequence of forms $G_1, G_2, G_3, \ldots$ is obtained.

Since the coefficient of y_n^r in the form G_1 has a non-zero value, the degree of each form G_s in the given sequence with respect to the variable y_n can be reduced below r, so one can multiply G_1 by a suitable form B_s and subtract the resulting product from G_s. Accordingly, for the arbitrary index s we set

$$G_s = B_s G_1 + g_{s1} y_n^{r-1} + g_{s2} y_n^{r-2} + \cdots + g_{sr},$$

where B_s is a form in the n variables $y_1, y_2, \ldots, y_n$, while the forms $g_{s1}, g_{s2}, \ldots, g_{sr}$ contain only the $n-1$ variables $y_1, y_2, \ldots, y_{n-1}$.

We now assume that our Theorem I. has already been proved for forms in $n-1$ variables, and apply it to the sequence of forms $g_{11}, g_{21}, g_{31}, \ldots$. It follows from Theorem I. that there is a number μ such that for every value of s there is a relation of the form

$$g_{s1} = b_{s1} g_{11} + b_{s2} g_{21} + \cdots + b_{s\mu} g_{\mu 1} = l_s(g_{11}, g_{21}, \ldots, g_{\mu 1}),$$

---

[4]This claim is odd: there is no terminating algorithm for finding the lowest power of x, although it is clearly $\leq r_1$.

where $b_{s1}, b_{s2}, \ldots, b_{s\mu}$ are forms in the $n-1$ variables $y_1, y_2, \ldots, y_{n-1}$. We now construct the form

$$g_{st}^{(1)} = g_{st} - l_s(g_{11}, g_{21}, \ldots, g_{\mu 1}), \quad (t = 1, 2, \ldots, r) \tag{25.1}$$

where in particular for $t = 1$

$$g_{s1}^{(1)} = 0.$$

We now apply Theorem I. again, in the case of $n-1$ variables, to the sequence of forms

$$g_{12}^{(1)}, g_{22}^{(1)}, g_{32}^{(1)}, \ldots.$$

It follows from Theorem I. that there is a number $\mu^{(1)}$ such that for every value of s there is a relation of the form

$$g_{s2}^{(1)} = b_{s1}^{(1)} g_{12}^{(1)} + b_{s2}^{(1)} g_{22}^{(1)} + \cdots + b_{s\mu^{(1)}}^{(1)} g_{\mu^{1(1)}}^{(1)} = l_s^{(1)}(g_{12}^{(1)}, g_{22}^{(1)}, \ldots, g_{\mu^{(1)}2})^{(1)},$$

where $b_{s1}^{(1)}, b_{s2}^{(1)}, \ldots + b_{s\mu^{(1)}}^{(1)}$ are forms in the $n-1$ variables $y_1, y_2, \ldots, y_{n-1}$.
We now set

$$g_{st}^{(2)} = g_{st}^{(1)} - l_s^{(1)}(g_{1t}^{(1)}, g_{2t}^{(1)}, \ldots, g_{\mu^{(1)}t}^{(1)}), \quad (t = 1, 2, \ldots, r) \tag{25.2}$$

where in particular for $t = 1, 2$

$$g_{s1}^{(2)} = 0, \quad g_{s2}^{(2)} = 0.$$

The application of Theorem I. to the sequence of forms $g_{13}^{(2)}, g_{23}^{(2)}, g_{33}^{(2)}, \ldots$ leads to the relation

$$g_{s3}^{(2)} = l_s^{(2)}(g_{13}^{(2)}, g_{23}^{(2)}, \ldots, g_{\mu^{(2)}3}^{(2)}),$$

and we set

$$g_{st}^{(3)} = g_{st}^{(2)} - l_s^{(2)}(g_{1t}^{(2)}, g_{2t}^{(2)}, \ldots, g_{\mu^{(2)}t}^{(2)}), \quad (t = 1, 2, \ldots, r) \tag{25.3}$$

from which it follows in particular that

$$g_{s1}^{(3)} = 0, \quad g_{s2}^{(3)} = 0, \quad g_{s3}^{(3)} = 0.$$

By repeated application of this method we obtain the relations

$$g_{st}^{(r-1)} = g_{st}^{(r-2)} - l_s^{(r-2)}(g_{1t}^{(r-2)}, g_{2t}^{(r-2)}, \ldots, g_{\mu^{(r-2)}t}^{(r-2)}), \quad (t = 1, 2, \ldots, r) \qquad (25.4)$$

$$g_{s1}^{(r-1)} = 0, \; g_{s2}^{(r-1)} = 0, \; g_{s,r-1}^{(r-1)} = 0,$$

and finally one obtains

$$g_{sr}^{(r-1)} = l_s^{(r-1)}(g_{1r}^{(r-1)}, g_{2r}^{(r-1)}, \ldots, g_{\mu^{(r-1)}r}^{(r-1)}), \quad (t = 1, 2, \ldots, r)$$

from which

$$0 = g_{st}^{(r-1)} - l_s^{(r-1)}(g_{1t}^{(r-1)}, g_{2t}^{(r-1)}, \ldots, g_{\mu^{(r-1)}t}^{(r-1)}), \quad (t = 1, 2, \ldots, r) \qquad (25.5)$$

follows. By addition of Eqs. (25.1), (25.2), (25.3), (25.4), (25.5) it arises that

$$g_{st} = l_s(g_{11}, g_{21}, \ldots, g_{\mu 1}) + l_s^{(1)}(g_{1t}^{(1)}, g_{2t}^{(1)}, \ldots, g_{\mu^{(1)}t}^{(1)}) + \cdots$$

$$+ l_s^{(r-1)}(g_{1t}^{(r-1)}, g_{2t}^{(r-1)}, \ldots, g_{\mu^{(r-1)}t}^{(r-1)}), \quad (t = 1, 2, \ldots, r).$$

We can replace the forms

$$g_{1t}^{(1)}, g_{2t}^{(1)}, \ldots, g_{\mu^{(1)}t}^{(1)}, \ldots g_{1t}^{(r-1)}, g_{2t}^{(r-1)}, g_{\mu^{(r-1)}t}^{(r-1)}$$

in the sequence through repeated application of Eqs. (25.1), (25.2), (25.3), …, (25.4) by linear combinations of the forms $g_{1t}, g_{2t}, \ldots, g_{mt}$, where m denotes the largest of the numbers $\mu, \mu^{(1)}, \ldots, \mu^{(r-1)}$. In this way we obtain from the last formula a system of equations of the form

$$g_{st} = c_{s1}g_{1t} + c_{s2}g_{2t} + \cdots + c_{sm}g_{mt} = k_s(g_{1t}, g_{2t}, \ldots, g_{mt}), \quad (t = 1, 2, \ldots, r).$$

If we multiply the last formula by y_n^{r-t} and these equations for $t = 1, 2, \ldots, r$, it follows, because

$$g_{s1} = y_n^{r-1} + g_{s2}y_n^{r-2} + \cdots + g_{sr} = G_s - B_s G_1,$$

that the equation

$$G_s - B_s G_1 = k_s(G_1 - B_1 G_1, G_2 - B_2 G_1, \ldots, G_m - B_m G_1)$$

holds, or, if C_s denotes a form in the n variables $y_1, y_2, \ldots, y_n$, that

$$G_s = C_s G_1 + k_s(G_1, G_2, \ldots, G_m) = L_s(G_1, G_2, \ldots, G_m),$$

i.e. the number m is such that Theorem I. holds for the sequence of forms $G_1, G_2,$ $G_3, \ldots$ and consequently also for the original sequence of forms $F_1, F_2, F_3, \ldots$. Consequently our Theorem I. holds in the case of n variables on the assumption that it is proved for forms in $n - 1$ variables. Since our Theorem I. is already known to be valid for forms in a single homogeneous variable, it holds in general.

Commentary

The proof of the Hilbert basis theorem is largely a matter of devising a good notation. The strategy is proof by induction, so we assume that the theorem is true for n variables, which we call $x_1, x_2, \ldots, x_n$, and we now have homogeneous forms in the $n + 1$ variables $x_1, x_2, \ldots, x_n, y$. Forms involving the xs and y will be written in capital letters, forms that do not involve the variable y in lower case. A form of degree d will be written as $F^{(d)}$ or $f^{(d)}$ if we need to indicate its degree, d.

We shall assume that, possibly after a change of variable, the first form in the sequence looks like

$$G_1^{(d)} = y^d + C_1,$$

where C_1 is a polynomial in y of degree at most $d - 1$ with coefficients that are polynomials in the xs. Hilbert put the form G_1 into his generating set for the sequence (G_s).

Consider the form $G_s^{(k)}$. If there is no power of y of degree d or more, we leave it alone. If

$$G_s^{(k)} = c_s y^r + D_s,$$

where $r > d$ is the highest power of y that occurs in G_s, and c is a form in the xs, then we reduce the degree of y in the term $c_s y^r$. To do so, we use the equation

$$G_1^{(d)} = y^d + C_1$$

to write

$$G_s^{(k)} = c_s y^{r-d} y^d + D_s = c_s y^{r-d}(G_1^{(d)} - C_1) + D_s =$$

$$c_s y^{r-d} G_1^{(d)} - c_s C_1 y^{r-d} + D_s.$$

We have written G_s in terms of G_1 and terms that are of degree at most $r - 1$ in y, because C_1 and D_s are polynomials in y of degree at most $d - 1$. Notice that the coefficient of G_1 in the new expression for G_s involves both xs and y.

By repeating this reduction as often as necessary, we write each form in this manner:

$$G_s^{(k)} = B_s G_1 + b_{s1} y^{r-1} + b_{s2} y^{r-2} + \cdots + b_{sr}.$$

Hilbert now looked at the column of coefficients of y^{r-1}:

$$b_{21}, b_{31}, \ldots, b_{s1}, \ldots.$$

By his inductive hypothesis, there is an integer k_1 such that every entry in this column is a linear expression (with coefficients that are polynomials in the xs) of

$$b_{21}, b_{31}, \ldots, b_{s_k 1}.$$

Hilbert now put the corresponding $G_2, G_3, \ldots, G_{s_k}$ into his generating set.

Consider now a G_t with $t > s_k$. By subtracting suitable combinations of the generators, here denoted H_t, it can be written as

$$G_t - H_t = b'_{t2} y^{r-2} + b'_{t3} y^{r-3} - \ldots,$$

and the same argument as before allows Hilbert to deal with the columns of b'_{t2}. He added the necessary $G_t - H_t$ to his generating set, and moved on to the next column in the same way, until all the columns were taken care of.

25.5 The Hilbert Basis Theorem and the *Nullstellensatz*

Hilbert's Theorem I. evolved into the Hilbert basis theorem, which says that any ideal in a polynomial ring has a finite basis, but this evolution was a complicated process that we cannot fully trace here. But we can see that the original formulation is rather different, and its proof in some respects inadequate. First of all, Hilbert worked with homogeneous polynomials, and they do not form a ring (the sum of two homogeneous expressions of different degrees is not homogeneous). McLarty in his (2012) speculates plausibly that Hilbert simply thought that one could drop the homogeneity requirement, but, as the editors of Hilbert's *Werke* note, this is not the case unless the forms satisfy some particular requirements. This is the error that marred his initial publications.

Second, Hilbert stated his theorem only for polynomial rings; how generally he thought it applied is a complicated historical matter. Third, and more importantly, Hilbert spoke of a countable sequence while plainly dealing with uncountable sets of forms, and then appealed to a procedure that even in the countable case is not algorithmic: there is no way of being sure one has found the expression of least degree of a given kind. But this, too, has a historical dimension. Gordan had no problem with these claims, and made similar ones himself. But Gordan did have ideas about how to express polynomials in ways that make Hilbert's procedure constructive if we grant that the expression of least degree can be found, and this was a contribution Hilbert adopted when he tried to make his approach more explicit. That said, even modern computers have not enabled mathematicians to deal effectively with forms of degree 5 or more.

In 1893 Hilbert returned to the theory of forms, and proved what he called his third general theorem, and which has become known as the *Nullstellensatz* or theorem on the places of the zeros (Hilbert 1893, pp. 320–325).

In the form he presented it, it says: Given m rational integral homogeneous functions[5] $f_1, f_2, \ldots, f_m$ of n variables $x_1, x_2, \ldots, x_n$, and let $F, F', F'', \ldots$ be any rational integral homogeneous functions of the same variables $x_1, x_2, \ldots, x_n$ with the property that they all vanish for those systems of variables for which the given m functions $f_1, f_2, \ldots, f_m$, then it is always possible to determine an integer r such that every product $\prod^{(r)}$ of r arbitrary functions from the sequence $F, F', F'', \ldots$ can be written in the form

$$\overset{(r)}{\prod} = a_1 f_1 + a_2 f_2 + \cdots + a_m f_m,$$

where the $a_1, a_2, \ldots a_m$ are suitable rational integral homogeneous functions in the variables $x_1, x_2, \ldots, x_n$.

We can give only a sketch of the proof. Hilbert derived it in two steps, on the further assumption that the sequence $F, F', F'', \ldots$ is finite. First, he assumed that the m forms $f_1, f_2, \ldots, f_m$ have only a finite number of common zeros. In this setting he argued by induction on the number of common zeros. The theorem is true when the forms have no common zero. To carry out the inductive step, Hilbert introduced a way of regarding the forms as forms in two variables (binary forms) and used the well-developed theory of binary forms to prove the theorem in that setting.

Second, to obtain the general proof, Hilbert argued by induction on the number of variables. The induction can begin with two variables.

The theorem entails, as he remarked, a result obtained earlier by Netto for inhomogeneous functions of two variables, that the rth power of any member of the sequence $F, F', F'', \ldots$ belongs to the module $(f_1, f_2, \ldots, f_m)$.

---

[5] That is, homogeneous polynomials.

Chapter 26
Hilbert's *Zahlbericht*

26.1 Introduction

Historians often note that two books in number theory open and close the nineteenth century in the theory of numbers: Gauss's *Disquisitiones Arithmeticae* at the start of the nineteenth century and Hilbert's *Zahlbericht*, or *Report on the Theory of Numbers*, at the end. Here we look at aspects of Hilbert's book, and hint at some of its influential choices, including its influence on the subsequent study of algebra.

26.2 An Overview of the *Zahlbericht*

In 1893 the DMV asked Hilbert and Hermann Minkowski to report on the theory of numbers. Minkowski was by then established as a world authority on the subject, but his friend Hilbert was known for his work on invariant theory and had done little more than express interest in number theory.

Hilbert was turning towards a subject that had become something of a German speciality and a source of pride within the field of mathematics ever since Gauss had published his *Disquisitiones Arithmeticae* in 1801. As we have seen, the subjects of quadratic forms, quadratic reciprocity and its generalization to higher powers, and cyclotomy took off with Gauss's work, and what sustained the interest of researchers was what Gauss had noted: the surprising difficulty in providing proofs, and the unexpected connections between topics that the search for proofs was likely to reveal. The cumulative production of such mathematicians as Eisenstein, Jacobi, Dirichlet, Dedekind, Kummer, and Kronecker was a daunting prospect and a challenge for the ambitious mathematician.

© Springer Nature Switzerland AG 2018
J. Gray, *A History of Abstract Algebra*, Springer Undergraduate Mathematics Series,
https://doi.org/10.1007/978-3-319-94773-0_26

Minkowski, however, soon withdrew from the project because he felt he was too busy with other things, especially his book *Geometrie der Zahlen* (*The Geometry of Numbers*). As a result, Hilbert's *Zahlbericht*, for all its 371 pages, is only a torso. A number of topics in number theory with a pedigree going back to Euler and Jacobi are under-represented, which may well have skewed the subsequent impact of Hilbert's problems on number theory. But Minkowski kept in touch with Hilbert, and performed the useful service of checking all the page proofs. Minkowski's opinion of the *Zahlbericht* was high. He wrote to Hilbert:

> I congratulate you now that finally the time has come, after the many years of labour, when your report will become the common property of all mathematicians, and I do not doubt that in the near future you yourself will be counted among the great classical figures of number theory.

He predicted correctly. Aside from the depth of insight Hilbert brought to the subject, and his careful organisation, the *Zahlbericht* is lucidly written. Hermann Weyl, one of the great writers of mathematics in German or English, regarded the preface to Hilbert's *Zahlbericht* as one of the great prose works in mathematical literature, and described in his obituary of Hilbert how as a young man he had heard "the sweet flute of the Pied Piper that Hilbert was, seducing so many rats to follow him into the deep river of mathematics". Hilbert also re-presented the history of number theory in a way that appealed to its practitioners, making it clear how it had emerged as a systematic science only during the past hundred years or so. More precisely, he re-wrote the subject so that knowledge of its history and familiarity with the older texts was no longer required. New readers could start here.

This fresh start partly came about from Hilbert's whole-hearted introduction of the ideas of Galois theory, which formed the second of the five parts of the Report.[1] Hilbert thought of Galois theory as a theory of algebraic numbers and therefore as an organising principle in algebraic number theory. This not only gave a clear conceptual shape to the subject, it explained why questions in number theory led naturally to and from questions about cyclotomy. The *Zahlbericht* is a paradigm example of the benefits of digging deep in pursuit of an explanation, and it put the seal on a century of stretching the definition of an integer until the general concept of an algebraic integer was paramount, and the ordinary integers had to travel with a qualifying adjective 'rational'.

Hilbert was particularly fond of analogies, which he hoped would suggest deep, if at times obscure, connections. He noted that a number of problems in number theory relied on complex function theory, and that there were numerous points in which the theory of algebraic functions and fields of algebraic numbers overlapped.

---

[1] An enthusiasm for Galois's work that he surely acquired from Klein, and which became a strongly held view in Göttingen that was shared by Hermann Weyl in the next generation.

(This had been the central thrust of paper by Dedekind and Weber in 1882.) As he put it in the preface (p. ix):

> Thus we see how far arithmetic, the Queen of mathematics, has conquered broad areas of algebra and function theory and has become their leader. The reason that this did not happen sooner and has not yet developed more extensively seems to me to lie in this, that number theory has only in recent years become known in its maturity .... Nowadays the erratic progress characteristic of the earliest stages of development of a subject has been replaced by steady and continuous progress through the systematic construction of the theory of algebraic number fields.

> The conclusion, if I am not mistaken, is that above all the modern development of pure mathematics takes place under the banner of number: the definitions given by Dedekind and Kronecker of the concept of number lead to an arithmetisation of function theory and serve to realise the principle that, even in function theory, a fact can be regarded as proven only when in the last instance it has been reduced to relations between rational integers. [...] The arithmeticization of geometry fulfils itself in the modern study of Non-Euclidean geometry, which it provides with a rigorous logical structure and gives the most direct possible and completely objection-free introduction of number into geometry

Hilbert's breakthrough with invariant theory had been achieved by adopting a strongly conceptual approach that minimised, but did not disdain, the computational side, and he now advocated taking the same approach in number theory:

> the fifth part develops the theory of those fields [the cyclotomic fields] which Kummer took as a basis for his researches into higher reciprocity laws and which on this account I have named after him. It is clear that the theory of these Kummer fields represents the highest peak attained today on the mountain of our knowledge of arithmetic; from it we look out on the wide panorama of the whole explored domain since almost all essential ideas and concepts of field theory, at least in a special setting, find an application in the proof of the higher reciprocity laws. I have tried to avoid Kummer's elaborate computational machinery, so that here, too, Riemann's principle may be realised and the proofs carried out not by calculations but purely by thought.

26.3 Ideal Classes and Quadratic Number Fields

As Lemmermeyer and Schappacher note in their introduction to the English translation of Hilbert's *Zahlbericht*, Hilbert did not use, and I would add did not possess, the modern way of thinking about algebraic number theory. He spoke of 'number rings', following Dedekind—meaning the ring of integers in a given number field—but did not think there was a subject called ring theory. More surprisingly, but in line with what he had learned from Weber, for someone who valued Galois theory so highly, he did not develop any tools from abstract group theory. In particular the notion of a quotient group is lacking in the *Zahlbericht* and so Hilbert was forced into some clumsy locutions. This says a lot about the state of abstract algebra in 1893–97.

That said, Hilbert was very happy to work with ideals in a given number ring, just as Dedekind had advocated. He defined two ideals **a** and **b** to be equivalent, written **a** $\sim$ **b**, if and only if there are algebraic integers α and β in the ring such that $\beta\mathbf{a} = \alpha\mathbf{b}$. The equivalence class of all ideals equivalent to a given one he called an ideal class, and he showed that this notion of equivalence respects products: if $\mathbf{a} \sim \mathbf{a}'$ and $\mathbf{b} \sim \mathbf{b}'$, then $\mathbf{ab} \sim \mathbf{a}'\mathbf{b}'$. All principal ideals are equivalent, they form the so-called principal class, which I shall denote $\mathbf{e}$, and $\mathbf{ea} = \mathbf{a}$.

There is a notion of quotient ideal: the ideal $\mathbf{b}$ is divisible by the ideal $\mathbf{a}$ if there is an ideal $\mathbf{c}$ such that $\mathbf{b} = \mathbf{ac}$. Hilbert showed that division, like multiplication, extends to ideal classes. He then showed, following Minkowski, and before him Dedekind and Kronecker, that the number of ideal classes is finite. The number of distinct ideal classes is called the class number of the field and its ring of integers.

Significantly, he did not say that the set of ideal classes has just been shown to form a finite abelian group, and so he could not appeal to any theory of such things.[2] He did note that if the class number is h then the hth power of every ideal is the ideal $\mathbf{e}$, which is trivial if one thinks of groups.

I have chosen to illustrate these general ideas with a selection of items from the third part of Hilbert's *Zahlbericht*, which is on quadratic fields. Hilbert began by recalling that in the field $\mathbb{Q}(\sqrt{m})$ obtained by adjoining $\sqrt{m}$ to $\mathbb{Q}$, where the positive square root is taken when $m > 0$ and the positive imaginary square root is taken when $m < 0$, has as its basis $\{1, \omega\}$, where

$$\omega = \frac{1 + \sqrt{m}}{2}, \text{ or } \omega = \sqrt{m},$$

according as m is congruent to 1 mod 4 or not.

The discriminant is $d = m$ in the first case and $d = 4m$ in all the others.

The discriminant is given as $(\omega - \omega')^2$.

How do prime ideals in $\mathbb{Q}$ fare in $\mathbb{Q}(\sqrt{m})$? Hilbert, following Dedekind, showed that the following happens:

- every rational prime that divides d is a square in $\mathbb{Q}(\sqrt{m})$;
- every odd rational prime that does not divide d splits into two conjugate primes in $\mathbb{Q}(\sqrt{m})$ if and only if d is a quadratic residue mod p;
- every odd rational prime that does not divide d remains prime in $\mathbb{Q}(\sqrt{m})$ if and only if d is a quadratic non-residue mod p;
- when $m \equiv 1 \pmod 4$ the prime 2 splits into two conjugate primes in $\mathbb{Q}(\sqrt{m})$ if $m \equiv 1 \pmod 8$ and remains prime if $m \equiv 5 \pmod 8$.

At this stage, Hilbert merely observed that it was now possible to exhibit all the ideal classes for any value of m, which is true but unenlightening. He could have investigated the group structure, as later mathematicians were to do, but he once again failed to draw maximum profit from the fact that the ideal classes form a finite abelian group—which may only serve to remind us that the history of mathematics, like the history of anything, is seldom tidy.

---

[2]Commutativity and associativity for ideals and ideal classes follows from commutativity and associativity in the ring itself.

Instead, and very importantly for later developments, Hilbert developed a theory of quadratic extensions of the Gaussian integers $Z[i]$, which he called Dirichlet's biquadratic field, with a view to raising it "in a purely arithmetic way to the same level that the theory of quadratic fields has since Gauss" in (Lemmermeyer 2007, pp. 528–529). To this end he computed the integral bases of the ideals and the decomposition into prime ideals. He then defined the genus of an ideal in terms of their characters—it would take us too far to give the definitions, but an interesting point can be made nonetheless. Hilbert put ideals in the same class in the same genus, and called the principal class that genus all of whose classes are trivial. He then stated the 'principal genus theorem' that every ideal in the principal genus is the square of some ideal class. To quote (Lemmermeyer 2007, p. 539):

> Hilbert then determines the number of genera, derives the quadratic reciprocity law, and finally gives an arithmetic proof of the class number formula for $\mathbb{Q}(i, \sqrt{m})$ and $m \in \mathbb{Z}$.

In the third section of the *Zahlbericht* Hilbert reworked this material, deduced the quadratic reciprocity law in this setting, and proved the 'principal genus theorem'.

More precisely, Hilbert showed that in a quadratic number field $\mathbb{Q}(\sqrt{d})$ there are precisely 2^{t-1} different genera, where t is the number of prime divisors of d, the square of any proper ideal class lies in the principal genus, and conversely every proper ideal class in the principal genus is the square of a proper ideal class.[3] The significance of the first remark is that it shows that precisely half the possible number of genera occur.

It is striking how closely this progression of ideas resembles Gauss's original path, transcribed from forms to ideals.[4]

26.4   Glimpses of the Influences of the *Zahlbericht*

Two books may serve to illustrate the state of algebraic number theory in the period shortly after Hilbert's *Zahlbericht*: Legh Wilber Reid's *The Elements of the Theory of Algebraic Numbers* (1910) and J. Sommer's *Vorlesungen über Zahlentheorie* (or, Lectures on Number Theory) (1907), which was also published in a French translation in 1911.

Reid, who was a professor at Haverford College in Pennsylvania, set himself the task of preparing people to study more advanced works, such as the *Zahlbericht*. He carefully introduced the key terms: integers, algebraic integers, and realms (his term for a field). He then studied the rational realm (which he denoted k), congruences among rational integers, and quadratic residues, and proved the theorem of quadratic reciprocity using Gauss's lemma. Then came several chapters of examples: the

---

[3] See (Frei 1979, p. 41).

[4] As (Lemmermeyer 2007, p. 541) observes, Weber's approach to the genus theory of ideals in quadratic fields in his *Lehrbuch der Algebra* (Sect. 22.4 above) follows Hilbert's closely.

realm $k(i)$, the realm $k(\sqrt{-3})$, the realm $k(\sqrt{2})$, and then the realm $k(\sqrt{-5})$, which allowed him to introduce the concept of ideals. He then considered the "general quadratic realm" $k(\sqrt{m})$ and its ideals, congruences modulo an ideal in these realms, units, and ideal classes.

His book could almost be read as an introduction to Sommer's, which is also billed as an introduction to contemporary number theory. It begins with a lengthy chapter on the general quadratic field and moves on to apply the theory to Fermat's last theorem and the relationship of the multiplication theory of ideals to the composition of forms. The book then concludes with a chapter on cubic fields and one on relative fields, topics that we cannot introduce here, and there is an appendix on the geometric theory of lattices that reflects the influence of Klein.

Both books make valuable contributions to their mathematical communities. Both start off much as Dirichlet had done in his *Lectures*, and then move into topics from Dedekind's various supplements. In view of the greater sophistication among his audience that Sommer felt he could assume, it is interesting to quote the introductory remarks that Hilbert contributed to Reid's book.

> The theory of numbers is a magnificent structure, created and developed by men who belong among the most brilliant investigators in the domain of the mathematical sciences: Fermat, Euler, Lagrange, Legendre, Gauss, Jacobi, Dirichlet, Hermite, Kummer, Dedekind and Kronecker. All these men have expressed their high opinion respecting the theory of numbers in the most enthusiastic words and up to the present there is indeed no science so highly praised by its devotees as is the theory of numbers. In the theory of numbers, we value the simplicity of its foundations, the exactness of its conceptions and the purity of its truths; we extol it as the pattern for the other sciences, as the deepest, the inexhaustible source of all mathematical knowledge, prodigal of incitements to investigation in other departments of mathematics, such as algebra, the theory of functions, analysis and geometry.

> Moreover, the theory of numbers is independent of the change of fashion and in it we do not see, as is often the case in other departments of knowledge, a conception or method at one time given undue prominence, at another suffering undeserved neglect; in the theory of numbers the oldest problem is often to-day modern, like a genuine work of art from the past. Nevertheless it is true now as formerly, a fact which Gauss and Dirichlet lamented, that only a small number of mathematicians busy themselves deeply with the theory of numbers and attain to a full enjoyment of its beauty. Especially outside of Germany and among the younger mathematicians arithmetical knowledge is little disseminated. Every devotee of the theory of numbers will desire that it shall be equally a possession of all nations and be cultivated and spread abroad, especially among the younger generation to whom the future belongs. Such is the aim of this book. May it reach this goal, not only by helping to make the elements of the theory of numbers the common property of all mathematicians, but also by serving as an introduction to the original works to which reference is made, and by inciting to independent activity in the field of the theory of numbers. On account of the devoted absorption of the author in the theory of numbers and the comprehensive understanding with which he has penetrated into its nature, we may rely upon the fulfilment of this wish.

Rhetoric, of course, but an indication of how Hilbert saw the importance of the subject he had done so much to redefine, and of the tradition that he was seeking to establish and extend.

Chapter 27
The Rise of Modern Algebra: Group Theory

27.1 Introduction

Here we look at how group theory also emerged "in its own right" in various books and papers around 1900, including the first monograph on abstract group theory, and then look at the influential work of the American mathematician Leonard Eugene Dickson.

27.2 The Emergence of Group Theory as an Independent Branch of Algebra

It makes sense for a modern mathematician to ask "What is group theory?" He or she, if a beginner, might simply want to know, and would presumably receive the answer that begins "A group is a set with a closed binary operation satisfying the following axioms: ...". Or the mathematician might be asking for an answer such as "Group theory is that branch of mathematics that deals with symmetry", or, if that's too vague, for an answer that indicates what role group theory plays, or is intended to play, in mathematics as a whole.

The historian of modern mathematics wants to know when mathematicians began asking such questions, and why. When was there a recognisable body of ideas that could be singled out and characterised, assessed for its importance, promoted, extended and otherwise used? There is usually no tidy answer to such a question, because it is about a social process of creation and recognition. Perhaps there is a moment when some definitive account was given, after which there was an identifiable subject—group theory—and people could say "I am a group theorist", "Today I proved a theorem in group theory". Usually there is a preceding period lasting quite some time when the subject was building, often in a subordinate but growing role within some other enquiry. It's quite common then for it to turn out

© Springer Nature Switzerland AG 2018
J. Gray, *A History of Abstract Algebra*, Springer Undergraduate Mathematics Series,
https://doi.org/10.1007/978-3-319-94773-0_27

that the subject has 'really', 'implicitly', 'tacitly' been around all along. In the case of group theory it has seriously been argued that Egyptian and later Islamic designs show an awareness of group theory, but that should not distract us from investigating when group theory emerged from the study of the solution of polynomial equations, and what happened as a result.

The first person to try to axiomatise the group concept was Cayley in 1849, responding to Cauchy's papers of a few years before, and Cayley may be the first person to use the term in its modern sense as applying to some symbols that obey the rules for multiplication and division. It is, however, unclear whether Cayley regarded his 'symbols of operation' as transformations, in which case he need not have said they obeyed the associative law, or if they are abstract, in which case he should couched his remarks about associative as an axiom and not as mere fact. In any case, historians of mathematics seem agreed that Cayley's ideas had little impact, perhaps because he did little with them to indicate the merit in and power of thinking group-theoretically.[1]

A more effective presentation of axioms for commutative groups is due to Kronecker in 1870, which he gave in his work as a number theorist. But matters changed significantly in the late 1870s. In those years Cayley raised the idea of studying groups for their own sakes, and called for an answer to the question of finding all groups of a given order. This is an absurdly difficult question, so much so that I doubt if Cayley had any idea how hard it was, but it is to his credit that he asked it at all, and it indicates that a subject called group theory had arrived (or at least was arriving). The paper also had an influence: Otto Hölder cited it specifically when writing his paper on groups of orders p^3, pq^2, pqr, p^4 in 1893.

Kronecker's ideas about commutative groups also had an impact in Berlin where they were picked up by Frobenius and Stickelberger. Frobenius was to mature into the most important Berlin mathematician of his generation; Stickelberger was more of a specialist. Together they hacked out a classification of finite abelian groups from its hidden place in the work of number theorists such as Gauss.

According to Wussing (1984, III.4), whose account we have been following, the year 1882 saw several crucial changes in the recognition of the importance of group theory within algebra. In that year Weber published a paper (1882) about prime numbers represented by quadratic forms in which he gave what may be the first proper definition of a group of order h. It is a system G of h elements, $\vartheta_1, \vartheta_2, \ldots \vartheta_h$ with a composition or multiplication such any two elements can be combined to make a third: $\vartheta_r \vartheta_2 = \vartheta_t$, where

$$(\vartheta_r \vartheta_s)\vartheta_t = \vartheta_r(\vartheta_s \vartheta_t),$$

---

[1] For a thorough study of definitions of a group and discussions in the early years of the twentieth century about the best definition, see (Hollings 2017).

and such that from either

$$\vartheta\vartheta_r = \vartheta\vartheta_s \quad \text{or from} \quad \vartheta_r\vartheta = \vartheta_s\vartheta$$

it follows that $\vartheta_r = \vartheta_s$.

From these assumptions Weber deduced that there is a unique element ϑ_0 in the system such that for every element ϑ_r

$$\vartheta_r\vartheta_0 = \vartheta_0\vartheta_r = \vartheta_r.$$

Furthermore, for every ϑ in G there is a unique ϑ' such that

$$\vartheta\vartheta' = \vartheta'\vartheta = \vartheta_0.$$

He added that the group is Abelian if for every pair of elements ϑ_r and ϑ_s it is the case that $\vartheta_r\vartheta_s = \vartheta_s\vartheta_r$.

Also in 1882 Eugen Netto published his book *Substitutionstheorie und ihre Anwendung an die Algebra* (Substitution theory and its applications in algebra) and Walther von Dyck, a pupil of Klein's published his paper *Gruppentheoretische Studien I* (Group-theoretic studies)—paper II followed the next year.

Netto's book is divided into two parts. The first develops group theory in the manner of Jordan's *Traité* with attention to the number of values of function may take, and, in Chap. 7, an interest in groups of various orders (p, pq, p^2). The second part applies this to the solution of algebraic equations, with chapters on cyclotomy, abelian equations, and algebraically solvable equations. In short, although Netto did not use the term, Galois theory. In the preface Netto acknowledged the importance of the books of Serret and Jordan and thanked his revered Professor Kronecker, regretting only that Kronecker's *Grundzüge* had been published too recently to have been any use in the preparation of this book.

His main claim to originality, said Netto of himself, was his recognition that the group formation, by its generality, has many advantages and this generality is clearest if the presentation is abstract. So although he called his book 'substitution theory' and not 'group theory', and tended to think of his groups as permutation groups, he did present finite commutative groups axiomatically (see p. 144, where he followed Kronecker's presentation). However, like his mentor, neither man saw any reason to develop group theory outside of a context where it could immediately be applied.

Netto's book was translated into English in 1892 and reviewed by Bolza the next year. His comments were divided between praise for the account of substitution groups in the first half of the book, and criticisms of the account of the Galois theory of equations (as we discussed in Sect. 15.2).

Von Dyck went much further. In his papers he included transformation groups in geometry and infinite groups generally in the range of objects he studied. He had learned from Klein how groups appeared in geometry, both projective and non-Euclidean geometry, and he tended to think of them as transformation groups with the transformations left out, as it were, and just the rules for combining the group elements left behind. So he can be read as offering the idea of a group as a free group on a certain number of generators which, perhaps, are then subject to some relations. It is indeed possible to argue that the combinatorial study of groups via their generators and relations represents a different way of thinking about groups to the axiomatic one, the approach via transformations and geometry, and the approach rooted in permutation groups.

The first person to act on the idea that groups are best thought about abstractly was Frobenius, who decided to rework the theory of finite groups without recourse to permutation-theoretic ideas. He recognised that every finite group is a subgroup of a permutation group (a result first proved by Cayley) and that Sylow's theorems are true (Sylow had proved them in 1872), but he re-derived them abstractly in a paper of 1887.

In 1889 Hölder published his proof of what has become known as the Jordan–Hölder theorem. He also proceeded entirely abstractly in his treatment of groups. Hölder's contribution differs from Jordan's contribution to the Jordan–Hölder theorem in that he proved that in an unrefinable composition series—a chain of maximal 'normal' subgroups, each one 'normal' in the one before—

$$G = G_0 \triangleright G_1 \triangleright \cdots \triangleright G_j \triangleright G_{j+1} \triangleright \cdots \triangleright G_n = \{e\},$$

it is not just the orders of the quotients $\frac{|G_j|}{|G_{j+1}|}$ that are independent of the choices made at each stage (about which maximal normal subgroup to pick) but the quotient *groups* themselves.

'The Group as a Fundamental Concept of Algebra' is a section heading in Wussing's book, and it is a good topic with which to conclude this section. Wussing finds that the role groups were made to play in Weber's investigations of Galois theory and algebraic number theory was decisive. In four lengthy papers collectively called *Theorie der Abel'schen Zahlkörper* (Theory of Abelian number fields) 1886–1887, Weber, relying on the paper by Frobenius and Stickelberger, showed how to use group theory in the study of algebraic number theory, and in a further paper of 1893 he advocated that groups be treated purely formally so that their theory can be applied indifferently to Galois theory and number theory, a view he repeated in the second volume of his *Lehrbuch*. That said, Weber otherwise showed little inclination to move beyond the permutation-theoretic presentation of groups; he recognised there was a general concept but in his hands it remained subordinate to other topics. This may be one reason why, as Wussing notes, the study of abstract groups for their own sake did not get going until the twentieth century. We should, however, add that in Volume 2 of his *Lehrbuch der Algebra* Weber repeated his earlier axioms for a finite group and then carefully noted that when infinite groups are involved one must also assume that every element has an inverse. This makes him the first person to give an abstract definition of an infinite group.

Wussing also noted that the first monograph to present group theory abstractly was de Séguier's *Théorie des Groupes Finis: Éléments de la Théorie des Groupes Abstraits* (1904), and it is indeed abstract. Whereas many earlier books and papers presented groups in specific settings—as groups of permutations or groups of geometric transformations—and were not altogether clear about the axioms that defined a group, de Séguier began with the idea that a group is at the very least a set of objects.[2] Moreover, he discussed what properties of sets (with reference to Cantor's (1895)) and what maps between sets are.

A group then emerges as a set with a product: the product of elements of the group is again an element of the group, and (Postulate 1) the product is associative, $(ab)c = a(bc)$. Furthermore (Postulate 2), the equation $ax = b$ always has a solution for every pair of elements a and b, as does the equation $xa = b$ (Postulate 3). De Séguier then deduced that there is a unique element u such that $ua = a = au$ for all elements a in the group. As a result, every element a has an inverse a^{-1} such that $aa^{-1} = u$. He then remarked that the three postulates are satisfied by numerous examples, so they are not self-contradictory, and they are independent, as is shown by various non-standard definitions of the product operation.

De Séguier then worked his way through a considerable amount of what was known in the emerging field of group theory. He wrote about subgroups or divisors and invariant or normal subgroups, and in some detail about the connection to Galois theory. From normal subgroups he passed to quotient groups, which he called factor groups, and their relation to homomorphisms between groups. He discussed Sylow's theorems and solvable groups, the structure of abelian groups, groups of prime power order, and several other topics.

Abstraction and the axiomatic presentation of various domains of mathematics were topical at the start of the twentieth century, promoted by the influence of Hilbert's *Grundlagen der Geometrie* (1899) and a growing feeling that one could get lost in studies that were too detailed. It offered both a consolidation of what had been discovered, and an opportunity to make new discoveries undistracted by accidental features coming from the origins of the subject. A group need not be seen as a permutation group, nor should a field necessarily belong to Galois theory. De Séguier's monograph may not have reshaped group theory, but it a marker that the subject had become autonomous, needing neither techniques nor motivation from other fields, pertinent though they may be.

There are several ways in which the history of an ongoing subject can bid farewell to its subject. In the present case, we can note the arrival of representation theory, specifically character theory, in the work of Frobenius and Burnside, and point to Burnside's $p^a q^b$ theorem: all groups of order $p^a q^b$ are solvable, where p and q are primes. In his introduction to Burnside's *Collected Papers* the group-theorist Walter Feit said that "I would guess that the elegance of both the statement and the proof have attracted more people to the study of characters than any other result in the

---

[2](Hollings 2017) concentrates on American postulation theorists such as Huntington and Moore, whom de Séguier cited.

subject". Put it together with Sylow's theorems, which show that groups of order p^a are solvable, and we have the result that the order of a nonabelian finite simple group must be divisible by at least three primes—as is the case for A_5, which is of order $60 = 2^2 \cdot 3 \cdot 5$.

27.3 Dickson's Classification of Finite Simple Groups

We can also note the publication of Dickson's *Linear Groups with an Exposition of the Galois Field Theory*, 1901. This is a study of matrix groups over finite fields, it having been recently shown that every finite field is of order p^n for some prime p (see Sect. 23.2)—these are what Dickson called Galois fields. Dickson's preface to his book may speak for itself, eloquent as it is not only about the amount of work that had by then been done on this relatively new branch of mathematics, but on how the theory is organised.

> Since the appearance in 1870 of the great work of Camille Jordan on substitutions and their applications, there have been many important additions to the theory of finite groups. The books of Netto, Weber and Burnside have brought up to date the theory of abstract and substitution groups. On the analytic side, the theory of linear groups has received much attention in view of their frequent occurrence in mathematical problems both of theory and of application. The theory of collineation groups will be treated in a forthcoming volume by Loewy. There remains the subject of linear groups in a finite field (including linear congruence groups) having immediate application in many problems of geometry and function-theory and furnishing a natural method for the investigation of extensive classes of important groups. The present volume is intended as an introduction to this subject. While the exposition is restricted to groups in a finite field (endliche Körper), the method of investigation is applicable to groups in an infinite field; corresponding theorems for continuous and collineation groups may often be enunciated without modification of the text.
>
> The earlier chapters of the text are devoted to an elementary exposition of the theory of Galois fields chiefly in their abstract form. The conception of an abstract field is introduced by means of the simplest example, that of the classes of residues with respect to a prime modulus. For any prime number p and positive integer n, there exists one and but one Galois field of order p^n. In view of the theorem of Moore that every finite field may be represented as a Galois field, our investigations acquire complete generality when we take as basis the general Galois field. It was found to be impracticable to attempt to indicate the sources of the individual theorems and conceptions of the theory. Aside from the independent discovery of theorems by different writers and a general lack of reference to earlier papers, the later writers have given wide generalizations of the results of earlier investigators. It will suffice to give the following list of references on Galois fields and higher irreducible congruences:
> Galois, "Sur la théorie des nombres", *Bulletin des sciences mathématiques* de M. Ferussac, 1830; *Journ. de mathématiques*, 1846.
> Schonemann, *Crelle*, vol. 31 (1846), pp. 269–325.
> Dedekind, *Crelle*, vol. 54 (1857), pp. 1–26.
> Serret, *Journ. de math.*, 1873, p. 301, p. 437; *Algèbre supérieure*.
> Jordan, *Traité des substitutions*, pp. 14–18, pp. 156–161.
> Pellet, *Comptes Rendus*, vol. 70, p. 328, vol. 86, p. 1071, vol. 90, p. 1339, vol. 93, p. 1065; *Bull Soc. Math, de France*, vol. 17, p. 156.
> Moore, *Bull. Amer. Math. Soc.*, Dec., 1893; Congress Mathematical Papers.

Dickson, *Bull. Amer. Math. Soc.*, vol. 3, pp. 381–389; vol. 6, pp. 203–204. *Annals of Math.*, vol. 11, pp. 65–120; Chicago Univ. *Record*, 1896, p. 318.

Borel et Drach, *Théorie des nombres et algèbre superiéure*, 1895.

The second part of the book is intended to give an elementary exposition of the more important results concerning linear groups in a Galois field. The linear groups investigated by Galois, Jordan and Serret were defined for the field of integers taken modulo p; the general Galois field enters only incidentally in their investigations. The linear fractional group in a general Galois field was partially investigated by Mathieu, and exhaustively by Moore, Burnside and Wiman. The work of Moore first emphasized the importance of employing in group problems the general Galois field in place of the special field of integers, the results being almost as simple and the investigations no more complicated. In this way the systems of linear groups studied by Jordan have all be [sic] generalized by the author and in the investigation of new systems the Galois field has been employed ab initio.

The method of presentation employed in the text often differs greatly from that of the original papers; the new proofs are believed to be much simpler than the old. For example, the structure of all linear homogeneous groups on six or fewer indices which are defined by a quadratic invariant is determined by setting up their isomorphism with groups of known structure. Then the structure of the corresponding groups on m indices, $m < 6$, follows without the difficult calculations of the published investigations. In view of the importance thus placed upon the isomorphisms holding between various linear groups, the theory of the compounds of a linear group has been developed at length and applied to the question of isomorphisms. Again, it was found practicable to treat together the two (generalized) hypoabelian groups. The identity from the group standpoint of the problem of the trisection of the periods of a hyperelliptic function of four periods and the problem of the determination of the 27 straight lines on a general cubic surface is developed in Chapter XIV by an analysis involving far less calculation than the proof by Jordan.

Chicago, November, 1900.

When the book was reprinted in 1958, the group-theorist and geometer Wilhelm Magnus added a short introduction, in which he praised the book as "a milestone in the development of modern algebra." The Jordan–Hölder Theorem, he went on, posed a fundamental problem in the theory of groups of finite order: Find all the simple groups, especially the non-trivial simple groups whose order is not a prime number. The way forward, Magnus suggested, had been through the study of finite fields (or as they were called later, Galois fields) and here Dickson's book had been especially useful. It was still, he said, the most extensive and thorough presentation of the theory of Galois fields available in the literature, and "Although many of his proofs would be shortened today, and although the same subject would be expounded differently now, the first five chapters of Dickson's book are still very readable, and contain an enormous wealth of both examples and theorems".

Magnus then went on to describe how Dickson, building on the work of such predecessors as Lie, Cartan, Killing, and Engel, had shown in two further papers that—the Mathieu and alternating groups aside—all the known simple groups lie in one of eight infinite systems that arise naturally from the analysis of linear groups with coefficients in a Galois field. He then added

It is a remarkable fact that after these two papers of Dickson's appeared no new simple groups of finite order were discovered for half a century. It was not until 1955 that a paper by C. Chevalley appeared presenting a new infinite system of finite simple groups. In this same paper Chevalley also announced that he had been able to obtain simple linear groups in a Galois field which are the analogue of the exceptional simple continuous groups not considered by Dickson.

Magnus also remarked that "I would only add that in addition to the remarkable enumeration of simple groups Dickson's book testifies to the geometric significance of many of these groups; I suspect that some good original historical work can be done on the geometry of finite groups".

Chapter 28
Emmy Noether

Fig. 28.1 Amalie Emmy Noether (1882–1935). Photographic portrait of Emmy Noether, Photo Archives, Special Collections Department, Bryn Mawr College Library

© Springer Nature Switzerland AG 2018
J. Gray, *A History of Abstract Algebra*, Springer Undergraduate Mathematics Series,
https://doi.org/10.1007/978-3-319-94773-0_28

28.1   Introduction

Emmy Noether is universally regarded as the greatest woman mathematician so far. Here we look at some of her work in the creation of modern algebra (Fig. 28.1).

Emmy Noether's life is reasonably well known, although there are good reasons for thinking that her mathematical contributions have not been properly understood by historians. She was born on 23 March 1882 in Erlangen in Germany, where her father Max Noether was Professor of Mathematics. She studied mathematics herself, taking a PhD under her father's friend Paul Gordan in 1907. The topic was computational invariant theory, in which Gordan was still the world expert, and which had been transformed by the more recent work of Hilbert.

In 1916 she was in Göttingen, where it was also easier to be, because the First World War was devastating the ranks of students and junior members of staff. She was there when Einstein came to discuss difficulties in formulating the general theory of relativity, a topic that was also of great interest to Hilbert at the time, who had a rival theory. Einstein's visit attracted the attention of Klein, and he turned to Emmy to clarify what a theory of conserved quantities and conservation laws might be in general relativity. She did this by re-writing conservation laws as symmetries, so conservation of momentum is a symmetry of the group of translational isometries, conservation of angular momentum a symmetry of the rotation group, and so on.

After the war Hilbert secured her a position, albeit unpaid, after facing down the opposition from members of the Philosophy Faculty, to which the Mathematics Department belonged, with the immortal words "We are not running a bath house".[1] Emmy Noether's attention shifted to abstract algebra, and by the end of the 1920s she had drawn a group of almost equally gifted young mathematicians around her, Emil Artin and Bartel van der Waerden among them, jocularly called the Noether boys. It is they who turned her way of thinking into 'Modern Algebra', while she did fundamental work on commutative and non-commutative algebra.

In 1933 the Nazis came to power in Germany, and set about barring Jews from the civil service and the universities. Emmy Noether was a Jew, and one with well-known left-wing sympathies,[2] but she was able to get a position at the women's university Bryn Mawr, in Pennsylvania, USA. She died there on 14 April 1935 from a post-operative infection.

Because of the three parts of her career there are, one might say, three Emmy Noethers known to history. There is the invariant theory one, and because she affected to despise her PhD work and because invariant theory is very technical, this Noether is often dismissed as a false start. But Colin McLarty is busy giving this part of her career a through examination and showing how deeply it was tied in to contemporary mathematics and how significant its implications were for Emmy Noether's later work, so this part of her reputation may well change in the next few years.

---

[1] She eventually obtained a modest salary.

[2] As was her brother Fritz, a physicist, who left for the Soviet Union where he was eventually murdered in the purges of the 1930s.

Then there is the Noether who turned physicists away from conservation laws and towards symmetries and symmetry groups.[3] This Noether is well-known to physicists who, however, do not always know the first and third ones. In fairness, many mathematicians do not know the second Noether.

Then there is the third Noether, the one presented here. Even this one is perhaps misunderstood. I shall give the traditional version, in which Noether's work on commutative rings is described, and her later work on non-commutative rings and number theory is barely hinted at. But McLarty has been arguing that Noether found maps between objects just as important, thus giving her work a category-theoretic spin.

And indeed there is a fourth Noether, Noether the great woman mathematician. There may yet be a fifth, because McLarty has been arguing for some time that too much has been taken over from the reminiscences and obituaries of two mathematicians who knew her well: the Soviet mathematician and topologist Pavel Alexandrov and Hermann Weyl, who many regard as Hilbert's true successor. They down-played her politics, and he has been arguing that it interestingly of a piece with her mathematics. For another integrative account of Emmy Noether's life and work, see (Koreuber 2015).

Influences

I have indicated that the influence of her father and of Paul Gordan on the young Emmy Noether is being re-examined. There can be no doubt that Hilbert's work on invariant theory and later number theory influenced her, because he was the dominant algebraist of his time. We should also note that the first abstract presentation of a ring was given in (Fraenkel 1914), which was based on the author's doctoral dissertation. The paper deliberately resembles Steinitz's account of fields, but, as its title indicates the paper is more concerned with the factorisation of zero divisors than the abstract properties of a ring per se. In it a ring is defined as what we would call a not necessarily commutative ring with a multiplicative unit.[4]

But the person she always singled out as the thinker it was worth coming back to was Dedekind: "*Es steht alles schon bei Dedekind*" ("Everything is already in Dedekind") she would say. There is a clear sense in which this is true. She worked on algebra, on rings and ideals in particular, leaning more to number theory than algebraic geometry, and very obviously chose Dedekind's approach over Kronecker's. But this should not blind us to a deeper commonality. Like Dedekind she looked hard for the right concepts around which to organise a theory. She wanted to know what was the deep part of an analogy that would take work forward, what concepts should guide the calculations. This is evident in one her most famous papers, to which we now turn.

---

[3]For this Noether, in an account written by someone who does know all three Noethers, see (Kosman-Schwarzbach 2011).

[4]Fraenkel added two further definitions to assist him in obtaining factorisation theorems; see (Corry 1996, p. 210).

28.2   Ideal Theory in Ring Domains

Emmy Noether began her paper *Idealtheorie in Ringbereichen* (Ideal theory in ring domains) by saying that it carried over of the factorisation theorem for rational integers and ideals in algebraic number fields to arbitrary integral domains and more general ring domains.[5]

She recalled the decomposition of an integer into products of prime powers,

$$a = p_1^{e_1} p_2^{e_2} \ldots p_s^{e_s} = q_1 q_2 \ldots q_s, \ q_j = p_j^{e_j}, j = 1, \ldots s,$$

and noted four facts about the prime powers, which she called the components of the decomposition:

1. No two have a common divisor, and no q is a product of numbers with no common divisor, so they are irreducible in that sense;
2. Any two components, q_j and q_k say, are relatively prime; that is, if q_k divides bq_j then b is divisible by q_k;
3. Every q_j is primary; that is if a product bc is divisible by some q then either b is divisible by q or some power of c is;
4. Every q is irreducible in the sense that it is not representable as the least common multiple of two proper divisors.

(We shall comment on her language below.)

She went on to remark that when the factorisation theorem for rational integers is stated in terms of primary factors it says that any primary decomposition has the same number of components, the same primes, and the same exponents. Formulated in terms of ideals one can even drop the remark about the ambiguous sign.

She now proposed to generalise this theorem to rings all of whose ideals have finite bases. This condition ensures that there are irreducible and prime ideals, which are lacking, she observed, in the ring of all algebraic integers, where there is no factorisation into prime ideals.

She proposed to show that the four facts mentioned above generalise to four versions of factorisation theory. The first fact leads to the idea of factorisation into divisor-free irreducible ideals, the remaining three generalise the idea of least common multiple. The first two (of the four) lead to a unique factorisation theorem, the third and fourth must allow for some components called isolated ideals.

Noether regarded the ring of polynomials in n variables with complex coefficients as the simplest ring that admitted all four versions. The theory of factorisation of these polynomial domains into maximal primary ideals had, she said, been worked out by Lasker and refined by Macaulay. But here the theory of irreducible ideals and relatively prime irreducible ideals still remained to be studied, although Macaulay had ventured some remarks on when the isolated primary components are

---

[5]There is an English translation of this paper by Daniel Berlyne, who took this course for pleasure in 2013–14; see arXiv:1401.2577 [math.RA].

unique. The theory of divisor-free irreducible ideals had been studied by Schmeidler, who had used the existence of a finite basis for these ideals. Here Noether promised a significant generalisation and consequent simplification of the theory.

In §1 of her paper Noether defined a ring R as a system of elements that one can add and multiply, and subtract, that obeys the associative and commutative laws for addition, the associative law for multiplication and the commutative law for multiplication, and the distributive law. She noted that she did not require that there be a multiplicative unit, and she did not exclude zero divisors.[6] A ring with no zero divisors was, she said, called an *integral domain*.

She defined an *ideal* I as a subset closed under subtraction and multiplication by arbitrary elements of the ring: $a, b \in I \Rightarrow a - b \in I$; $a \in I, r \in R \Rightarrow ar \in I$. She said an ideal J is *divisible* by an ideal I if J was a subset of I. This makes sense: in the ring $\mathbb{Z}$ the ideal (6) is divisible by the ideal (3) because, indeed, the integer 6 is divisible by the integer 3. Be warned: the ideal doing the dividing is the bigger ideal!

She restricted her attention to those rings for which every ideal has a finite basis, and observed that if $I, I_1, I_2, \ldots$ is a chain of ideals each of which is divisible by the next one—notice that this gives us an ascending chain of ideals—then the chain is finite: there is an integer n such that $j \geq n \Rightarrow I_j = I_n$. She also noted the converse: if every ascending chain of ideals is finite then every ideal has a finite basis.[7]

The extent to which her paper reworks Lasker's gradually becomes clear. In §2 of her paper Noether showed that every ideal can be represented as the least common multiple (lcm) of finitely many irreducible ideals. She defined the lcm of the ideals $I_1, \ldots, I_m$ as the set of elements common to all ideals, that is, as their intersection $\bigcap_j I_j$. (So, just as divisors get bigger, multiples get smaller!) She observed that one can always assume that when an ideal is presented as an lcm it is the lcm of ideals none of which divide any of the others.

Now she set about obtaining a decomposition of an ideal into irreducibles. She defined an ideal to be irreducible if it cannot be presented as the lcm of two other ideals (other than trivially, when the ideals are the same). Then she proved that every ideal is an lcm of finitely many irreducible ideals. When this is done by successive division (i.e., successive intersection) that reduces the ideal every time, the decomposition is said to be the *shortest*. In §3 she showed that the number of components in two such decompositions is always the same provided they are as short as possible—I gloss over this point.

In §4 Noether came to primary ideals. An ideal I is *primary* if $ab \in I, a \notin I \Rightarrow b^n \in I$ for some integer n. If the integer $n = 1$ in every case the ideal is said to be *prime*. The significant theorem here is this one:
If Q is a primary ideal there is a unique prime ideal P that is a divisor of Q and such that Q is a divisor of some power of P. This is easier to understand in the

---

[6] A ring R has zero divisors $a, b \in R$ if $a \neq 0, b \neq 0, ab = 0$.

[7] As Corry points out (1996, p. 239), Noether's use of the ascending chain condition is much more explicit and central in her paper (1926), which unfortunately I cannot consider here.

language of subsets: we have $P \supseteq Q \supseteq P^n$ for some integer n. Recall that $P \supseteq P^2 \supseteq \ldots P^n \supseteq \ldots$, but note that there need be no value of the integer n such that $Q = P^n$.

Noether showed that irreducible ideals are primary. Moreover, two shortest representations of an ideal as an lcm of irreducible ideals give rise to the same primary ideals.

In §5 she showed that when an ideal is represented as the lcm of maximal primary ideals the number of components and the number of primary ideals are unique.

The paper continued, but I shall leave it here, noting that she studied the decomposition into relatively prime irreducible ideals, the possible uniqueness of the so-called isolated components (which I have not yet defined), and quite a bit more. Nor, indeed, was this Noether's only paper on the subject, but it is perhaps enough for present purposes.

28.3 Structural Thinking

One way to measure Emmy Noether's achievements in abstract algebra is simply to pick up any advanced undergraduate or graduate text book in the subject. Mathematicians simply do ring theory her way. They may stipulate that a ring has a multiplicative unit, but that's the only change they make. That is why rings that satisfy an ascending chain condition are called Noetherian rings, in her honour. They form a large class of rings that very neatly and naturally captures what is happening in the rings of interest in number theory and algebraic geometry. Of course much has happened since 1935, but homological algebra, category theory, and even non-commutative algebra grow out of her work – they do not replace it or diminish it.

It is a rare thing to redefine a field so completely. It means that two or three generations of mathematicians have found the topic of rings important and her way of formulating the theory helpful. A detailed study of how this came about would take us too far afield, but one aspect can be singled out. It is known as the structural approach.

To over-simplify, Emmy Noether directed attention to a class of rings large enough to include most of the interesting known examples, and she provided a way of thinking about them which seldom said anything about properties, still less the forms, of typical elements. Everything was kept at the level of subsets. She was explicit about this, and people liked the simplifications and clarity it brought to the subject. Arguably, it created the subject of abstract (commutative, Noetherian) rings.

One way to see that this is an over-simplification is to see that it over-estimates the set-theoretic way of thinking. If you look back at her work she did not talk about subsets, but divisors, not about intersections but least common multiples. She could have done, the language had been available since the time of Cantor, and Krull, another 'Noether boy', was to introduce it, but she kept it because the links to number theory were important to her. Her work on non-commutative algebra and number theory, which is much harder, demonstrates her priorities.

Nonetheless, her work eclipsed a whole school of computational algebraists for as much as two generations. She had a high opinion of the work of Lasker and Macaulay, with its heavy use of explicit polynomials and the methods of elimination theory (resultants, resolvents, ...) but the Noether boys did not; that kind of mathematics became deeply unfashionable and people who worked in the area felt unfairly neglected. Modern computing has restored the computational side somewhat, but no-one would deny that the conceptual framework of modern algebra was laid down by Noether.

It is worth concluding by observing that neither she, nor Dedekind before her, nor Hilbert, took up abstract mathematics before they felt they had to. The ideas of modern abstract algebra did not fall from the sky, they did not seduce mathematicians away from good honest work. They grew from the ground, and they accomplish things. Nowhere is this clearer than in the very concept of a ring. The ring of integers is the paradigm mathematical object; formulating geometry and number theory in the language of rings is currently a massive mathematical operation, and Noether's work is a turning point in that endeavour.

Chapter 29
From Weber to van der Waerden

29.1 Introduction

This section is a commentary on the opening chapter of Corry's *Modern algebra and the rise of mathematical structures*, and more particularly his paper 'From *Algebra* (1895) to *Moderne Algebra* (1930): Changing Conceptions of a Discipline. A Guided Tour Using the *Jahrbuch über die Fortschritte der Mathematik*', in (Gray and Parshall (eds.) 2006, pp. 221–243).[1]

The story that Corry traces through the pages of the abstracting and reviewing journal of the time start with the books of Serret, Jordan, and especially Weber on algebra, and ends with the book that gave modern algebra its name: van der Waerden's *Moderne Algebra* (first edition 1930). Corry calls van der Waerden's approach 'structural': his book presents groups, ideals, rings, and fields (as many a textbook has done since). Various other objects are relegated to the status of mere examples (the rational and the real numbers, for example). The once fundamental theorem of algebra ceases to be fundamental. Key concepts are introduced axiomatically.

This gets particularly interesting when Corry points to texts written between Weber's and van der Waerden's: one by Dickson, one by Hasse (both major number theorists) one by the by-then elderly Fricke and another by Otto Haupt. These books offer much the same selection of topics (what Corry, following Elkana, calls the body of knowledge) but in a way that reflects old-fashioned priorities (the image of the knowledge, Corry's term for the things one says about mathematics).

To trace the evolution of the body and the change in the image, Corry uses the *Jahrbuch*, both for the wealth of information it contains and because its structure (division into major topics and sub-topics) displays a shifting image of mathematics. So he contrasts Weber's major paper of 1893 on Galois theory with his three-volume *Lehrbuch der Algebra* to show that Weber had a more modern image in that paper

---

[1] See http://www.tau.ac.il/~corry/publications/articles/jahrbuch.html for the final draft of the paper.

© Springer Nature Switzerland AG 2018   297
J. Gray, *A History of Abstract Algebra*, Springer Undergraduate Mathematics Series,
https://doi.org/10.1007/978-3-319-94773-0_29

than he chose to give the subject in his textbook. Corry next shows that while group theory was a recognised part of algebra by 1900, it was presented as a technique subordinate to the study of substitutions and determinants, and Galois theory is part of the theory of equations. At this stage the classifications of the *Jahrbuch* are close to those that Weber used in his textbook. Algebra has at its core the solution of polynomial equations in one or several variables.

Quaternions and vector analysis make their appearance in a new classification scheme in 1905. The background here is an international investigation into the merit of vectors and vector methods in teaching and research (and a resolution of a controversy as to the related role of quaternions). More importantly for us, what is called 'universal algebra' appears, which was the name for such things as Wedderburn's investigations of what we call 'algebras'. But another major paper, Lasker's on modular systems, was forced into the section on the theory of forms— not wrong, exactly, but not comfortable.

Corry moves on to note that Steinitz's paper on the abstract theory of fields, which was published in 1910 and is seminal in its importance for the subject and for structural algebra generally, did not even make it to the algebra pages of the *Jahrbuch* but appeared under the rubric of 'Function Theory: General'. Indeed, he pointed out that, strange though this is, the paper could hardly have gone anywhere else at the time.

I pass over Corry's discussion of a different classification used by American mathematicians at the time, which largely confirms his analysis, and come to the next major change in the *Jahrbuch*, the one in 1916. For the first time, group theory and the abstract theory of fields and modules has a place of its of own, and the theory of rings was reviewed in due course in this section. But Galois theory was still confined within the theory of equations; only in 1925 did group theory finally make it to algebra.

Finally, what of van der Waerden's *Moderne Algebra*? It was reviewed under the heading of Algebra and Arithmetic, which was appropriate, alongside Hasse's essay on modern algebraic methods and the re-edition of Steinitz's monograph on the abstract theory of fields.

From Gauss and Galois to van der Waerden

Does Corry's account of the slow, undirected but concept-driven emergence of modern algebra from what might be called the Weberian view extend backwards? Does it offer a good way of looking at developments from, say, the 1830s to the 1890s? I shall leave this as an issue for you to address, but some methodological lessons can be drawn from his paper.

Corry's analysis reminds us that many significant changes happen only slowly, or force themselves on everyone's attention only after a long period under the surface. There are many reasons why this might be, or have been, the case.

One is inertia: there are many papers in the sections of the *Jahrbuch* relevant to the emergence of modern algebra that are not discussed because they did not promote the changes we (and Corry) are interested in. These papers are of varying merit: they may all have new things to say but not all of them have equal claims on everyone's attention. Many of them do not contest the body or the image of algebra at the time they were written. The mathematicians who wrote them have to be brought along, and their contributions fairly reflected in the *Jahrbuch*. If one supposes that the leading mathematicians of the day teach the new material and the new approaches, it will take time for their weaker but nonetheless productive students to start writing the papers that will, in their turn, be conventional for their day. Indeed, in 1900 quite a few authors were professional engineers or school teachers but only avocational mathematicians; it is seldom the case that these people write a truly original paper (Macaulay was one such).

Another, perhaps more significant, reason might be called complexity, or the emergence of a need for a division of labour. There will often be major mathematicians who do not share what historians later find to be an emerging orthodoxy. There is a curious shift at work here. What, after all, should one say is the major purpose of mathematics? A perfectly good answer is that it solves problems. Consider the analogy with work in science. Even a hazy acquaintance with modern science shows that quite massive and complicated machinery is needed to make the discoveries (the genome project, the large hadron collider, ...) and consequently there is a large collection of people whose job it is to understand this or that aspect of the machinery—not the genome or hadrons, particularly. So is the real job of the mathematician to solve equations, a task that, to be sure, would involve group theory. Or is it to build a 'machine', such as group theory, remembering that an application might be to the solution of polynomial equations?

This is a perfectly responsible question of the kind that every mathematician designing a syllabus or planning the course of their own career has to confront. There may well be no unique answer. Historians of mathematics can observe that different answers are given at different times for different reasons. For example, van der Waerden wrote his *Moderne Algebra* as a young man who had recently joined the group around Emmy Noether. She was a profound and forceful mathematician who gathered around her a number of gifted young mathematicians—Emil Artin was another—and who, by her vision of how to do mathematics, was able to bring along a number of other powerful mathematicians such as Hasse. She was also well established in Göttingen, which had resumed its place as the leading centre for mathematics quite quickly after the First World War, and this gave her extra influence. Noether and the 'Noether boys', as they were known, were in an ideal place as a group to make explicit the rise of structural algebra and the relocation (and relative demotion) of problem solving.

We have come to a third factor, which might be called taste, or priorities. One might say that certain new problems were now the 'right' sort of problems to tackle. Noether promoted a vision of mathematics in which half the task is to get the context right in which a concrete problem can be best posed, the other half then being to solve the problem. When context building is so elaborate it can be easy to forget that there are old-fashioned problems too; people who know Noether's work well say that she did not forget.

A taste for the abstract singles out certain key players in the story of the arrival of modern algebra. Whatever *modern* mathematics is taken to mean—other than anything and everything happening recently—it means abstract and structural. This is the most obvious feature of the Noether approach. So we can look for abstract, structural aspects in, for example, the theory of groups. We would expect their introduction to be a new phase in the study and use of groups, one that would happen for identifiable reasons and one that would not carry everyone along. Dickson, for example, evidently thought that abstract finite groups are interesting, and that it would be valuable to know all the infinite families of finite simple groups. This is a very hard problem, but not so big that it would have to be divided up among members of a team. The historian sees that it was done, that it has roots going back at least to Jordan, and that Jordan was in turn responding to the work of Galois. The historian notes that a few people responded to Galois's work, Jordan among them, and that these people also provoked others. By the time Dickson set himself this task there was quite a body of knowledge about group theory, quite a few facts known and techniques developed for him to use, and quite a few people who could read his work and comment intelligently on it. Within the community of mathematicians gathered at Chicago, Dickson probably did not have to defend his decision to do group theory at all: there were quite a few group theorists about, and a problem within group theory now seemed a worthwhile thing to do, one that Dickson but not many others could be expected to tackle successfully. Of course, it helped that it had an ancestry in Galois theory: this gave it an importance that not every problem can have.

Jordan, on the other hand, as the very title of his book makes clear, was doing two things, both of them different from Dickson: solving polynomial equations and developing the theory of substitution groups. One curious question here when abstraction is at stake is the status of the concept of a quotient group. Historians disagree over whether Jordan possessed the concept, or whether it came in only with the rise of abstraction and the work of people like Otto Hölder. It has been argued, for example in (Schlimm 2008), that if not in 1870 then by 1873 Jordan could have proved the full Jordan–Hölder theorem. That he did not is because, it is suggested, he saw no need for a fully abstract concept and therefore had no interest in regarding the composition factors as stand-ins for quotient groups.

Yet another factor might be that, like greatness, some are born abstract, some achieve it, and some have it thrust upon them. Abstraction is 'achieved', on this metaphor, as mathematicians responded to the suggestion of Jordan (and its echo in the work of Cayley) that it might be worthwhile finding all (finite) groups. Jordan, as we saw, set out many of the tools needed to get started, and identified and partially classified interesting classes of groups.

Wussing's argument is that abstract group theory was forced upon mathematicians inasmuch as it emerged from the coalescing of three strands of work: permutation theory, implicitly group-theoretic thinking in geometry, and number theory. If it is the case that geometrical transformations can usefully be thought of as forming groups, it is nonetheless far from obvious what these infinite groups permute. The common features of groups in the two settings are only recognisable when one sheds the permutation-theoretic clothing and sees the transformations underneath.

The number-theoretic strand that Wussing considered leads to the other strand of modern algebra: rings and fields. The move to stretch the concept of 'integer' as far as it will go is in tension with the classical (should one say 'core'?) concern of number theory: to solve polynomial equations in integers or rational numbers, otherwise known as Diophantine analysis. Gauss's cyclotomic integers are a generalisation of the Gaussian integers that certainly play a number of roles in Diophantine analysis. Kummer's ideal numbers, on the other hand, are an abstraction, brought in to improve the theory, the 'machine'. The largely German version of Galois theory replaced adjunction of roots with extension of fields, and within an extension of the field of rational numbers the mathematicians looked for the corresponding algebraic integers.

The extent to which embracing this programme is staying with or moving away from Diophantine analysis differs from the question of whether Galois theory is or is not helping anyone to solve polynomial equations, because one might argue that once a given polynomial equation cannot be solved by radicals it has to be handed over to the analysts for further treatment and so ceases to be interesting to an algebraist, whereas a number theorist cannot abandon a Diophantine equation.

In this respect it is interesting to note that the 10th of Hilbert's famous list of Mathematical Problems, presented at the International Congress of Mathematicians in Paris in 1900 was:

> Given a Diophantine equation with any number of unknown quantities and with rational integral numerical coefficients: To devise a process according to which it can be determined in a finite number of operations whether the equation is solvable in rational integers.

In the end it was shown in the 1960s by Julia Robinson in America and Yuri Matijasevich in the Soviet Union that this problem cannot be solved. Hilbert would have been quite happy with an algorithm which would correctly say of any Diophantine equation whether it had a solution; in particular, exactly in analogy with Galois theory, he would have accepted a negative answer. What Robinson and Matijasevich showed was that there is no algorithm of this kind.

There are several strands to the history of number theory. Here I have concentrated on the emergence of a concept of algebraic integer and attempts to generalise the theory of prime factorisation to the new setting, concentrating on the theory of binary quadratic forms and quadratic integers. This topic mixes with the wider topic of polynomial rings over a field, which interests geometers, and with the abstract theory of fields. There are other strands, which I have not attempted to cover: there are intimate connections with the theory of elliptic functions, there are other heavily analytical approaches (for example, associated with the Riemann zeta function), tests for prime numbers and searches for prime factors, and there are Diophantine equations at every level of difficulty from the merely recreational to the currently impossible. The strand I have followed leads with seeming inevitability from Kummer to Kronecker, Kronecker to Dedekind and Weber, from them to Hilbert, and from him to Emmy Noether, who brought in the geometrical side and created the essentials of the modern theory of commutative rings.

Even this is a contrived story, inasmuch as Hilbert played down a major aspect that went back to Kummer and was revived after Hilbert's major work on number theory was done. A 'full' story, whatever that might be, would have to take on board all the strands omitted here. But even the edited story displays some of the features of the earlier story about the emergence of abstract group theory. There is a small group of people who want to bring about abstraction, others who oppose it, yet others who have an eclectic attitude to it, and quite probably a majority of so-called minor figures who were content to ignore this work.

29.2   van der Waerden on the Origins of *Moderne Algebra*

In 1975, prompted by the interest of Garrett Birkhoff, van der Waerden published a short paper entitled 'On the sources of my book *Moderne Algebra*'. However much forty tumultuous years may have affected van der Waerden's memories for events and influences, the comments are still very interesting.

Under Emmy Noether's famous motto "Everything is already in Dedekind", he explained that he had studied mathematics and physics at Amsterdam from 1919 to 1924, supplementing his study of classical algebra by reading about Galois theory and other subjects in Weber's "admirable" *Lehrbuch der Algebra*. He had also begun to ask questions in what was always his favourite subject, algebraic geometry. However

> When I came to Göttingen in 1924, a new world opened up before me. I learned from Emmy Noether that the tools by which my questions could be handled had already been developed by Dedekind and Weber, by Hilbert, Lasker and Macaulay, by Steinitz and by Emmy Noether herself.

Taking advantage of the possibility of borrowing books from the mathematical library at Göttingen (unusual in those days) he first caught up with (Steinitz 1910). Chapter 5 of *Moderne Algebra*, he tells us, essentially followed Steinitz, but the proof of the existence of a primitive element "is due to Galois. It was Emmy Noether who drew my attention to this proof".

Group theory van der Waerden said he learned from Emmy Noether's lectures and later conversations with Artin and Schreier in Hamburg, and he wrote little about the topic in his own book because of the excellent textbooks (Speiser 1923) and (Burnside 1911).

As for the theory of ideals

> When I came to Göttingen one of my main problems was the generalization of Max Noether's "fundamental theorem" $F = Af + B\phi$ to n dimensions. The conditions F has to satisfy are "local conditions" in the neighbourhood of the single points of intersection of the curves $f = 0$ and $\phi = 0$. If P is a point of intersection, the local conditions define a "primary ideal" Q, and the original ideal $M = (f, \phi)$ is the intersection of these "primary ideals". The terminology is modern, but the ideas are those of Max Noether and Bertini.

It seems, he went on, that Hilbert was the first to realize that an n-dimensional generalization of Noether's theorem would be desirable (this is incorrect, as the work of König shows).

> Emmanuel Lasker, the chess champion, who took his Ph.D. degree under Hilbert's guidance in 1905, was the first to solve this problem. He proved that, quite generally, every polynomial ideal $(f_1, \ldots, f_r)$ is an intersection of primary ideals.

> In her 1921 paper "Idealtheorie in Ringbereichen" [...] Emmy Noether generalized Lasker's theorem to arbitrary commutative rings satisfying an 'ascending chain condition'. Chapter 12 of my book, Allgemeine Idealtheorie der kommutativen Ringe, is based on this paper of Emmy Noether. The proof of Hilbert's Finite Basis Theorem in §80 is due to Artin; he presented it in a seminar lecture at Hamburg in 1926. The ascending chain condition is very weak; it is satisfied in all polynomial domains over any field and in many other cases. If stronger assumptions are made concerning the ring, one can even prove that the primary ideals are powers of prime ideals and that every ideal is a product of prime ideals. In Emmy Noether's paper [...] (1926), five axioms were formulated which ensure that every ideal is a product of prime ideals. Rings satisfying these axioms are now called 'Dedekind Rings'. In these rings Dedekind's theory of ideals in algebraic number fields and fields of algebraic functions of one variable is valid.

In a brief indication of how inter-active the group around Emmy Noether was, van der Waerden then went on

> The theory of Dedekind fields was presented in Chapter 14 of my book. Emmy Noether's proofs were simplified, making use of an idea of W. Krull contained in §3 of Krull's paper [...] (1927). Emmy Noether was a referee for this paper, and she told Artin about it. Artin simplified Krull's proof and presented it in a seminar in Hamburg, in which I participated. Artin's simplified version of Krull's proof was reproduced in §100.

Chapters 11 and 13 on elimination theory and polynomial ideals, we learn, drew on Macaulay's Cambridge Tract *Modular systems*, 1916, and so "are based on the work of the school of Kronecker" while also being "closely connected with Emmy Noether's work". In Chap. 15, on linear algebra, van der Waerden used Châtelet's *Leçons sur la Théorie des Nombres* (1913) on the advice again of Emmy Noether.

In 1926, having obtained his PhD, he used a Rockefeller Fellowship to study with Hecke, Artin, and Schreier in Hamburg. He met Artin and Schreier nearly every day, and followed Artin's "marvellous" lectures on algebra.[2] Artin already had a contract with Springer for a book on the subject, but first he and van der Waerden decided to write it together on the basis of the notes van der Waerden was taking, and then Artin apparently professed himself so pleased with the notes that he suggested van der Waerden write the book on his own. So was born *Moderne Algebra*.

---

[2]Corry has pointed out to me that Artin and Schreier's work at the time on a theory of purely real fields would have shown van der Waerden the extent to which the real numbers can be made a specific example of a structure that is hierarchically built up in a purely abstract and algebraic manner, thus removing them from their hitherto privileged place in most textbooks.

Chapter 30
Revision and Final Assignment

The final chapter is devoted to revising the topics and a discussion of the final assignment.

The opening two chapters of the last part of the book point towards the mathematics of the early twentieth century, and begin to raise the question of what was significantly different about it. Chapter 22 offers a look at what algebra was taken to be in the hands of one of its leading exponents around 1900 and Chap. 23 looks at the origins of the modern concept of a field.

Chapter 24 offered an introduction to ideas that later Emmy Noether was to find important, if also to rewrite. The original idea is that if two plane algebraic curves, with equations $f(x, y) = 0$ and $g(x, y) = 0$, meet in a finite number of points then any curve through those points should lie in the ideal defined by the polynomials $f(x, y)$ and $g(x, y)$ in the ring $\mathbb{C}[x, y]$. Suitably refined, this turns out to be true, and the refinements are a nice exercise in projective geometry. But the theorem does not generalise to higher dimensions, and that means that the ideal theory of polynomial rings is not exactly that of number rings. The mathematicians who first sorted this out are Lasker and Macaulay, and the solution is the introduction of the concept of a primary ideal. On the way we meet ideas that Hilbert was later to generalise in his Basis Theorem.

Chapter 25 introduces Hilbert, who came to dominate mathematics around the start of the twentieth century. His invariant theory came as a shock.[1] What was the new viewpoint from which it all seemed easy (or, according to Gordan, "Not mathematics but theology".)? By the way, how scrupulously precise was it?

Hilbert went on to do many things in his *Zahlbericht*, not all of which we could discuss in Chap. 26. In particular, his account of the work of Kummer and Kronecker, which he seems to have found uncongenial, has been omitted entirely from this book. But one of the things Hilbert did do was to bring number theory and

---

[1] It is a generalisation of much that had gone before; $b^2 - 4ac$ has very little to do with it.

© Springer Nature Switzerland AG 2018
J. Gray, *A History of Abstract Algebra*, Springer Undergraduate Mathematics Series,
https://doi.org/10.1007/978-3-319-94773-0_30

Galois theory together, and another was to sloganise. How important was this for the development of modern algebra?

Chapter 27 is a story about the rise of group theory. It is selective, and potentially misleading, but enough to indicate that at some stage, and notwithstanding all its rich connections to other branches of mathematics, group theory became a subject in its own right, with its own priorities. Very little could be said about its particular methods, in particular the search for groups of particular orders (p^2, pqr, etc.), Sylow's theorems, the representation theories of Burnside and Frobenius, and Burnside's magnificent theorem that groups of order $p^a q^b$ are solvable. You should consult the resources section for some remarks about groups and geometry in the late nineteenth century.

Chapter 28 tells a similar story about the rise of ring theory, concentrating on the first of Emmy Noether's contributions. It would be perverse to claim that what she did was merely a continuation of what had gone on before, but to what extent and in what ways was it significantly new?

Chapter 29 follows this up by taking a look at the change from Hilbert's idea of algebra to van der Waerden's. Leo Corry's discussion is very helpful here.

Between them, these chapters look at several different aspects of algebra and allow us to study the emergence of the modern concepts of group, ring, and field. They are framed by Hilbert's influential account and van der Waerden's even more influential replacement, in an analysis by the historian Leo Corry.

The final assessment counted for 60% of the marks. Over the years I set several different final essays, trying to push the students to use their reflections on what they had learned in 3 years of algebra, all of it very different from anything they had done at school, while constraining them to address a historical question.

The first two questions I used were straight-forward:

- Describe in detail at least three mathematical advances drawn from this book (preferably from the list on page xv). On the basis of these examples, and any others you see fit to introduce, consider the claim that during the 'long' nineteenth century (1789–1914) algebra became modern algebra for good reasons. What do you take this claim to mean and how valid do you take it to be?
- Describe three key advances that contributed to the creation of modern algebra, making clear what was important about them. To what extent do you think modern algebra was created by accident or by design? Note that your three examples should be well spaced throughout the subject and the period.

But it proved more interesting to ask the students to respond to two more striking opinions.

- Emmy Noether once said that "Everything is already in Dedekind". As regards modern algebra, what do you think she could have meant by that, and how correct do you think she was?

- Consider the following extract from Alfred North Whitehead, *Process and Reality*, 1929 (Free Press ed. 1979, 39):

> ... So far as concerns philosophy only a selected group can be explicitly mentioned. There is no point in endeavouring to force the interpretations of divergent philosophers into a vague agreement. What is important is that the scheme of interpretation here adopted can claim for each of its main positions the express authority of one, or the other, of some supreme master of thought – Plato, Aristotle, Descartes, Locke, Hume, Kant. But ultimately nothing rests on authority; the final court of appeal is intrinsic reasonableness.
>
> *The safest general characterization of the European philosophical tradition is that it consists of a series of footnotes to Plato.* I do not mean the systematic scheme of thought which scholars have doubtfully extracted from his writings. I allude to the wealth of general ideas scattered through them. His personal endowments, his wide opportunities for experience at a great period of civilization, his inheritance of an intellectual tradition not yet stiffened by excessive systematization, have made his writing an inexhaustible mine of suggestion.

I would have added the name of Leibniz to the list, but in any case the sentence I have italicised has become famous. I then asked:

How reasonable would it be to say that modern algebra consists of a series of footnotes to Gauss?

For the Noether question I offered the following advice.

This famous quotations is one of those aphorisms that makes you think a variety of things at once, such as: It's a brilliant insight; it's plain wrong; it's true in these ways but not in those; it would have seemed more true then than now; and even 'Huh?'. Spend some time sorting these and other responses out in your mind before you start to write. When you do:

- Start by making a list of what you take to be the main sources or examples of modern algebra in the nineteenth century, and then select three that you will use to assess her remark.
- Look for moments when the nature of the subject changed, or the criteria for its importance shifted.
- Who else might she have considered, and how different would her answer have been then?
- Who was she possibly ignoring; why? Do remember her actual opinions.
- You are not being asked to write about Noether's own work, but before writing your essay you should assume that she thought that she too had important things to say; don't write an essay that leaves no room for work after Dedekind.

I varied this advice slightly for the Whitehead question. I asked them to think about why Whitehead, an eminent figure in his day, might have thought it worth while making a remark of this kind, given that it is plainly intended to be provocative, and think about how he interpreted it so that it becomes more plausible. I asked them to consider if, as applied to modern algebra, the aphorism might have seemed more (or less) true in 1900 or 1930 than today, and if it was more likely to be believed by some people than others (and if so, who?). More specifically:

- it might seem clear that for present purposes Dirichlet was – and saw himself as – a footnote to Gauss. But Dedekind and Weber? Emmy Noether? Artin and van der Waerden?
- when, if at all, does this claim lose its force?
- or would you prefer to argue that the claim is plain silly, and simply false to the history as you see it.
- look for moments when the nature of the subject changed, or the criteria for its importance shifted.
- which other mathematicians might have been so inspirational? The obvious alternative is Galois, but don't be afraid to look widely. Be prepared to argue well, and to give other people their due, and remember that a footnote (in Whitehead's sense) can be as good as the original, or better.

For each essay I also recommended that

- the essay should show how and to what extent judgements of this kind are grounded in the mathematics;
- wealth of mathematical detail is not necessary, but mathematical accuracy is;
- the response is grounded in evidence presented in your essay;
- the essay should not end up with a final paragraph full of superlatives that don't stand up to scrutiny.

Finally, I urged that all manner of conclusions are OK—if well argued! These questions do not have a right answer; enough exciting things have happened in a century of mathematics (your chosen subject at university, after all) to make the question worth discussing. I asked students to draw on their years as mathematics students to offer an informed opinion about the growth of modern algebra.

The best essays would put forward an argument that is convincing, supported with relevant facts, and well organised. The judgements it reaches are, when necessary, subtle and balanced. The coverage should be broad, with no potentially damaging omissions, and no unnecessary digressions. Ideally, and without being cranky, the essay would be original in its emphases, or its conclusions, and will make good use of quotations. Accuracy is important: the historical facts must be correct and to the point, the mathematics correct, clear, and relevant.

Appendix A
Polynomial Equations in the Eighteenth Century

A.1 Introduction

We look first at attempts in the eighteenth century to formulate and then prove a theorem about the roots of polynomial equations. Then we look at Gauss's criticisms of these attempts, and at one of his own, for what they can tell us about what was (perhaps tacitly) understood in algebra at the time and what was demanded of a proof.

A.2 The Fundamental Theorem of Algebra Before Gauss

As the calculus became established in the early eighteenth century, people became interested in evaluating integrals, among them

$$\int \frac{dx}{x+a}, \int \frac{dx}{x^2+a^2}, \int \frac{dx}{x^2-a^2}, \int \frac{(x+a)dx}{x^2+bx+c}.$$

This led naturally to the question of evaluating integrals of the form

$$\int \frac{P(x)dx}{Q(x)},$$

where P and Q are polynomials. If the denominator, Q, could be factored into linear and quadratic factors the question would reduce to one already solved. The existence of such a factorisation is one of the forms of the fundamental theorem of algebra:

Theorem. *The Fundamental Theorem of Algebra: A polynomial of degree n can always be factored into n linear factors, possibly with repetitions.*

© Springer Nature Switzerland AG 2018
J. Gray, *A History of Abstract Algebra*, Springer Undergraduate Mathematics Series,
https://doi.org/10.1007/978-3-319-94773-0

I have made the above statement deliberately vague, and not stated whether the coefficients are real or complex or whether the factors are to be real or complex. Eighteenth century mathematicians of the period were not sure what to make of complex numbers, and always sought real answers, so they often spoke of proving that every polynomial can be factored into terms of degree one or two.[1]

Some mathematicians disputed that the theorem was even true, as apparently Leibniz had in the seventeenth century. It was well known, of course, that there was a formula expressing the roots of a quadratic equation in terms of the coefficients. The equation

$$ax^2 + bx + c = 0$$

has roots

$$x = \frac{-b \pm \sqrt{b^2 - 4ac}}{2a},$$

—that, or an equivalent result, had been known for over 3000 years by then. It was also known that in the sixteenth century Tartaglia and Cardano had solved the general cubic equation and given the result as a formula involving both cube roots and square roots. The paradoxical result that when the cubic equation has three real roots the formula seems inevitably to require working with complex numbers had been somewhat illuminated by Bombelli a generation later. People also knew that Ferrari, a colleague of Cardano's, had shown how to solve equations of degree 4, but crucially, no-one knew of a general method for solving equations of degree 5 or more.

Euler did believe in the fundamental theorem of algebra. He wrote to Johann Bernoulli on 15 September 1739 to say that he believed that every polynomial equation could be resolved into the right number of linear factors and quadratic factors, although he admitted he did not have a proof. He needed this result in order to complete his solution of the problem of solving linear differential equation s of any order with constant coefficients. He repeated the claim in a letter he wrote to Johann's nephew, Nicolaus (I) Bernoulli on 1 September 1742, but Nicolaus was not convinced, and offered as a counter-example

$$x^4 - 4x^3 + 2x^2 + 4x + 4 = 0,$$

which has these roots

$$1 + \sqrt{2 + \sqrt{-3}}, \ 1 - \sqrt{2 + \sqrt{-3}}, \ 1 + \sqrt{2 - \sqrt{-3}}, \ 1 - \sqrt{2 - \sqrt{-3}}.$$

---

[1]For much of the information in this Appendix, see Gilain (1991).

Nicolaus found the roots, but denied that they could be written in the form $a +$ $b\sqrt{-1}$.[2]

Euler replied on 10 November. He factored Nicolaus's example as a product of two quadratics, and claimed that he could now prove the fundamental theorem of algebra for all polynomials of degree at most four. He also discussed the problem with Christian Goldbach, who queried the theorem for polynomials of the form $x^4 + px + q$; Euler corrected him.[3] Then Nicolaus Bernoulli replied, now agreeing with Euler that the fundamental theorem of algebra was true. He also claimed that he could prove the fundamental theorem of algebra, provided every imaginary quantity is an elementary function of quantities of the form $a + b\sqrt{-1}$, a view, he added, that nobody denies (evidently he had changed his own position).

Euler's *Introductio in Analysin Infinitorum* on the theory and use of infinite series is one the first great books of mathematics. The later-to-be-famous zeta function appears here, as does Euler's unification of the logarithmic, exponential, and trigonometric functions. Euler began by defining what a function is (a curious mixture of the input-output definition we use today and a confident belief that any function could be represented as an infinite series). In Chapter II Euler discussed how functions could be treated, and made many assertions about polynomial functions (it is understood that their coefficients are real): they have either linear or quadratic factors; the number of complex factors is always even (§30); the complex factors can always be paired in such a way that the product of these pairs is real (§32); a polynomial of odd degree always has a real factor (§34); a polynomial of odd degree has an odd number of real factors (§35); a polynomial of even degree has an even number of real factors (§36).

Euler returned to this theme in Chapter IX when he offered partial proofs of these claims—but he never attempted to justify them fully, and for this his great rival, Jean le Rond d'Alembert, criticised him. Euler replied that the book had been finished in 1745, since when he had made great progress "although I freely confess that I do not yet have a solid demonstration that every algebraic expression can be resolved into real trinomial factors".[4]

As one might suspect, a priority dispute was underway, for by then d'Alembert had published his own proof of the fundamental theorem of algebra, which he had submitted to the Berlin Academy in 1746 (it was published in 1748). It grew out of an earlier account (from 1745) , where d'Alembert had argued that any algebraic function of a complex number takes values that are complex numbers, and so, by regarding a root of a polynomial equation as such a function, a polynomial equation has complex roots. Because the equation has real coefficients it follows that if $a +$ $b\sqrt{-1}$ is one root then $a - b\sqrt{-1}$ is another. It seems that he soon came to regard this account altogether too naive—it is little more than assuming what has to be proved—in 1746 d'Alembert tried to prove the existence of complex roots directly.

---

[2]Curiously, Maple gives the roots in the form Bernoulli found them.

[3]See Euler (2015), nos. 62 and 63, 5 February 1743 and 26 February 1743.

[4]Quoted in Gilain (1991, p. 111).

Now he took the polynomial equation

$$x^m + ax^{m-1} + \cdots + fx + g = 0$$

and looked at the curve with equation

$$G(x, y) = x^m + ax^{m-1} + \cdots + fx + y = 0.$$

He rewrote this as a power series for x:

$$x = a^{m/n} + by^{r/s} + \cdots,$$

using the methods of the calculus and infinite series, and deduced that when y is small x is necessarily complex. He then deduced that x will be complex for any value of y, and so found a complex value for x when $y = g$.

This is generally considered to be the first serious attempt on the fundamental theorem of algebra, although it contains no attempt to prove the convergence of the series for x, or to show that claim is valid for all values of y and not just a bounded set.[5]

Then in 1749 Euler wrote his great memoir on the subject (it was published in 1751). In it, he aimed to show that every root of a polynomial equation is of the form $a + b\sqrt{-1}$, where a and b are real.

Euler dealt with polynomials of odd degree by arguing that when $|x|$ is very large the value of the polynomial is large and of one sign when x is negative and the other when x is positive, and so it has at least one real root. Similarly, a polynomial of even degree with a negative constant term has at least two real roots. He then claimed that a polynomial of degree $2^k m$, m odd, can be treated as one of degree 2^{k+1} by multiplying it by powers of x. That left polynomials of degree exactly a power of 2, and he dealt with the cases where the polynomial is of degree 4, 8, 16 and finally 2^k.

To solve a polynomial equation of degree 4, Euler wrote it, as it was well-known that one could, in the form

$$x^4 + bx^2 + cx + d = 0.$$

If this can be factorised, it can be written as

$$(x^2 + ux + \lambda)(x^2 - ux + \mu) = 0.$$

Euler deduced from that that u satisfies this equation:

$$u^6 + 2bu^4 + (b^2 - 4d)u^2 - c^2 = 0.$$

---

[5]In his (1799) Gauss was to criticise d'Alembert's use of infinite series in his proof, and showed by means of an example that it was unsound, while politely admitting that it was perhaps capable of being re-cast in a more reliable form.

The point is not so much that this equation is a cubic in u^2 and so can be solved, as to hint at a pattern upon which Euler was to rely when he got to the general case. Here, he observed that the equation for u can be solved according to the second of his basic assumptions, so u is known, and finally he showed that λ and μ can be found as rational functions of b, c, d, and u.

Euler dealt with the equations of degree 8 and 16 similarly. When it came to general case of a polynomial equation of degree 2^k Euler invoked the theory of the symmetric functions that went back to Newton. If a polynomial expression of degree n can be written as a product of n linear factors, say

$$x^n + a_1 x^{n-1} + \cdots a_{n-1} x + a_n = (x - \alpha_1)(x - \alpha_2) \ldots (x - \alpha_n),$$

then

$$a_1 = -(\alpha_1 + \alpha_2 + \ldots \alpha_n) \tag{A.1}$$

$$\cdots$$

$$a_n = (-1)^n \alpha_1 \cdot \alpha_2 \cdots \alpha_n. \tag{A.2}$$

The expressions on the right are the various symmetric functions of the αs. The fundamental theorems about them are:

1. that every rational function in the αs that is symmetric (i.e. invariant under all permutations of the αs) is a rational function of the symmetric functions.
2. a rational function of the αs that takes k distinct values as the αs undergo their $n!$ permutations satisfy an equation of degree k whose coefficients are rational functions of the coefficients of the original equation.

When the polynomial equation has degree 4, Euler argued that u must be the sum of two of the roots, because from the formal identity

$$x^4 + bx^2 + cx + d = (x^2 + ux + \lambda)(x^2 - ux + \mu) = (x - \alpha_1)(x - \alpha_2)(x - \alpha_3)(x - \alpha_4)$$

some expression of the form $u = \alpha_1 + \alpha_2$ must be true and so u takes on 6 values as the roots are permuted, which is why it satisfies an equation of degree 6. In fact, the possibilities are

$$u_1 = \alpha_1 + \alpha_2 = p = -u_4 = -(\alpha_3 + \alpha_4),$$

$$u_2 = \alpha_1 + \alpha_3 = q = -u_5 = -(\alpha_2 + \alpha_4),$$

$$u_3 = \alpha_1 + \alpha_4 = r = -u_6 = -(\alpha_2 + \alpha_3),$$

so the equation for u is

$$(u^2 - p^2)(u^2 - q^2)(u^2 - r^2) = 0.$$

This is indeed an equation of even degree with a negative constant term $-p^2q^2r^2$, and Euler did check that pqr is real. He showed in fact that

$$pqr = (\alpha_1 + \alpha_2)(\alpha_1 + \alpha_3)(\alpha_1 + \alpha_4),$$

which is invariant under all permutations of the αs, and so is expressible rationally in terms of the coefficients of the original equation, and hence real.

The proof in the case of a polynomial equation of degree 2^k was only sketched by Euler. He supposed the equation had no term of degree $2^k - 1$, which is trivial, and investigated the consequences of formally factorising it into two factors of degree 2^{k-1}. This gave him an equation of the form

$$x^{2^k} + bx^{2^k-2} + \cdots = (x^{2^{k-1}} + ux^{2^{k-1}-1} + \cdots)(x^{2^{k-1}} - ux^{2^{k-1}-1} + \cdots).$$

The so-far-undetermined coefficients are $2^k - 1$ in number, equal to the number of coefficients in the original polynomial.

The coefficient u must be a sum of half the roots of the equation, so 2^{k-1} in number, so u takes $^{2^k}C_{2^{k-1}} = 2N$ values, where N is odd, as Euler showed. So, he said, u satisfies an equation of degree $2N$ with real coefficients that, moreover, is of the form

$$(u^2 - p_1^2)(u^2 - p_2^2) \cdots (u^2 - p_N^2) = 0,$$

with a negative, real, constant term. So u^2 is a known real quantity and the remaining coefficients in each of the two factors can then be written as rational functions of u and the coefficients of the original polynomial.

Whatever the weaknesses of this argument, it acquires a certain amount more force when the equations of degrees 8 and 16 are done explicitly first. The serious attempt to show *why* the computations work in these cases gives weight to the feeling that they can be trusted to work in general. That said, Euler's approach founders on the confusion of two questions: is every imaginary number (for which read: number-like object that is not real) of the form $a + b\sqrt{-1}$ for real a and b? and does every polynomial equation (with real coefficients) have exactly as many roots as its degree.

In his (1772) Lagrange set himself the task of completing Euler's proof of the fundamental theorem of algebra, and the resulting argument is reasonably convincing, but not completely so, and it was left to the twenty-year-old Carl Friedrich Gauss in his doctoral dissertation of 1797 to make the decisive criticism. Gauss addressed it chiefly to d'Alembert's attempts, but it applies to all the others, as he said elsewhere.

Gauss objected that d'Alembert had not proved that the roots actually exist; instead he had assumed the existence of the roots and showed only that, in that case, they had to be complex. The possibilities that the roots did exist but could not be manipulated like numbers; or that the roots did not even exist, had not been considered. That said, Gauss admitted that "the true strength of the proof seems to me not to have been weakened at all by all the objections", and remarked that one could build a rigorous proof on that foundation (in *Werke* 3, 11).[6]

Gauss then turned to the second of Euler's arguments, that reduced the problem to the factorisation of polynomials of degree a power of 2; once factorisation is assured the fundamental theorem of algebra follows immediately by induction. Here Gauss objected that not only did Euler's approach tacitly assume that polynomial equations have roots, but the proof of factorisation replaced the original equation with a system of quadratic equations for which there is no guarantee that solutions exist, for it produces $2m - 1$ equations in $2m - 2$ unknowns. Gauss then commented that Lagrange (1772) had thoroughly resolved some of the objections to Euler's argument, but gaps in the proof remained, notably the assumptions that polynomial equations have roots and that the only problem is to show that they are complex numbers.[7]

From Gauss's Critique

Euler tacitly supposes that the equation $X = 0$ has $2m$ roots, of which he determines the sum to be $= 0$ because the second term in X is missing. What I think of this licence I have already declared in art. 3. The proposition that the sum of all the roots of an equation is equal to the first coefficient with the sign changed, does not seem applicable to other equations unless they have roots; now although it ought to be proved by this same demonstration that the equation $X = 0$ really does have roots, it does not seem permissible to suppose the existence of these. No doubt those people who have not yet penetrated the fallacy of this expression will reply, "Here it has not been demonstrated that the equation $X = 0$ can be satisfied (for this expression means that the equation has roots) but it has only been demonstrated that the equation can be satisfied by values of x of the form $a + b\sqrt{-1}$; and indeed that is taken as axiomatic". But although types of quantities other than real and imaginary $a + b\sqrt{-1}$ cannot be conceived of, it does not seem sufficiently clear how the proposition awaiting demonstration differs from that supposed as axiomatic. [...] Therefore that axiom can have no other meaning than this: Any equation can be satisfied either by the real value of an unknown, or by an imaginary value expressed in the form $a + b\sqrt{-1}$, or perhaps by a value in some other form which we do not know, or by a value which is not totally contained in any form. But how such quantities which are shadowy and inconceivable can be added or multiplied is certainly not understood with the clarity which is required in mathematics. [...]

Finally, Lagrange has dealt with our theorem in the commentary *Sur la Forme des Racines Imaginaires des Equations*, 1772. This great geometer handed his work to the printers when he was worn out with completing Euler's first demonstration [...]. However, he does not touch upon the third objection at all, for all his investigation is built upon the supposition that an equation of the mth degree does in fact have roots.

---

[6]For an evaluation of D'Alembert's proof and Gauss's criticism of it, see Gilain's introduction in (D'Alembert 1746, lxxvi–xcii).

[7]Remmert has aptly remarked that "the Gaussian objection against the attempts of Euler–Lagrange was invalidated as soon as Algebra was able to guarantee the existence of a splitting field for every polynomial." (in Ebbinghaus et al. 1990, 105). Remmert went on to point out that this had been already observed by Kneser in 1888.

Gauss then offered his own proof, which dealt systematically with polynomial equations of any degree. He took a polynomial of the form

$$f(z) = z^m + a_1 z^{m-1} + \ldots + a_m$$

with real coefficients, and looked for its roots in an infinite plane whose points are specified by polar coordinates r, ϕ. Such roots will be the common points of the equations

$$U = r^m \cos m\phi + a_1 r^{m-1} \cos(m-1)\phi + \ldots + a_{m-1} r \cos \phi + a_m = 0,$$

and

$$T = r^m \sin m\phi + a_1 r^{m-1} \sin(m-1)\phi + \ldots + a_{m-1} r \sin \phi = 0 ,$$

which are the curves defined by the real and imaginary parts of the equation $f(z) = 0$: $\mathrm{Re}\, f(z) = 0$ and $\mathrm{Im}\, f(z) = 0$, as z varies in the plane of complex numbers. Gauss now argued as follows. Outside a suitably large circle of radius R centred on the origin each of these curves meets a concentric circle of radius $r > R$ in two disjoint sets of $2m$ distinct points, and therefore these curves consist of $2m$ arcs going off to infinity in the plane (let r increase indefinitely). Moreover, for suitably large values of R the z^m is dominant and therefore the curves $\mathrm{Re}\, f(z) = 0$ and $\mathrm{Im}\, f(z) = 0$ meet the circle of radius R alternately. Gauss now endeavoured to show that inside the circle of radius R these curves are each made up of m disjoint 'parabola-shaped' pieces, but they can only join up if a curve $\mathrm{Re}\, f(z) = 0$ crosses a curve $\mathrm{Im}\, f(z) = 0$. At such a crossing point, the equation has a root, and so the fundamental theorem of algebra is proved.

He argued that the curves are real algebraic curves in m pieces that come from and go to infinity; they cannot stop, break apart, or spiral to a point in the fashion of some transcendental curves (such as $y = 1/\log(x)$). Rather, as he put it, an algebraic curve that enters a bounded region of the plane also leaves it (§ 21, footnote). Insofar as this is no easier to prove than the fundamental theorem of algebra itself, Gauss's proof is defective. That said, the topological nature of Gauss's proof is attractive, and Gauss saw it as the heart of the matter. This was a remarkable insight at a time when neither Gauss nor any one else could have provided a proper distinction between the real numbers and the rational numbers.

Gauss's second proof (dated late 1815) is a *tour de force* in the use of symmetric functions and does not involve complex variables.[8] His third proof (dated January 1816) in many ways returns to the ideas of his first. On this occasion, Gauss wanted to show that there is a root of the polynomial $f(z)$ within a circle of suitably large radius R by considering the quantity the square of the modulus of $f(z)$, which is $t^2 + u^2$, where $t = \mathrm{Re}\, f(z)$ and $u = \mathrm{Im}\, f(z)$. He was able to define a double integral

---

[8] An English translation can be found in Smith's *Sourcebook*.

over the region $0 \le r \le R$, $0 \le \varphi \le 2\pi$ that—on the assumption that t and u do not simultaneously vanish—took the value 0 when integrated first with respect to φ and then with respect to r, but a non-zero value 0 when integrated first with respect to r and then with respect to φ. The only way out of this apparent contradiction was to deny its fundamental assumption that the integral may be evaluated in both orders and give the same result, and therefore Gauss concluded that there were points where $t^2 + u^2 = 0$, and so t and u simultaneously vanish, and so the polynomial has a root. The proof therefore rests on the insight that when a double integral is replaced by a repeated integral the order of integration may matter when the integrand becomes infinite (in later language, has a simple pole).

The fourth and last of Gauss's proofs was published in 1849 and was produced for the 50th anniversary of his first proof, the occasion for a celebration of Gauss's distinguished career. Gauss dropped his criticisms of the eighteenth century proofs, noted that Cayley had given a proof more recently, and then took up his first proof again. He now dealt with the case of multiple roots and the possible configurations of curves that can arise, and concluded that all roots occur in the way described.

Appendix B
Gauss and Composition of Forms

B.1 Composition Theory

The table of forms of discriminant -161 in Sect. 4.5 shows that of the eight possibilities a priori, exactly half occur. It illustrates a deep but important theorem that Gauss was able to establish: the combinations that occur are those for which the product of the characters (scoring Y as 1 and N as -1) is $+1$. To prove this theorem, Gauss developed his theory of composition of forms, which is a way of combining two forms of a given discriminant to produce a third of the same discriminant. He also required a theory of ternary forms (quadratic forms in three variables, x, y, and z), but it was later shown by Arndt (1859) that this can be reduced to an application of a theorem of Legendre's.[1]

We have already seen in Sect. 3.2 that if a quadratic form Q represents a product mm' then there are quadratic forms of the same discriminant as Q that represent m and m'. The aim of the composition of forms is the reverse statement: given two quadratic forms, one that represents m and one that represents m', to produce a quadratic form that represents the product mm'.

To be precise, Gauss's composition of forms can be carried out for forms of different discriminants that satisfy certain requirements, but we shall just consider the case where the discriminants are the same—in this case the 'certain requirements' are automatically met.

Given two quadratic forms

$$Q_1(x, y) = a_1 x_1^2 + 2b_1 x_1 y_1 + c_1 y_1^2 \quad \text{and} \quad Q_2(x, y) = a_2 x_2^2 + 2b_2 x_2 y_2 + c_2 y_2^2$$

---

[1]Legendre's theorem says that the ternary form $ax^2 + by^2 + cz^2$ has solutions in positive integers if and only if a, b, c are not all of the same sign and $-bc, -ca, -ab$ are squares modulo a, b, c respectively. Gauss had used his theory of ternary forms to deduce this result.

© Springer Nature Switzerland AG 2018
J. Gray, *A History of Abstract Algebra*, Springer Undergraduate Mathematics Series,
https://doi.org/10.1007/978-3-319-94773-0

with the same discriminant, their composite should be a form

$$Q(X, Y) = AX^2 + 2BXY + CY^2$$

of the same discriminant, where

$$X = p_0 x_1 x_2 + p_1 x_1 y_2 + p_2 x_2 y_1 + p_3 y_1 y_2, \quad Y = q_0 x_1 x_2 + q_1 x_1 y_2 + q_2 x_2 y_1 + q_3 y_1 y_2,$$

where $p_0, \ldots, p_3, q_0, \ldots, q_3$ are all integers. These conditions can be met, as will be shown below.

We also require that if the form $Q_1(x, y)$ is equivalent to a form $Q_1'(x, y)$, and the form $Q_2(x, y)$ is equivalent to a form $Q_2'(x, y)$ then the composite of $Q_1(x, y)$ and $Q_2(x, y)$ is equivalent to the composite of $Q_1'(x, y)$ and $Q_2'(x, y)$. This extends composition to classes of forms. A significant problem with the theory, however, is that the composition of two forms is not unique: it is only determined up to proper equivalence. So every theorem about the composition of forms must be extended to show that any choices made along the way result in equivalent outcomes.

It cannot be said that composition theory is pleasant. In his book (Weil 1984, pp. 332–335) André Weil sketched Legendre's approach, which we will look at below, and then remarked

> No doubt the Gaussian theory, as Gauss chose to describe it (DA.Art.234–260), is far more elaborate; so much so, indeed, that it remained a stumbling-block for all readers of the *Disquisitiones* until Dirichlet restored its simplicity by going back very nearly to Legendre's original construction (Dir. II. 107–114, 1851).

Weil then speculated that Legendre's ideas may have influenced Gauss, contrary to Gauss's claim that he first saw Legendre's book after "the greater part" of the *Disquisitiones* was in print; Gauss started work on composition of forms late in 1798, and might have seen Legendre's *Essai* on a visit to Pfaff around that time.

Be that as it may, as Weil went on to remark

> But there is at the same time an essential difference between his treatment and that of Gauss. Legendre never made the distinction between proper and improper equivalence; his classes of quadratic forms are the Lagrangian classes [...]. Thus it was left to Gauss to discover that the "Gaussian" classes, under the operation of composition, make up a finite commutative group, and to draw the consequences of this fact.

Gauss for his part established that composition is well-defined on classes (and indeed orders and genera) and that the classes of a given discriminant form what we would call (but he could not!) a finite abelian group. The principal class, which contains the form $(1, 0, d)$, plays the role of the identity element. The principal genus consists exactly of the squares of classes of forms. Moreover, the characters are a group homomorphism onto the group $\{1, -1\}$.

As we have seen Gauss's theory of composition of forms was regarded as important in its day because it delivers insights into what quadratic forms of a given discriminant can represent a given number, even though it has always been regarded as very hard.

Using his composition theory, Gauss showed (§252) that for a given discriminant there are the same number of classes in every genus of the same order. Moreover (§261) half of the assignable characters for a positive nonsquare discriminant correspond to no properly primitive genus, and if the discriminant is negative to no properly primitive positive genus. From this he deduced a new proof of the theorem of quadratic reciprocity in §262. As for the half of the assignable characters that are not excluded, he showed after a lengthy 'digression' into the theory of forms of the form $ax^2 + by^2 + cz^2$ that they all do occur (§287). He was understandably pleased with this result, and wrote

> We believe that these theorems are among the most beautiful in the theory of binary forms, especially because, despite their extreme simplicity, they are so profound that a rigorous demonstration requires the help of many other investigations.

This is, when you think about, a very revealing statement about what he took good mathematics to be.

Gauss's Route to the Above Results

Although we cannot follow Gauss's account of composition of forms in full detail, we can indicate how it was able to deliver the results that it did in Gauss's hands, and this will show some of the 'group-theoretic' ideas at work underneath.[2] To do so we shall employ some notation that he did not provide. We write $\chi(f)$ for the genus of a form f and $O(f)$ for its order, ff' for the composite of the forms f and f', and $f \leadsto a$ to indicate that the form f represents the number a. We shall also restrict our attention to forms of a fixed discriminant.

In §246 Gauss investigated how the genus of a composite of two forms is connected with the genera of the individual forms. In our restricted setting, the forms f, f', and $F = ff'$ all have the same determinant, say D. Suppose that p is an odd prime dividing D. There are fixed relationships for the forms f and f', and if $f \leadsto a$ and $f' \leadsto a'$ then $F = ff' \leadsto aa'$. The fixed relationships determine the values of $\left(\frac{a}{p}\right)$ and $\left(\frac{a'}{p}\right)$ and therefore of $\left(\frac{aa'}{p}\right)$, so the genus of $F = ff'$ is determined from the genera of f and f'.

Similar arguments deal with the case when 4 enters the total character of F, and when D is congruent to 2 or 6 mod 8; I omit these.

In §247 Gauss deduced that if f, g, f', g' are primitive forms, $O(f) = O(g)$, $\chi(f) = \chi(g)$, $O(f') = O(g')$, and $\chi(f') = \chi(g')$, then $\chi(ff') = \chi(gg')$.

If moreover f belongs to the principal genus, so all its characters are $+1$, then $\chi(ff') = \chi(f')$. And if f and f' are properly primitive forms of the same genus then ff' is in the principal genus, which implies in particular that the form $f^2 = ff$ is in the principal genus.

---

[2] Among modern accounts, see §2.2 and §2.3 of Lemmermeyer (2000).

In §248 Gauss deduced that if $\chi(f) = \chi(g)$ and $\chi(f') = \chi(g')$ then $\chi(ff') = \chi(gg')$.

In §252 Gauss showed that for a given determinant there are the same number of classes in every genus of the same order (and in §253 he observed that this result is false if the condition on the order is dropped). The proof is as follows. Let G be a genus composed of the classes $K_0, K_1, \ldots K_{n-1}$ and H another genus in the same order. Let L be a class in the genus H. There is a properly primitive class M such that $MK_0 = L$. Consider the classes $MK_0, MK_1, \ldots MK_{n-1}$; by §249 they are all distinct, by §248 they are all of the same genus, and all the classes in H arise in this way. Therefore the genera contain the same number of classes.

A crucial role is played by the ambiguous classes. Gauss enumerated them in §257, and by a long argument in §258 he showed that the number of properly primitive ambiguous classes of a given determinant is half the number of assignable characters.

In §260 he considered the properly primitive ambiguous classes of a given determinant, say $H_0, H_1, \ldots H_{n-1}$, where H is the principal class. He took a form k and considered the classes $kH_0, kH_1, \ldots kH_{n-1}$; they are distinct and properly primitive. If $k^2 = K$ then also $(kH_j)^2 = K$, and conversely, if $k_1^2 = K$ then $k_1 = kH_j$ for some $j = 0, 1, \ldots n - 1$ (again I omit the argument). Therefore, if the number of properly primitive classes is r and the number of properly primitive ambiguous classes is n then the number of properly primitive classes that are squares is r/n.

Gauss then returned to the example of forms of discriminant -161 and commented

> Thus, e.g., for $D = -161$, the number of all positive properly primitive classes is 16, the number of ambiguous classes 4, so the number of classes that can arise from the duplication of any class must be 4. As a matter of fact we find that all classes contained in the principal genus are endowed with this property; thus the principal class $(1, 0, 161)$ results from the duplication of the four ambiguous classes; $(2, 1, 81)$ from duplicating the classes $(9, 1, 18)$, $(9, -1, 18)$, $(11, 2, 15)$, $(11, -2, 15)$; $(9, 1, 18)$ from duplicating the classes $(3, 1, 54)$, $(6, 1, 27)$, $(5, -2, 33)$, $(10, 3, 17)$; finally $(9, -1, 18)$ from duplicating the classes $(3, -1, 54)$, $(6, -1, 27)$, $(5, 2, 33)$, $(10, -3, 17)$.

In §261 Gauss put all this together to deduce that "Half of all the characters assignable for a given determinant cannot belong to any properly primitive genus". As he noted for clarity, the result should be interpreted to mean that at least half the assignable characters do not occur.

To prove this, Gauss denoted the number of properly primitive genera by m and the number of classes in each genus by k, so that the number of classes is km. He denoted the number of characters by n, and therefore the number of properly primitive ambiguous classes is $n/2$ (by §258). Therefore (by §260) the number of properly primitive squares is $2km/n$, and (by §247) these classes belong to the principal genus. Therefore if (as Gauss promised to show later) all classes in the principal genus are squares then $2km/n = k$ and so $m = n/2$. In any case it is impossible that $2km/n > k$, so at least half the assignable characters do not occur.

To finish his analysis, Gauss developed a theory of ternary forms, which will not be discussed. Using it, he could deduce (§286) that every form in the principal genus arises as a square—a result that came to be called the principal genus theorem—and then (§287) that exactly half the assignable characters do occur. In this way he established he had earlier called (§261) "the truth of this profound proposition concerning the most deeply hidden properties of numbers".

This result has struck generations of later mathematicians as important, and they have explored the connections that lead to its proof. Among the most important of these has been Dirichlet's class number formula, which Dedekind used to prove these results of Gauss's in Supplement 4 of Dirichlet's *Lectures*. To quote Lemmermeyer (2007, p. 538)

> Dedekind returns to genus theory of binary quadratic forms in his supplement X: §153 gives the first inequality, §154 the quadratic reciprocity law, and in §155 he observes that the second inequality of genus theory (the existence of half of all the possible genera) is essentially identical with the principal genus theorem: "Every class of the principal genus arises from duplication." He then adds:
>
>> It is impossible for us to communicate the proof, which Gauss has based on the theory of ternary quadratic forms; but since this deep theorem is the most beautiful conclusion of the theory of composition, we cannot help but derive this result, without the use of Dirichlet's principles, in a second way, which will also form the basis for other important investigations.[3]

B.2   Gaussian Composition of Forms

Consider the simplest case, the product of the forms $x^2 + y^2$ and $u^2 + v^2$, given by

$$(x^2 + y^2)(u^2 + v^2) = (xu - yv)^2 + (xv + yu)^2 = X^2 + Y^2.$$

The product is of degree four in the original variables x, y, u, v but of degree two in the new variables

$$X = xu - yv, Y = xv + yu.$$

The idea behind composition of forms is to recognise in the product of two forms $ax^2 + 2bxy + cy^2$ and $a'u^2 + 2b'uv + c'v^2$, which is of degree four in the variables x, y, u, v, two expressions X and Y of degree two in those variables such that the product is of degree two in X and Y. So in §234 of the *Disquisitiones Arithmeticae* Gauss looked for integer values of the constants

$$p_1, p_2, p_3, p_4, \text{ and } q_1, q_2, q_3, q_4$$

---

[3]See (Dedekind 1871, 407).

such that on setting

$$X = (x \ y) \begin{pmatrix} p_1 & p_2 \\ p_3 & p_4 \end{pmatrix} \begin{pmatrix} u \\ v \end{pmatrix}$$

and

$$Y = (x \ y) \begin{pmatrix} q_1 & q_2 \\ q_3 & q_4 \end{pmatrix} \begin{pmatrix} u \\ v \end{pmatrix}$$

one has

$$AX^2 + 2BXY + CY^2 = (ax^2 + 2bxy + cy^2)(a'u^2 + 2b'uv + c'v^2) \qquad \text{(B.1)}$$

for some A, B, C that are determined by $a, b, c, a', b',$ and c'. If moreover the six quantities

$$p_1q_2 - p_2q_1, \ p_1q_3 - p_3q_1, \ p_1q_4 - p_4q_1, \ p_2q_3 - p_3q_2, \ p_2q_4 - p_4q_2, \ p_3q_4 - p_4q_3$$

have no common divisor, then Gauss said that the form in X and Y was the composite of the other forms. It may help to notice that these products are the determinants in the 2 by 2 minors of this matrix

$$\begin{pmatrix} p_1 & p_2 & p_3 & p_4 \\ q_1 & q_2 & q_3 & q_4 \end{pmatrix}$$

When Eq. (B.1) is expanded, it becomes a complicated set of nine conditions for the eight unknown ps and qs, from which it follows, after a considerable amount of work, that the ps and qs can be found only under certain conditions, such as that the discriminant of the forms are the same. In this case, the discriminant of the product is the same as the discriminants of the original two forms. But Gauss could not finish here, because when the conditions are met it does not follow that the ps and qs are uniquely defined.

To resolve the ambiguity in the composition, Gauss had to find a way of fixing the signs. It turns out that on any definition of a composite,

$$p_1q_2 - p_2q_1 = \pm a, \ \text{and} \ p_1q_3 - p_3q_1 = \pm a'.$$

So Gauss added these condition to fix uniqueness, that the above signs are both $+$:

$$p_1q_2 - p_2q_1 = a, \ \text{and} \ p_1q_3 - p_3q_1 = a'.$$

We can see that some skill is required to produce a useful definition of the composite of two forms, because we have the identity

$$(ax^2 + 2bxy + cy^2)(a'u^2 + 2b'uv + c'v^2) = \frac{1}{aa'}(X^2 + R^2Y^2), \qquad (B.2)$$

where $b^2 - ac = -R^2 = b'^2 - a'c'$ and

$$X = aa'xu + ab'xv + ba'yu + (bb' + nyv)yv, \quad Y = (ax + by)v + (a'u + b'v)y.$$

If we took this as the correct definition the composite of any two forms with the same discriminant would be the trivial form.

In this case, we find that

$$p_1q_2 - p_2q_1 = a, \ p_1q_3 - p_3q_1 = a', \ p_1q_4 - p_4q_1 = b + b',$$

$$p_2q_3 - p_3q_2 = b' - b, \ p_2q_4 - p_4q_2 = c', \ p_3q_4 - p_4q_3 = c.$$

All these difficulties motivate my decision not to explain Gauss's theory of how two forms are composed, but I do note that in fact the definition of the composite of the form $2x^2 + 2xy + 3y^2$ with itself is $x^2 + 5y^2$. This goes a long way to explaining what Fermat had noticed about the divisors of $x^2 + 5y^2$.

Exercise. Derive Brahmagupta's identity (B.2) by writing

$$ax^2 + 2bxy + cy^2 = (1/a)(A + iRy)(A - iRy),$$

where $A = ax + by$ and $R^2 = ac - b^2$, and

$$a'u^2 + 2b'uv + c'v^2 = (1/a')(A' + iRv)(A' - iRv),$$

where $A' = a'u + b'v$ and $R^2 = a'c' - b'^2$, and multiplying out. This generalises the strategy at the start of this section.

B.3   Dirichlet on Composition of Forms

Smith's account of Dirichlet's method of composition of forms is very helpful.[4] To compose two forms with the same discriminant D, say

$$ax^2 + 2bxy + cy^2 \text{ and } a'x^2 + 2b'xy + c'y^2,$$

---

[4]See (Smith 1859, §111). Dirichlet's account was originally a short, Latin, academic dissertation, later published in Crelle's *Journal* in 1851.

Dirichlet chose two representable numbers M and M', both prime to $2D$, such that

$$am^2 + 2bmn + cn^2 = M \text{ and } a'm'^2 + 2b'm'n' + c'n'^2 = M',$$

and two numbers ω and ω' such that

$$\omega^2 \equiv D \bmod M \text{ and } \omega'^2 \equiv D \bmod M'.$$

This allowed him to observe that $ax^2 + 2bxy + cy^2$ was equivalent to $Mx^2 + 2\omega xy + \frac{\omega^2 - D}{M}$ and $a'x^2 + 2b'xy + c'y^2$ was equivalent to $M'x^2 + 2\omega'xy + \frac{\omega'^2 - D'}{M'}$. So now he investigated when these forms can be combined.

If it is possible to find a number Ω such that

$$\Omega \equiv \omega \bmod M, \quad \Omega \equiv \omega' \bmod M', \text{ and } \Omega^2 \equiv D \bmod MM'$$

then the composite can be defined as

$$MM'x^2 + 2\Omega xy + \frac{\Omega^2 - D}{MM'}y^2.$$

There is such an Ω, moreover, if the values of M and M' are suitably chosen. The result is that equivalence classes of forms of the same discriminant can always be defined (the resulting equivalence class is well-defined and independent of all the choices made). So, to reiterate, equivalence classes of quadratic form s form a group, but the calculation of the composite still depends on making certain choices.

It is instructive to compute the Dirichlet composite of the form $2x^2 + 2xy + 3y^2$ with itself. We write the form in the form $(2, 1, 3)$. When we try to use only the second stage of Dirichlet's procedure, we have to try to solve the congruence $\Omega^2 \equiv -5 \bmod 4$, which has no solutions. We therefore go back to the start, and transform one of the forms into a more appropriate form. With $m = 0$ and $n = 1$ we get $M = 3$ and we replace $(2, 1, 3)$ with $(3, 1, 2)$. The composite of this with $(2, 1, 3)$ is found by solving the congruences

$$\Omega \equiv 1 \bmod 3, \quad \Omega \equiv 1 \bmod 2, \text{ and } \Omega^2 \equiv -5 \bmod 6,$$

for which the solution is $\Omega = 1$. The resulting form is $(6, 1, 1)$, which is equivalent to $(1, 0, 5)$, as we expected.

Exercise. Consider the forms that Legendre attempted to compose, and the ones that must also be considered because they complete the list of positive forms of discriminant $b^2 - ac = -41$. They are

$$A = x^2 + 41y^2, \text{ or } (1, 0, 41)$$

$$B = 2x^2 + 2xy + 21y^2, \text{ or } (2, 1, 21)$$

$$C_+ = 5x^2 + 4xy + 9y^2, \text{ or } (5, 2, 9)$$

$$C_- = 5x^2 - 4xy + 9y^2, \text{ or } (5, -2, 9)$$

$$D_+ = 3x^2 + 2xy + 14y^2, \text{ or } (3, 1, 14)$$

$$D_- = 3x^2 - 2xy + 14y^2, \text{ or } (3, -1, 14)$$

$$E_+ = 6x^2 + 2xy + 7y^2, \text{ or } (6, 1, 7)$$

$$E_- = 6x^2 - 2xy + 7y^2, \text{ or } (6, -1, 7).$$

In each case they are ready to be composed. One has to find Ω and then reduce the resulting form to one of these eight reduced forms. For example, to compose C_+ and D_+ we have to solve the congruences

$$\Omega \equiv 2 \bmod 5, \quad \Omega \equiv 1 \bmod 6, \text{ and } \Omega^2 \equiv -41 \bmod 15.$$

The solution is $\Omega = 7$, so the composite is $CD = (17, 7, 6)$ or $15x^2 + 14xy + 6y^2$. This reduces to $(6, -1, 7)$, which is E_-.

1. Check that the composite C_+E_- is equivalent to D_-.
2. Compute enough composites to be sure you are comfortable with the procedure.
3. What is the group these eight reduced forms make up? [Hint: it is enough to compute the squares, AA, BB etc.]

As we have seen, Legendre came close to defining a useful concept of composition that met the above requirements. But it was subtly flawed, as we shall now see, following the discussion in Cox (1989, p. 42) of a passage in Legendre's *Essai sur la théorie des nombres*, vol. II, pp. 39–40. Legendre regarded the following five forms of discriminant -41 as the reduced forms[5]:

$$A = x^2 + 41y^2$$

$$B = 2x^2 + 2xy + 21y^2 \quad C = 5x^2 + 4xy + 9y^2$$

$$D = 3x^2 + 2xy + 14y^2 \quad E = 6x^2 + 2xy + 7y^2$$

He then worked through his method for finding composites, and came up with these results:

$$AA = A, \quad AB = B, \quad AC = C, \quad AD = D, \quad AE = E,$$

---

[5]This definition of C replaces the one given by Cox, $C = 5x^2 + 6xy + 10y^2$, which is not reduced and reduces to $5x^2 - 4xy + 10y^2$.

$$BB = A, \ BC = C, \ BD = E, \ BE = D,$$

$$CC = A \text{ or } B, CD = D \text{ or } E, CE = D \text{ or } E,$$

$$DD = A \text{ or } C, DE = B \text{ or } C,$$

$$EE = A \text{ or } C.$$

Cox observes that the ambiguity must be resolved before there is any hope of these forms forming a group. This was not a problem for Legendre, who did not have that idea, but it was for Gauss. It is exacerbated by the fact that Legendre's concept of equivalence is not that of proper equivalence, so some ambiguities of sign are inevitable. With proper equivalence there are eight inequivalent reduced forms, not five, because each of C, D, and E splits into two (in each case the corresponding value of $2b$, the coefficient of xy, may be replaced by its negative).

B.4   Kummer's Observations

Kummer published two major papers on his ideal numbers in 1847. While they will not be discussed, the following remarks in the shorter of those papers on the use of ideals and complex numbers in arithmetic are interesting. They provide evidence that Gauss's composition of forms was found to be very difficult, and they suggest that another approach might work. But we must be careful: Kummer's ideas about ideals are not the same as Dedekind's later ideas, and Kummer was not claiming that his work has resolved, or reformulated Gauss's theory—only that perhaps it could.

> The ideal factors of complex numbers enter, as shown, as factors of genuine complex numbers; so ideal factors multiplied with suitable other ones must always give rise to genuine complex numbers. Now this question of the connection between ideal factors and genuine complex numbers is, as I shall show, using the results I have already obtained, of great interest because it stands in close relationship with the most important section of number theory. The two most important results concerning this question are the following:
>
> • There is always a definite, finite number of ideal complex multipliers which are necessary and sufficient to make all possible ideal complex numbers into genuine ones;
> • Every ideal complex number has the property that a definite integer power of it makes a genuine complex number.
>
> I shall now go into a more detailed investigation of these two theorems. Two ideal complex numbers that, multiplied by one and the same ideal number, give the same genuine complex number I call *equivalent* or say belong to the same class because the investigation into genuine and ideal complex numbers is identical with the classification of certain compound forms of degree $\lambda - 1$ in $\lambda - 1$ variables, concerning which Dirichlet has found the principal result, although he has not yet published it, so that I do not exactly know if his principle of classification coincides with this, drawn from the theory of complex numbers. As a particular case the theory of forms of degree two in two variables, albeit only in the case when the determinant λ is a prime number, is comprehended in these researches, and

here our classification agrees with Gauss's and not with Legendre's. This sheds a bright light on Gauss's classification of forms of the second degree and on the true basis for the distinction between proper and improper equivalence, which it cannot be denied enters in the *Disquisitiones Arithmeticae* in a seemingly inappropriate way. If indeed two forms, such as $ax^2 + 2bx + cy^2$ and as $ax^2 - 2bx + cy^2$, or as $ax^2 + 2bx + cy^2$ and as $cx^2 + 2bx + ay^2$, are to be considered as belonging to different classes when in truth an essential difference between these two is not to be found, and yet on the other hand the Gaussian classification must be recognised as mostly corresponding to the nature of things; then one must if one is to take $ax^2 + 2bx + cy^2$ and as $ax^2 - 2bx + cy^2$ merely as representative of two other, entirely distinct ideas in number theory. But in truth these are nothing other that two distinct ideal factors belonging to one and the same number.

The entire theory of forms of the second degree in two variables can indeed be presented as the theory of complex numbers of the form $x + y\sqrt{D}$ and then necessarily leads to ideal complex numbers of the same kind. These are classified in just the same way as ideal multipliers which are necessary and sufficient to make genuine complex numbers of the form $x + y\sqrt{D}$. These coinciding with the Gaussian classification, this opens up the true basis for them.

The general study of ideal complex numbers has the closest analogy with Gauss's very difficult section on the composition of forms [which we discuss below], and the principal result, which Gauss proved for these forms on §§291 et seq. also holds for the composition of general ideal complex numbers. Here, to each class of ideal numbers there belongs another class that, multiplied by this one, brings about a genuine complex number (the genuine complex numbers are here the analogue of the principal class). Here there are also classes that, multiplied by themselves, give genuine complex numbers (the principal class) – thus, the class anceps. If one takes an ideal class $f(\alpha)$ and raises it to powers one always finds, by the second of the results above, a power which makes it a genuine complex number; if h is the smallest number for which $f(\alpha)^h$ is a genuine complex number the $f(\alpha)$, $f(\alpha)^2$, $f(\alpha)^3$, ... $f(\alpha)^h$ all belong to different classes.

At this point, said Kummer, he declined to go further into the domain of complex numbers. He would not attempt to find the true number of classes because he had been told that Dirichlet had already done this using similar principles. Instead, he brought this paper to a close with a short paragraph about cyclotomy.

Appendix C
Gauss's Fourth and Sixth Proofs
of Quadratic Reciprocity

The proofs are interesting because they draw on the theory of cyclotomy.

C.1 Gauss's Fourth Proof

In the survey of Gauss's proofs of quadratic reciprocity it would have been best to include his fourth proof (for reasons which I shall give shortly). Unfortunately, the original version, which he discovered in 1805 and published in 1811, is long and, if only for that reason, difficult to follow, although it has been translated into English (see Nagell 1951, pp. 74–80), so we shall have to content ourselves with a brief summary. I have chosen the one in Lemmermeyer (2000, pp. 95–97).

The proof is worth including because it too connects to the theory of cyclotomy and because it proved to be influential. Fix an odd prime p and let $\zeta = e^{2\pi i/p}$. It is trivial that $\sum_{k=0}^{p-1} \zeta^k = 0$, but Gauss was led to consider the quadratic sum $G = \sum_{k=0}^{p-1} \zeta^{k^2}$. It is a matter of calculation that

$$G^2 = \left(\frac{-1}{p}\right) p,$$

but Gauss wanted to evaluate G, and the determination of the sign of a square root that baffled Gauss from 1801, when he first had the idea of the proof, to late in 1805 when he wrote to his friend Olbers (3 September 1805):

> The determination of the sign of the root has vexed me for many years. This deficiency overshadowed everything that I found; over the last four years, there was rarely a week that I did not make one or another attempt, unsuccessfully, to untie the knot. Finally, a few days ago, I succeeded – but not as a result of my search but rather, I should say, through the mercy of God. As lightning strikes, the riddle has solved itself.

© Springer Nature Switzerland AG 2018
J. Gray, *A History of Abstract Algebra*, Springer Undergraduate Mathematics Series,
https://doi.org/10.1007/978-3-319-94773-0

Gauss was able to show that the positive square root was to be taken in each case:

$$\tau = \begin{cases} \sqrt{p} & \text{if } p \equiv 1 \bmod 4 \\ i\sqrt{p} & \text{if } p \equiv 3 \bmod 4. \end{cases}$$

Gauss then defined $\tau_m(k) = \sum_{t=0}^{m-1} \exp(\frac{2\pi i k t^2}{m})$; note that $\tau_p(1) = \tau$. He then showed that

$$\tau_{mn}(k) = \tau_m(an)\tau_n(am).$$

So, if p and q are odd primes

$$\tau_{pq}(1) = \tau_p(q)\tau_q(p) = \left(\frac{p}{q}\right)\left(\frac{q}{p}\right)\tau_p(1)\tau_q(1),$$

and the proof of quadratic reciprocity follows.

The day before Gauss made his celebrated discovery that the regular 17-gon is constructible by straight edge and circle, he had already had an important insight into the cyclotomic equation which he later described to a former student of his, Christian Gerling (6 January 1819).

> Already earlier I had found everything related to the separation of the roots of the equation $\frac{x^p-1}{x-1}$ into two groups on which the beautiful theorem in the D. A. on p. 637 depends, in the winter of 1796 (during my first semester in Göttingen), without having recorded the day. By thinking with great effort about the relation of all the roots to each other with respect to their arithmetic properties, I succeeded, while I was on a vacation in Braunschweig, on that day (before I got out of bed) in seeing this relation with utmost clarity, so that I was able to make on the spot the special application to the 17-gon and to verify it numerically (Gauss, *Werke* X.1, 125; translation in (Frei 2006, 161)).

The theorem Gauss was referring to occurs in §357 of the *Disquisitiones Arithmeticae*, where he shows how to use two (Gaussian) periods of the cyclotomic equation to factorise it (four times) as

$$4(x^{p-1} + x^{p-2} + \cdots + 1) = G(x)^2 - p^* H(x)^2,$$

where $G(x)$ and $H(x)$ are polynomials with integer coefficients and $p^* = +p$ if p is of the form $4n + 1$ and $p^* = -p$ if p is of the form $4n - 1$. In other words, $p^* = \left(\frac{-1}{p}\right)p$. In the previous paragraph, §356, Gauss had already shown that the periods are the roots of the polynomial $x^2 + x + \frac{1}{4}(1 - p^*)$, which makes it clear that another proof of quadratic reciprocity could be at hand. Indeed, he had mentioned this result much earlier, in §124, when discussing the quadratic characters of $+7$ and -7.

In other words, the pth cyclotomic polynomial factorises as the difference or the sum of two squares over the field $\mathbb{Q}(\sqrt{p^*})$ according as -1 is or is not a square modulo p. Or, to be even more modern in our expression of the result, the field

$\mathbb{Q}(\sqrt{p^*})$ is a subfield of the cyclotomic field $\mathbb{Q}(\zeta)$, where ζ is a primitive pth root of unity.

Here are two small examples of this factorisation.

- When $p = 5$, $G(x) = 2x^2 + x + 2$ and $H(x) = x$;
- When $p = 7$, $G(x) = 2x^3 + x^2 - x - 2$ and $H(x) = x^2 + x$.

C.2 Gauss's Sixth Proof

In this section I give an English translation of Gauss's sixth proof of quadratic reciprocity. The proof is not easy, and it is followed by a commentary that should help. It is interesting because it illustrates how Gauss showed that ideas in one part of mathematics are connected, often in mysterious ways, to ideas in other parts. It is also interesting for the unusually personal introduction that Gauss gave to it. This proof also uses quadratic sums, but avoids the difficult determination of the sign.

From Gauss, 'New proofs and extensions of the fundamental theorem in the study of quadratic residues', (1818b).

The fundamental theorem of quadratic residues, which belongs to one of the most beautiful truths of higher arithmetic, is indeed easily found by induction but it was only to be proved after extensive difficulties. In such research it is often to be seen that the proofs of the simplest truths, which the researcher can certainly discover by induction for himself, lie buried very deeply and can only be brought to the light of day first after many vain attempts and in a different way from the one originally sought. Further, it not seldom appears that as soon as one way is found many other ways are found that lead to the same goal, some shorter and more direct, others at the same time coming from the side and proceeding on quite different principles, so that between these and the preceding researches one can scarcely conjecture any connection. Such a wonderful connection between hidden truths give these considerations not only a certain appropriate richness, they should be diligently researched and clearly presented because they not seldom lead to new means and extensions of science.

Although the arithmetical theorem dealt with here has, by earlier efforts, been provided with four entirely different proofs, and can seem to be completely finished, I return here to some new observations and present two new proofs, which shed a new light on these facts. The first is in a certain way related to the third, because it derives from the same lemma, but later it takes a new path so that it can with justice be counted as a new proof that does not coincide with the third and should not be considered inferior. But the sixth proof calls upon a completely different and most subtle principle, and gives a new example of the wonderful connection between arithmetic truths that at first glance seem to lie very far from one another. These two proofs provide us with a new and very simple algorithm for deciding if a given integer is a quadratic residue or nonresidue of a given prime number.

Still another reason was provided, which led me to publish a new proof now that I had already spoken of 9 years earlier. In fact, as I was occupied in 1805 with the theory of cubic and biquadratic residues, where a very difficult circumstance that I had begun to work through, happened to draw on almost the same skill as originally in the theory of quadratic residues. Without further notice, indeed, were those theorems—which had handled those questions completely, and had presented a wonderful analogy with the corresponding theorems about quadratic residues— found by induction as soon as they were only sought for in a proper way, and all remaining researches led on all sides to a complete proof that for a long time I had sought in vain. This was indeed the impulse that I so much sought to add more and more proofs of the already-known theorems on quadratic residues, in the hope that from these many different methods, one or another could illuminate something in the related circumstances. This hope was in no way idle, and untiring efforts were finally crowned with success. Shortly, I will be in a position to publish the fruits of my studies, but before I undertake that difficult work I have decided to return once more to the theory of quadratic residues, and to say what is still to be said, and so in a certain sense to bid farewell to this part of the higher arithmetic.

Gauss's Sixth Proof of Quadratic Reciprocity

§1

Theorem. *1. Let p denote a positive odd prime number, n a positive integer not divisible by p, x an indefinite quantity, then the function*

$$1 + x^n + x^{2n} + \cdots + x^{n(p-1)}$$

is divisible by the function

$$1 + x + x^2 + \cdots + x^{p-1}.$$

Proof. Let g be a positive integer such that $gn \equiv 1(p)$ and suppose $gn = 1 + hp$, then

$$\frac{1 + x^n + x^{2n} + \cdots + x^{n(p-1)}}{1 + x + x^2 + \cdots + x^{p-1}} = \frac{(1 - x^{np})(1 - x)}{(1 - x^n)(1 - x^p)}$$

$$= \frac{(1 - x^{np})(1 - x^{gn} - x + x^{hp+1})}{(1 - x^n)(1 - x^p)}$$

$$= \frac{1 - x^{np}}{1 - x^p} \cdot \frac{1 - x^{gn}}{1 - x^n} - \frac{x(1 - x^{np})}{1 - x^n} \cdot \frac{1 - x^{hn}}{1 - x^p},$$

which is obviously an integral function. Q.E.D.

So every integral function of x that is divisible by $\frac{1-x^{np}}{1-x^n}$ is divisible by $\frac{1-x^p}{1-x}$.

§2

Let α denote a positive primitive root for the modulus p, i.e. let α be a positive integer with the property that the smallest positive residues of the powers $1, \alpha, \alpha^2, \ldots \alpha^{p-2}$ modulo p are, ignoring the sequence in which they occur, identical with the numbers $1, 2, 3, \ldots, p - 1$. Further, denote by $f(x)$ the function

$$x + x^\alpha + x^{\alpha^2} + \cdots + x^{\alpha^{p-2}} + 1,$$

so obviously

$$f(x) - 1 - x - x^2 - \cdots - x^{p-1}$$

will be divisible by $1 - x^p$ and a fortiori it will also be divisible by $\frac{1-x^p}{1-x} = 1 + x + x^2 + \cdots + x^{p-1}$; so $f(x)$ will also be divisible by this last function. But it follows from this, if x represents an indefinite quantity, that also $f(x^n)$ is divisible by $\frac{1-x^{np}}{1-x^n}$ and (by the previous paragraph) by $\frac{1-x^p}{1-x}$, at least whenever n is an integer that is not divisible by p. If, however, n is divisible by p then each member of the function $f(x^n)$ diminished by 1 will be divisible by $1 - x^p$, and in this case $f(x^n) - p$ will be divisible by $1 - x^p$ and therefore also by $\frac{1-x^p}{1-x}$.

§3

Theorem. 2. *If one sets*

$$x - x^\alpha + x^{\alpha^2} + \cdots - x^{\alpha^{p-2}} = \xi,$$

then $\xi^2 \mp p$ *will be divisible by* $\frac{1-x^p}{1-x}$ *if one takes the upper sign whenever* p *is of the form* $4k + 1$ *and the lower sign whenever* p *is of the form* $4k + 3$.

Proof. It is easy to see that of the following $p - 1$ functions,[1] where $1 \leq j \leq p-1$,

$$(-1)^{j-1}x^{\alpha^{j-1}}\xi + \sum_{k=2}^{p}(-1)^{k-1}x^{\alpha^{j+k-3}+\alpha^{j-1}}$$

---

[1] Gauss wrote these expressions out explicitly.

the first equals 0 and the others are divisible by $1 - x^p$. Therefore the sum of them all is divisible by $1 - x^p$, which is:

$$\xi^2 - \left(f(x^2) - 1\right) + \left(f(x^{\alpha+1}) - 1\right) - \left(f(x^{\alpha^2+1}) - 1\right) + \cdots + \left(f(x^{\alpha^{p-2}+1}) - 1\right)$$

$$= \xi^2 - f(x^2) + f(x^{\alpha+1}) - f(x^{\alpha^2+1}) + \cdots + f(x^{\alpha^{p-2}+1}) = \Omega.$$

So this expression Ω is also divisible by $\frac{1-x^p}{1-x}$. Now, among the exponents $2, \alpha + 1, \alpha^2 + 1, \ldots, \alpha^{p-2}$ only one is divisible by p, namely $\alpha^{\frac{1}{2}(p-1)} + 1$, so, by the previous paragraph, the following individual parts of the expression Ω

$$f(x^2), f(x^{\alpha+1}), f(x^{\alpha^2+1}), \ldots$$

with the exception of the term $f(x^{\frac{1}{2}\alpha(p-1)+1})$ are divisible by $\frac{1-x^p}{1-x}$. One can therefore set each term aside so that the function

$$\xi^2 \mp f(x^{\frac{1}{2}\alpha(p-1)+1})$$

remains divisible by $\frac{1-x^p}{1-x}$, where the upper or lower sign is chosen according as p is of the form $4k + 1$ or $4k + 3$. And because moreover $f(x^{\frac{1}{2}\alpha(p-1)+1}) - p$ is divisible by $\frac{1-x^p}{1-x}$, so also $\xi^2 \mp p$ will be divisible by $\frac{1-x^p}{1-x}$, as was to be shown.

So that the double sign cannot have an ambiguous value, we will denote by ε the number $+1$ or -1 according as p is of the form $4k + 1$ or $4k + 3$. Therefore $\frac{(1-x)(\xi^2-\varepsilon p)}{1-x^p}$ will be an integral function of x, which we shall denote by Z.

§4

Let q be a positive odd number and therefore $\frac{1}{2}(q - 1)$ an integer. Then accordingly $(\xi^2)^{\frac{1}{2}(q-1)} - (\varepsilon p)^{\frac{1}{2}(q-1)}$ will be divisible by $\xi^2 - \varepsilon p$ and therefore also by $\frac{1-x^p}{1-x}$. If we set

$$\varepsilon^{\frac{1}{2}(q-1)} = \delta, \text{ and } \xi^{q-1} - \delta p^{\frac{1}{2}(q-1)} = \frac{1 - x^p}{1 - x} Y,$$

then Y will be an integer function of x and $\delta = +1$ if either or both of p, q are of the form $4k + 1$, and with $\delta = -1$ if both p, q are of the form $4k + 3$.

§5

If we now assume that q is likewise a prime number (different from p), then, from the theorem proved in the D.A. §51, we have that

$$\xi^q - (x^q - x^{q\alpha} - x^{q\alpha^2} - \cdots - x^{q\alpha^{p-2}})$$

is divisible by q, or is of the form qX, where X is an integral function of the numerical coefficients of x (which is true of the other integral functions Z, Y, W that are involved). If we denote the index of the number q with respect to the modulus p and the primitive root α by μ, so $q \equiv \alpha^\mu \pmod{p}$, then the numbers $q, q\alpha, q\alpha^2, \ldots, q\alpha^{p-2}$ are congruent modulo p to the numbers $\alpha^\mu, \alpha^{\mu+1}, \alpha^{\mu+2}, \ldots, \alpha^{p-2}, 1, \alpha, \alpha^2, \ldots \alpha^{\mu-1}$, and therefore[2] each of $x^{q\alpha^j} - x^{\alpha^{\mu+j}}$, $j = 0, 1, \ldots, p - 2 - \mu$, and each of $x^{q\alpha^j} - x^{\alpha^k}$, $j = p - 1 - \mu, \ldots p - 2$, $k = j - p + 1 + \mu$, is divisible by $1 - x^p$. If one takes these quantities apart from their sign and adds them, it is clear that the function

$$x^q - x^{q\alpha} - x^{q\alpha^2} - \cdots - x^{q\alpha^{p-2}} \mp \xi$$

is divisible by $1 - x^p$ and indeed the upper or lower sign is valid according as μ is even or odd, i.e. according as q is a quadratic residue or nonresidue modulo p. If we therefore set

$$x^q - x^{q\alpha} - x^{q\alpha^2} - \cdots - x^{q\alpha^{p-2}} - \gamma\xi = (1 - x^p)W,$$

where $\gamma = +1$ or $= -1$ according as q is a quadratic residue or nonresidue modulo p, then clearly W is an integral function.

§6

After these preparations we derive from the combination of the preceding equations that

$$q\xi X = \varepsilon p(\delta p^{\frac{1}{2}(q-1)} - \gamma) + \frac{1 - x^p}{1 - x}\left(Z(\delta p^{\frac{1}{2}(q-1)} - \gamma) + Y\xi^2 - W\xi(1 - x)\right).$$

We now assume that on dividing the function ξX by

$$x^{p-1} + x^{p-2} + \cdots + x + 1$$

the quotient is U and the remainder T, or

$$\xi X = \frac{1 - x^p}{1 - x}U + T,$$

so that U, T are integral functions with respect to the numerical coefficients and T is of a lower degree than the divisor. ;Then it is the case that

$$qT - \varepsilon p(\delta p^{\frac{1}{2}(q-1)} - \gamma) = \frac{1 - x^p}{1 - x}\left(Z(\delta p^{\frac{1}{2}(q-1)} - \gamma) + Y\xi^2 - W\xi(1 - x) - qU\right)$$

---

[2]Gauss wrote these expressions out explicitly.

and obviously this equation can hold only when the left and right sides both vanish. For this, $\varepsilon p(\delta p^{\frac{1}{2}(q-1)} - \gamma)$ is divisible by q, as likewise is $\delta p^{\frac{1}{2}(q-1)} - \gamma$, and because $\delta^2 = 1$ the number $p^{\frac{1}{2}(q-1)} - \gamma\delta$ is divisible by q.

If now we denote by β the positive or negative unit, according as p is a quadratic residue or nonresidue modulo q, then $p^{\frac{1}{2}(q-1)} - \beta$ will be divisible by q and therefore also $\beta - \gamma\delta$; this can only happen when $\beta = \gamma\delta$. From this the fundamental theorem follows immediately. Indeed:

I: Whenever one, the other, or both of the numbers p, q are of the form $4k + 1$ and therefore $\delta = +1$ and so $\beta = \gamma$ then so simultaneously is q a quadratic residue for p and p a quadratic residue for q, or simultaneously q is a nonresidue for p and p a nonresidue for q.

II: Whenever both numbers p, q are of the form $4k + 3$ and therefore $\delta = -1$, then $\beta = -\gamma$ and either simultaneously is q a quadratic residue for p and p a nonresidue for q or simultaneously is q a nonresidue for p and p a residue for q. Q.E.D.

C.3   Commentary

We proceed in two steps. First we introduce some ad hoc notation designed to make Gauss's argument easier to follow. Then we bring this notation into line with current notation, so as to make the connection to quadratic reciprocity easier to see.

An Outline of Gauss's Proof

Let us define

$$c(x) = 1 + x + x^2 + \cdots + x^{p-1} = \frac{1 - x^p}{1 - x}.$$

In §1 Gauss showed that

$$c(x)|c(x^n).$$

In §2 he showed that

$$c(x)|f(x),$$

where

$$f(x) = 1 + x + x^\alpha + x^{\alpha^2} + \cdots + x^{\alpha^{p-2}},$$

and α is the smallest primitive root of p, and deduced that if $p \nmid n$ then $c(x^n) \mid f(x^n)$, but if $p \mid n$ then $c(x) \mid f(x^n) - p$.

In reading §3 it helps to note that ξ can be written as this function of x:

$$g(x) = 1 - x^\alpha + x^{\alpha^2} + \cdots - x^{\alpha^{p-2}}.$$

So the $p - 1$ functions Gauss considered can be written as

$$x^{\alpha^{j-1}} \left(\xi - g(x^{\alpha^{j-1}}) \right).$$

This makes it easier to see what the sum of these expressions is and why $c(x) \mid \Omega$. Gauss then deduced that

$$c(x) \mid \xi^2 \mp f(x^{\alpha^{\frac{1}{2}(p-1)}+1}).$$

In §4 Gauss introduced the odd number q, and in §5 he set $q \equiv \alpha^\mu \pmod{p}$, considered $\xi^q - g(x^q)$ and observed that $1 - x^p$ divides $x^{q\alpha^j} - x^{\alpha^{p+j}}$ for $j = 0, 1, 2, \ldots, p - 2$. From this he deduced that

$$1 - x^p \mid g(x^q) \mp xi$$

with the rule for the sign being that the minus sign is taken if and only if μ is even. So he wrote

$$g(x^q) - \gamma\xi = (1 - x^p)W,$$

where γ is $+1$ if q is a quadratic residue modulo p and -1 otherwise.

Finally, in §6 let us set $A = \delta p^{\frac{1}{2}(q-1)} - \gamma$, so the first equation becomes

$$q\xi X = \varepsilon p A + c(x) \left(ZA + Y\xi^2 - W\xi(1 - x) \right),$$

which Gauss rewrote, using $\xi X = c(x)U + T$, as

$$qT - \varepsilon p A = c(x) \left(ZA + Y\xi^2 - W(1 - x) - qU \right).$$

For both sides to vanish it is necessary that $q \mid \varepsilon p A$ and therefore that $q \mid A$, and therefore that

$$q \mid p^{\frac{1}{2}(q-1)} - \gamma\delta.$$

Moving Towards Modern Notation

The first thing to notice is an implication of the use of the primitive root. It is clear that the residues of the form α^{2k} are squares mod p and the residues of the form α^{2k+1} are nonresidues. So we can rewrite

$$\xi = g(x) = 1 - x^{\alpha} + x^{\alpha^2} + \cdots - x^{\alpha^{p-2}}$$

as

$$g(x) = 1 - \left(\left(\frac{1}{p}\right) x^{\alpha} + \left(\frac{2}{p}\right) x^{\alpha^2} + \cdots + \left(\frac{p-2}{p}\right) x^{\alpha^{p-2}} \right).$$

Similarly, we can write

$$g(x^j) = 1 - x^{j\alpha} + x^{j\alpha^2} + \cdots - x^{j\alpha^{p-2}}$$

so

$$g(x^j) = 1 - \left(\left(\frac{j}{p}\right) x^{\alpha} + \left(\frac{2j}{p}\right) x^{\alpha^2} + \cdots + \left(\frac{j(p-2)}{p}\right) x^{\alpha^{p-2}} \right).$$

Now in any case $c(x)$ divides $f(x^{\alpha^{\frac{1}{2}(p-1)+1}}) - p$ (by §1), so once Gauss deduced that $c(x)$ divides $\xi^2 \pm f(x^{\alpha^{\frac{1}{2}(p-1)+1}})$ if follows that $c(x)$ divides $\xi^2 \pm p$. Furthermore, if $p = 4k + 1$ then $\alpha^{\frac{1}{2}(p-1)+1} = \alpha^{2k+1}$, and if $p = 4k + 3$ then $\alpha^{\frac{1}{2}(p-1)+1} = \alpha^{2k+2}$, so in §3 Gauss showed that $c(x)$ divides $\xi^2 - p$ if $p = 4k + 1$ and $c(x)$ divides $\xi^2 + p$ if $p = 4k + 3$. So in the first case $+p$ is a square modulo $c(x)$, and in the second case $-p$ is a square modulo $c(x)$. Notice that the rule for ε means that $\varepsilon = \left(\frac{-1}{p}\right)$.

It will help to put a modern gloss on this. We can follow what Gauss did by working in the ring $\mathbb{Z}[x]$. We need to know, what is certainly the case, that an integer that is prime in $\mathbb{Z}$ remains prime in this ring, and that when p is a prime the polynomial $c(x)$ is irreducible, so the quotient $\mathbb{Q}[x]/\langle c(x)\rangle$ is a field, and we have just seen that Gauss proved that $\pm p$ are squares in this field, so the fields $\mathbb{Q}(\sqrt{\pm p})$ are subfields of the cyclotomic field $\mathbb{Q}(\zeta)$, where $\zeta = e^{2\pi i/p}$ is a primitive pth root of unity.

It remains to see how this helps to prove quadratic reciprocity. The first clue, in §4, is that $\delta = 1$ if either or both of p, q are of the form $4k + 1$ and $\delta = -1$ if both p and q are of the form $4k - 1$. So, $\delta = (-1)^{\frac{1}{2}(p-1)\frac{1}{2}(q-1)}$, and *if* we knew that quadratic reciprocity was true, we would know that $\delta = \left(\frac{p}{q}\right)\left(\frac{q}{p}\right)$.

Next, in §5, Gauss introduced $\gamma = \left(\frac{q}{p}\right)$, and in §6 showed that

$$q\xi X = \varepsilon p(\delta p^{\frac{1}{2}(q-1)} - \gamma) + \frac{1-x^p}{1-x}\left(Z(\delta p^{\frac{1}{2}(q-1)} - \gamma) + Y\xi^2 - W\xi(1-x)\right).$$

The quantity $\varepsilon p(\delta p^{\frac{1}{2}(q-1)} - \gamma)$ is $\left(\frac{-1}{p}\right)p\left((-1)^{\frac{1}{2}(p-1)\frac{1}{2}(q-1)}p^{\frac{1}{2}(q-1)} - \left(\frac{q}{p}\right)\right)$, and modulo q we have $p^{\frac{1}{2}(q-1)} \equiv \left(\frac{p}{q}\right)$.

Gauss now introduced $\beta = \left(\frac{p}{q}\right)$, so his conclusion, tacitly working modulo $c(x)$ and q, is that $\beta = \gamma\delta$ says that

$$\left(\frac{p}{q}\right) = \left(\frac{q}{p}\right)(-1)^{\frac{1}{2}(p-1)\frac{1}{2}(q-1)},$$

which is indeed the statement of quadratic reciprocity for odd primes p and q.

Appendix D
From Jordan's *Traité*

D.1 Jordan, Preface to the *Traité*

The problem of the algebraic solution of equations is one of the first that imposed itself on the research of mathematicians. From the beginnings of modern algebra, several procedures have been proposed for solving equations of the first four degrees: but these diverse methods, isolated one from another and founded on artifices of calculation, constituted facts rather than a theory until the day when Lagrange submitted them to a profound analysis, found the common foundation on which each rested and reduced each to a truly analytic method, taking as his starting point the theory of substitutions.

The failure of Lagrange's method for equations of degree greater than four gave rise to the belief that it was impossible to solve them by radicals. In fact, Abel proved this fundamental proposition; then, studying the particular equations that are susceptible to this method of resolution obtained the remarkable class of equations that bear his name. He undertook this work with great enthusiasm until death struck him down; the fragments which he left to us allow us to judge the importance of his unfinished edifice.

These beautiful results were, however, only the prelude to a great discovery. It was reserved to Galois to place the theory of equations on its definitive base, by showing that to each equation there corresponded a group of substitutions, in which its essential characteristics are reflected, notably all those that deal with its resolution by means of other, auxiliary, equations. According to this principle, being given an arbitrary equation it suffices to know one of its characteristic properties to determine its group, from which, reciprocally one deduces its other properties.

From this elevated point of view, the problem of the solution by radicals, which yesterday seemed to form the sole object of the theory of equations, appears as only the first link in a long chain of questions relative to the transformation of irrationals and their classification. Galois, applying his general methods to this particular problem, found without difficulty the characteristic property of the groups of

© Springer Nature Switzerland AG 2018
J. Gray, *A History of Abstract Algebra*, Springer Undergraduate Mathematics Series,
https://doi.org/10.1007/978-3-319-94773-0

equations that are solvable by radicals, the explicit form of the groups of equations of prime degree, and two important theorems relative to the case of composite degrees. But in the preparation of his publications he left several fundamental propositions without proof. M. Betti did not delay in filling in this gap with an important memoir, where the complete series of Galois's theorems was established rigorously for the first time.

The study of the division of the transcendental functions offered Galois a new and brilliant application of his method. A long time before, Gauss had proved that the equations for the division of the circle are solvable by radicals; Abel had established the same result for the division equations of elliptic functions, supposing that the division of the periods had been carried out, a proposition that M. Hermite was to extend to abelian functions. But it remained to study the modular equations on which the division of the periods depends. Galois determined their groups and remarked that those of degrees 6, 8, or 12 can be reduced by a degree. M. Hermite, carrying out this reduction, showed that it is enough to solve the equations of the first four degrees to identify the reduced equation obtained in the case of quinquisection with the general equation of the fifth degree, which furnishes a solution of the latter equation by means of elliptic functions. M. Kronecker came at the same time to the same result by almost the inverse method, which M. Brioschi then took up and developed in some remarkable pages.

Another fertile path was opened to analysts by the celebrated memoirs of M. Hesse on the inflection points of curves of the third order. In fact, the problems of analytic geometry provide a slew of other remarkable equations whose properties, studied by the most famous geometers, principally MM. Cayley, Clebsch, Hesse, Kummer, Salmon, Steiner, are well known today and to which the methods of Galois can be applied without difficulty.

The theory of substitutions, which thus became the foundation of all questions about equations, is still only a little advanced. Lagrange only made it open; Cayley attacked it on several occasions. MM. Bertrand, Brioschi, Hermite, Kronecker, J.-A. Serret, E. Mathieu have alike been occupied; but, despite the importance of their work, the question, as advanced as it is difficult, remains today still almost intact. Three fundamental notions have however begun to emerge: that of primitivity, which as already indicated was found in the works of Gauss and Abel; that of transitivity, which belongs to Cayley; and finally the distinction between simple and compound group. It is to Galois that this last notion, the most important of them all, is due.

The purpose of this work is to develop the methods of Galois and to constitute them into a body of doctrine by showing how easily they allow us to solve all the principal problems in the theory of equations. To make them more readily known, we have taken the elements as our point of departure, and we have expounded, as well as our own research, all the principal results obtained by the geometers who have preceded us. But we have often modified quite profoundly the statement of these propositions and their method of proof, while deriving everything from uniform principles that are as general as possible. The abundance of material has constrained us, besides, to suppress all historical development. This is why we have left on one side, not without regret, the celebrated demonstration given by Abel of the impossibility of solving the fifth degree equation by radicals, this beautiful theorem that can be established today by much simpler considerations.

Among the works that we have consulted, we must cite, in particular, besides the Works of Galois, to which all this is only a commentary, the *Cours d'Algèbre Supérieure* of M. J.-A. Serret. It is the assiduous reading of this book which introduced us to algebra and inspired in us the desire to contribute to its progress.

We must equally thank here MM. Clebsch and Kronecker for the precious information that they provided us. It is thanks to the liberal communications of M. Clebsch that we could tackle the geometrical problems in Book III, Chapter III, the study of the groups of Steiner, and the trisection of hyperelliptic functions. We owe to M. Kronecker the notion of the group of the division equation of the latter functions. We would have liked to draw more than we have done from the works of this illustrious author concerning these equations. Several causes prevented us: the thoroughly arithmetic nature of the methods, so different from our own; the difficulty in reconstructing whole a sequence of demonstrations that most often are barely indicated; finally the hope to see one day grouped into a consecutive and complete body of doctrine these beautiful theorems that presently are the envy and the despair of geometers.

This Work is divided into four Books:

The first book is devoted to indispensable notions in the theory of congruences;

Book II is divided into two Chapters, the first devoted to the study of substitutions in general, the second to those substitutions defined analytically and principally to linear substitutions.

Book III has four Chapters. In the first, we state the principles of the general theory of equations. The following three contain applications to algebra, to geometry, and to the theory of transcendentals.

Finally, in Book IV, divided into seven Chapters, we determine the various types of equations solvable by radicals, and we obtain for these types a complete system of classification.

D.2   Jordan, *General Theory of Irrationals*

Introduction

What follows is a fairly literal translation of Book III, Chapter 1 of Jordan's *Traité* of 1870, the part where he introduced his version of Galois theory. The intention is to make an accurate impression of this very important work available to an English audience. Three more technical passages are presently omitted.

I have kept the word 'substitution', where I could have used the word 'element', because it reminds us that the groups involved act by permuting roots of an equation. But I replaced the phrase that a subgroup 'commutes with all the elements' of a group with the modern term 'normal subgroup', leaving 'normal' in inverted commas to signal that I have introduced this word. Jordan's long sentences and nineteenth-century constructions have been intermittently modernised when I thought it helped.

A number of more modern turns of phrase have, however, been avoided. Quantities are called 'functions' that might have been called 'expressions'; a quantity is 'rationally expressible' if it is a member of the base field, which is usually the rational field enlarged, perhaps, by some roots of unity.

The notation is almost entirely Jordan's, but I could not write I for a subgroup and turned therefore to K, and I replaced expressions like $1 \cdot 2 \cdots m$ by $m!$. Set-theoretic notation and phrases were not introduced into mathematics until later and they have been avoided here. I chose not to replace Jordan's slightly cumbersome presentation of cosets by a modern one in order to keep closer to the historical flavour of the original.

Jordan numbered his theorems and corollaries in roman; the arabic numbering is provided by TEX. Otherwise I have tried to follow the layout of the original, including the use of italics.

§348 Let $F(x) = 0$ be an arbitrary algebraic equation of degree m. It will have m roots: and one knows that every symmetric function of these roots can be written rationally in terms of the coefficients of the equation.

In general these functions are the only ones that enjoy this property. Suppose in fact that one has $\varphi = \psi$, φ being a non-symmetric function of the roots, and ψ a rational function of the coefficients. Let us replace in ψ each of the coefficients by its value as a function of the roots; the equation $\varphi = \psi$ will become a relation between these roots, a relation that cannot reduce to an identity because the second member is symmetric and the first is not. But one cannot admit that there is in general and of necessity any relation of this kind between the roots: because one can create an equation of degree m having as its m roots entirely arbitrary quantities $x_1, \ldots, x_m$. It is therefore only in certain particular cases that such relations can exist between the roots that a non-symmetric function of these roots is expressible by means of the coefficients.

But one can generalise the problem and look for what are, for every given equation $F(x) = 0$, the functions of the roots that are capable of being expressed rationally as a function of the coefficients and certain irrationals given arbitrarily a priori, irrationals that we will call *adjoined to the equation*.

We consider as *rational* nonetheless every quantity expressible rationally by means of the coefficients of the equation and the adjoined quantities.

An equation with rational coefficients is said to be *irreducible* when it has no root in common with an equation of lower degree and rational coefficients.

§349 Lemma I If one of the roots of an irreducible equation $f(x) = 0$ satisfies another equation with rational coefficients $\varphi(x) = 0$ then all its roots satisfy it.

In fact, let us look for the greatest common divisor of $\varphi(x)$ and $f(x)$; it cannot reduce to a constant the equations $\varphi(x) = 0$ and $f(x) = 0$ having roots in common. It will therefore be a function of x, $\psi(x)$, and equating it to zero one will have an equation whose roots evidently satisfy each of the two equations $\varphi(x) = 0$ and

$f(x) = 0$. This last equation being irreducible, the degree of $\psi(x)$ cannot be less than that of $f(x)$; therefore $\psi(x)$ is equal to $f(x)$ up to a constant factor.

Lemma 1 *If all the roots of the equation $\varphi(x) = 0$ satisfy the equation $f(x) = 0$ then $\varphi(x)$ will be an exact factor of $f(x)$ up to a constant factor.*

In fact, $\varphi(x)$ is divisible by $f(x)$ (Lemma I). If the quotient of this division does not reduce to a constant, it will be divisible by $f(x)$, etc.

§350 Lemma II Let $F(x) = 0$ be an equation all of whose roots $x_1, \ldots x_m$ are distinct. One can find a function V of these roots such that the $m!$ expressions that one obtains by permuting the roots in every possible way have distinct numerical values.

In fact, let us set

$$V = M_1 x_1 + M_2 x_2 + \cdots,$$

$M_1, M_2, \ldots$ being undetermined integers. By equating arbitrarily two of the functions that one obtains among them from V by substitutions of the roots, one obtains a conditional equation that the $M_1, M_2, \ldots$ must satisfy. None of these equations is an identity; because the coefficients of $M_1, M_2, \ldots$ in each of them are the differences of the roots $x_1, \ldots x_m$ which by hypothesis do not vanish. Moreover these equations are finite in number. It is therefore easy to determine integers $M_1, M_2, \ldots$ in such a way that they satisfy none of these equations.

In what follows we denote by V_1 one of the values of the function V, chosen arbitrarily; by $V_a, V_b, \ldots$ the values which one deduces from it when one makes substitutions represented by $a, b, \ldots$ between the roots.

§351 Lemma Let G be any group of substitutions between the roots $x_1, \ldots, x_m$. One can form a function of these roots the numerical value of which is unaltered by the substitutions of G and changes with all other substitutions.

In fact, let $1, a, b, \ldots$ be the substitutions of G. Let us set

$$W_1 = (X - V_1)(X - V_a)(X - V_b) \ldots,$$

X being an undetermined constant. A substitution of G, such as a, transforms W_1 into

$$W_1 = (X - V_a)(X - V_{a^2})(X - V_{ba}) \ldots = W_a;$$

but the substitutions of G, forming a group, are the same as $a, a^2, ba, \ldots$ up to order. The binomial factors that compose W_a are therefore the same, up to order, as those that compose W_1; this function is therefore not altered by the substitutions of G.

On the contrary, let α be a substitution foreign to this group, it transforms W_1 into

$$W_1 = (X - V_\alpha)(X - V_{a\alpha})(X - V_{b\alpha})\ldots = W_\alpha.$$

The binomial factors that compose W_α being essentially different from those that compose W_1 these two expressions are not identical, and cannot take equal values except for certain particular values of the quantity X that it will be easy to avoid.

§352 Lemma III The function V being chosen as in Lemma II, one can express each of the roots $x_1, \ldots, x_m$ as a rational function of V and the coefficients of $F(x)$.

Let $V_1, \ldots V_\mu$ be the $\mu = (m-1)!$ values taken by V as one permutes the $m-1$ roots $x_2, \ldots, x_m$ without changing the position of x_1. One can form an equation in V of degree μ, namely

$$(V - V_1)(V - V_2)\ldots(V - V_\mu) = 0, \tag{D.1}$$

whose roots $V_1, V_2, \ldots$ are all distinct and of which the coefficients, which are symmetric functions of the roots $x_2, x_3, \ldots x_m$ of the equation

$$\frac{F(x)}{x - x_1} = 0$$

are expressible rationally in the coefficients of this equation, that is to say as a function of x_1 and the coefficients of $F(x)$. It follows that Eq. (D.1) can be put in the form

$$f(V, x_1) = 0,$$

f denoting a rational function of V and x_1. Now, this equation is satisfied by $V = V_1$. One therefore has identically

$$f(V_1, x_1) = 0,$$

whence it follows that the equation

$$f(V_1, x) = 0$$

will be satisfied by $x = x_1$. Consequently the equations $F(x) = 0$ and $f(V_1, x) = 0$ have a root in common, x_1. On the other hand, the equations cannot have any other root in common. For, if they had another, x_2, the equation

$$f(V, x_1) = 0$$

will be satisfied by $V = V_1$. Now this equation follows from the equation

$$f(V, x_1) = m_0, \text{ or } (V - V_1)(V - V_2)\ldots(V - V_\mu) = 0$$

by changing x_1 and x_2 into each other. On the other hand by this exchange the quantities $V_1, V_2, \ldots, V_\mu$ change into others, $V_1', V_2', \ldots, V_\mu'$ all distinct from the first ones, by hypothesis; the equation $f(V, x_2) = 0$ can therefore be put in the form

$$(V - V_1')(V - V_2')\ldots(V - V_\mu') = 0,$$

and one sees that it cannot have V_1 as a root.

The equations $F(x) = 0$ and $f(V_1, x) = 0$ having only a single root in common, one easily finds this root. To do this, one looks for the greatest common divisor of $F(x)$ and $f(V_1, x)$ and one continues this operation until one obtains a remainder of degree one; in equating this remainder to zero one obtains an equation that reveals the value of x_1, and this value will evidently be rational in V_1 because the operation of greatest common divisor can never introduce radicals.

One can operate in the same way to find the other roots, and one will thus obtain rational expressions for all the roots, such as

$$x_1 = \psi_1(V_1), \quad x_2 = \psi_2(V_1), \ldots (*).$$

§353 Lemma [Fundamental theorem, Theorem I] Let $F(x) = 0$ be an equation with distinct roots $x_1, \ldots, x_m$ with certain auxiliary quantities $y, z, \ldots$ adjoined. Then there is always a group of substitutions on the roots such that every function of the roots that is numerically unaltered by the substitutions of this group is rationally expressible and conversely.

Proof. Let V be a function of the roots altered by every substitution, and denote the substitutions by $1, a, b, c, \ldots$. Then the quantity V_1 is a root of the equation

$$(X - V_1)(X - V_a)\ldots = 0,$$

whose coefficients, being symmetric in the roots $x_1, \ldots, x_m$, are rational. If this equation is not irreducible the left-hand side will break into several irreducible factors. Let $Y = (X - V_1)(X - V_a)\ldots$ be the one of these that vanishes for $X = V_1$, so V_1 is a root of the irreducible equation $Y = 0$.

Then every function φ of the roots that is invariant under the substitutions $1, a, b, \ldots$ is expressible rationally. For, each of the roots $x_1, \ldots, x_m$ being a rational function of V_1 and the coefficients of $F(x)$, φ itself will be a rational function of V_1 and its coefficients. Let $\psi(V_1)$ be this function. But these substitutions change V_1 respectively into $V_a, V_b, \ldots$ and do not change the coefficients of $F(x)$, so one will have

$$\psi(V_1) = \psi(V_a) = \psi(V_b) = \cdots = \frac{1}{\mu}(\psi(V_1) + \psi(V_a) + \psi(V_b) + \cdots),$$

where μ is the degree of the polynomial Y. This function, being symmetric with respect to the roots of the equation $Y = 0$, can be expressed rationally by the coefficients of this equation, which are themselves rational.

Conversely, every function expressible rationally is invariant under the substitutions $1, a, b, \ldots$. Indeed, let $\varphi = \psi(V_1)$ be such a function. V_1 satisfies the equation $\varphi = \psi(V_1)$ which has rational coefficients and so all the roots of the irreducible equation $Y = 0$ must satisfy it. Therefore the function $\psi(V_1)$ does not alter when V_1 is replaced successively by $V_a, V_b, \ldots$ which is the same as operating on the roots $x_1, \ldots, x_m$ by the substitutions $a, b, \ldots$,

§354 It only remains to show that *the substitutions* $1, a, b, \ldots$ *form a group*, which presents no difficulty.

The polynomial Y being a function of the indeterminate X, whose coefficients are rational, cannot alter under any of the substitutions $1, a, b, \ldots$. Let us apply, for example, the substitution a. This polynomial becomes

$$(X - V_a)(X - V_{a^2})(X - V_{ab}) \ldots$$

For this new polynomial to be identical with Y, whatever X is, it is necessary that the quantities $V_a, V_{a^2}, V_{ab}, \ldots$ be no other than the quantities $V_1, V_a, V_b, \ldots$ up to order. But by assumption two distinct substitutions give essentially different values to the function V. It is therefore necessary that the substitutions $a, a^2, ab, \ldots$ are identical, up to order, with the substitutions $1, a, b, \ldots$. Thus, a and b being two arbitrary substitutions in the sequence $1, a, b, \ldots$, the substitution ba must equally belong to the sequence: the substitutions of this sequence therefore form a group.

§355 The group defined by the preceding theorem can be called the *group of the equation relative to the adjoined quantities* $y, z, \ldots$. The group can vary according to the nature of the adjoined quantities. Among all the groups that one can obtain in this way, there is one particularly remarkable G and we can call it in an absolute manner the *group of the equation*. It is the one obtained when no quantity at all is adjoined.

This group contains all the others, for, if H is the group obtained by adjoining the arbitrarily chosen quantities $y, z, \ldots$, then a function invariant under the substitutions of G and variable for all other substitutions is expressible rationally before, and *a fortiori* after the adjunction of the $y, z, \ldots$. Therefore it will be unaltered under all the substitutions of H; therefore all these substitutions are contained in G.

§356 Lemma If two functions φ_1, ψ_1 of the roots of the given equation are numerically equal, the same equality will hold between the functions φ_a, ψ_a obtained by applying to each of them an arbitrary substitution of G.

Proof. For, the function $\varphi_1 - \psi_1$, being zero, is expressible rationally, therefore it is not altered by the substitution a, therefore $\varphi_a - \psi_a = 0$.

§357 Theorem II *Every irreducible equation $F(x) = 0$ has a transitive group, and conversely.*

Proof. For, suppose that the group is not transitive. Let x_1 be an arbitrary one of the roots; $x_1, \ldots, x_m$ the roots with which it is permuted by the substitutions of G. These substitutions do not alter the symmetric functions in $x_1, \ldots, x_m$, therefore these symmetric functions are rational. Therefore $F(x)$ admits a rational divisor $(x - x_1) \ldots (x - x_m)$.

Conversely, suppose that G is transitive, then $F(x)$ cannot admit a rational divisor such as $(x - x_1) \ldots (x - x_m)$. For, let x_{m+1} be a root of the equation other than $x_1, \ldots, x_m$. G has a substitution that replaces x_1 by x_{m+1}. It transforms $(x - x_1) \ldots (x - x_m)$ into a new product different from this one because it admits the factor $(x - x_{m+1})$. Therefore the product $(x - x_1) \ldots (x - x_m)$ not being unaltered by all the substitutions of G cannot be rational.

§358 Theorem III *The order of the group of an irreducible equation of degree n all of whose roots are rational function of one of them, is also n.*

Proof. For, let $x_1, \ldots, x_n$ be the roots of the equation, V_1 a function of these roots that is altered by every substitution; it can be written as a function of x_1 alone. Let $V_1 = f(x_1)$, this function satisfies an equation of degree n:

$$[V - f(x_1)] \ldots [V - f(x_n)] = 0,$$

the coefficients of which, being symmetric in $x_1, \ldots, x_n$, are rational. Therefore the order of the group of the equation (which is the degree of the irreducible equation of which V_1 is a root) cannot be greater than n. But on the other hand this group being transitive its order is divisible by n; therefore it is equal to n.

§359 Theorem IV *The group of the equation of degree mn obtained by eliminating y between the equations*

$$a_m y^m + \cdots + a_0 = 0, \quad \text{and} \quad b_n(y)x^n + \cdots + b_0(y) = 0$$

is not primitive. Conversely, every equation whose group is not primitive results from such an elimination.

Proof. Presently omitted.

§360 We shall say henceforth that an equation is *primitive* or *not primitive*, *simple* or *composed* according as its group is primitive or not primitive, simple or composed; the *order* of the equation and its *composition factors* are the order and the composition factors of its group, two equations are isomorphic if their groups are isomorphic, etc.

§361 The group of an equation being known, one can propose to reduce it progressively by the successive adjunction of auxiliary quantities. With each of these adjunctions two cases can present themselves. If the irreducible equation $Y = 0$ remains irreducible, it is clear that the group undergoes no change. If on the contrary, thanks to the new adjunction, the polynomial $(X - V_1)(X - V_a)(X - V_b) \ldots$ decomposes into simpler factors $(X - V_1)(X - V_a), \ldots, (X - V_b) \ldots,$ one finds a new group H, smaller than G, and formed from only the substitutions $1, a, \ldots$.

We shall first examine what happens when one adjoins to the given equation certain functions of its roots; then we shall pass to the case where the adjoined quantities are roots of other equations.

§362 Theorem V *Let G be the group of the equation $F(x) = 0$, φ_1 a rational function of its roots. Define H_1 to be the elements of G that do not alter the numerical value of φ_1, then (1) H_1 is a subgroup of G and (2) adjoining the value of φ_1 reduces the group of the equation to H_1.*

Proof.

(1) In fact, let a and a_1 be two substitutions of G that do not alter φ_1: one has $\varphi_a = \varphi_1$, $\varphi_{a_1} = \varphi_1$, from which one deduces (§356) that $\varphi_{a_1 a} = \varphi_a = \varphi_1$. So the substitution $a_1 a$ does not alter φ_1, which proves the first of our propositions.

(2) Let us adjoin the value of φ_1 to the equation. After this operation, the reduced group of the equation can only contain substitutions that belonged initially to G, and besides it can only contain substitutions that do not alter φ_1, the value of this quantity being supposed rationally known. Therefore all these substitutions are contained in H_1. Conversely, it contains all the substitutions of H_1. In fact, let a be one of the latter substitutions, ψ_1 a function of the roots expressible rationally as a function of ϕ_1 and quantities already known, and let

$$\psi_1 = \chi(\varphi_1),$$

χ denoting a rational function. From this equality one deduces the following (§356): $\psi_a = \chi(\varphi_a)$, and as $\varphi_a = \varphi_1$ it follows that $\psi_a = \psi_1$. Thus every rationally expressible function ψ_1 is unaltered by the substitution a and this substitution therefore belongs to the reduced group.

Lemma I *The adjunction of several functions of the roots, $\varphi_1, \varphi_1', \ldots$ reduces the group of the equation to a group H' formed of the substitutions that alter neither φ_1 nor φ_1', etc.*

Lemma II *Two functions φ_1 and ψ_1 invariant under the same elements of G are expressible rationally in terms of each other.*

Proof. For, on adjoining φ_1 to the equation one reduces its group to those of its substitutions that do not alter φ_1; and ψ_1, being invariant under these substitutions, becomes rational.

[§§363, 364 show explicitly how to calculate ψ_1 as a function of φ_1, and are presently omitted.]

§365 The method of calculation which we are going to expound is due originally to Lagrange. It allows us to establish the following proposition:

Let $F(x) = 0$ be an algebraic equation, τ_1 a function of its roots the algebraic expression of which remains invariant under a certain group of substitutions $(1, a, b, \ldots)$ and all the values which are algebraically distinct are also numerically distinct. Then every function t_1 of the roots x_i which is unaltered algebraically by the group $(1, a, b, \ldots)$ is a rational function of τ_1 and the coefficients of $F(x)$.

Proof. For, let $1, a, b \ldots; \beta, a\beta, b\beta \ldots; \gamma, a\gamma, b\gamma \ldots$ be the various possible substitutions possible among the roots, and let n be an arbitrary integer. The function $\tau_1^n t_1 + \tau_\beta^n t_\beta + \cdots = \theta_1^{(n)}$ will be unaltered by every substitution and accordingly will be expressible rationally as a function of the coefficients. This done, t_1 is expressed rationally by [the explicit formula given at the end of §364].

Theorem VI *With the same definitions as in Theorem V, let $a_0, a_1, a_2, \ldots$ be the substitutions of H_1 (a_0 reducing to 1), $a_0, a_1, a_2, \ldots;$ $a_0 b, a_1 b, a_2 b, \ldots;$ $a_0 c, a_1 c, a_2 c, \ldots;$ those of G. Then the equation*

$$(Y - \varphi_1)(Y - \varphi_b)(Y - \varphi_c) \ldots \tag{D.2}$$

of degree equal to the ratio of the orders of G and H, will have rational coefficients and be irreducible.

Proof.

1: Let σ be an arbitrary substitution of G; it transforms the terms $\varphi_1, \varphi_b, \varphi_c$ into each other (§§61 and 363). The coefficients of Eq. (D.2), being symmetric in $\varphi_1, \varphi_b, \varphi_c$ will therefore not be altered by σ, but σ is an arbitrary substitution in G, therefore they are rational.

2: Equation (D.2) is irreducible, because if it had, for example, the factor $(Y - \varphi_1)(Y - \varphi_b)$ this factor would be unaltered by the substitution c, but this substitution transforms it into $(Y - \varphi_c)(Y - \varphi_{bc})$. For this to be unchanged, Y remaining indeterminate, it must be the case that φ_c and φ_{bc} must be equal in some order to φ_1 and φ_b. Suppose, for example, that $\varphi_c = \varphi_b$. One concludes that $\varphi_{cb^{-1}} = \varphi_1$. Therefore cb^{-1} must be one of the substitutions $a_0, a_1, a_2, \ldots$ and c will be of the form $a_\rho b$, which it is not.

§367 Remark: The substitutions of G that do not alter φ_1 being $a_0, a_1, a_2, \ldots,$ those that do not alter φ_b will be $b^{-1}a_0 b, b^{-1}a_1 b, b^{-1}a_2 b, \ldots,$ those which do not alter φ_c will be $c^{-1}a_0 c, c^{-1}a_1 c, c^{-1}a_2 c, \ldots;$ etc. For, let σ be a substitution of G that does not alter φ_b, then one will have

$$\varphi_b = \varphi_{b\sigma}, \text{ whence } \varphi_1 = \varphi_{b\sigma b^{-1}}, \text{ whence } b\sigma b^{-1} = a_\rho, \sigma = b^{-1}a_\rho b.$$

§368 Theorem VII *Adjoining $\varphi_1, \varphi_b, \ldots, \varphi_c$ simultaneously reduces the group of the equation to K, the largest subgroup of H_1 that is 'normal' in G.*

Proof. In fact, the group is reduced to the substitutions common to the groups

$$H_1 = (a_0, a_1, \ldots), \quad H_b = (b^{-1}a_0b, b^{-1}a_1b, \ldots), \quad H_c = (c^{-1}a_0c^{-1}, {}^{-1}a_1c, \ldots),$$

formed respectively by the substitutions that do not alter $\varphi_1, \varphi_b, \varphi_c, \ldots$. Now let J be the group formed by these common substitutions, s an arbitrary one of the substitutions, σ an arbitrary substitution of G. The substitution $\sigma^{-1}s\sigma$ will be common to the transform of the groups $H_1, H_b, H_c, \ldots$ by σ. But up to order these transformed groups are identical with $H_1, H_b, H_c, \ldots$, for each of the substitutions $\sigma, b\sigma, c\sigma, \ldots$, belonging to G, can be put in one of the forms $a_\rho, a_\rho b, a_\rho c, \ldots$. For example, let $b\sigma = a_\rho d$. The transform of the group H_b by σ will be composed of the substitutions

$$\sigma^{-1}b^{-1}a_0b\sigma = d^{-1}a_\rho^{-1}a_0a_\rho d, \quad \sigma^{-1}b^{-1}a_1b\sigma = d^{-1}a_\rho^{-1}a_1a_\rho d, \ldots,$$

which are none other than the substitutions of H_d. Therefore $\sigma^{-1}s\sigma$ belongs to J, this group therefore commutes with σ, therefore it is contained in K.

Conversely, the group K being contained in H_1, the transforms of its substitutions by $b, c, \ldots$ will be contained in $H_b, H_c, \ldots$. But these transforms reproduce, up to order, the substitutions of K, therefore all the substitutions of K are common to $H_1, H_b, H_c, \ldots$.

§369 Remark: In the particular case where the group H is already 'normal' in G then H and its conjugates are not only equal but equal K, and the functions $\varphi_1, \varphi_b, \varphi_c, \ldots$, invariant under the same substitutions of G, will be expressible rationally as functions of any one of them. Conversely, if the functions $\varphi_1, \varphi_b, \varphi_c, \ldots$ are expressible rationally as functions of any one of them, they will be invariant under the same substitutions of G; one will therefore have $H_1 = H_b = H_c = \ldots$ and this group will be transformed into itself by all the substitutions of G

Theorem VIII *Let the order of G be N, and the order of K be $\frac{N}{v}$, then the order of G', the group of the Eq. (D.2), is v.*

Proof. In fact, set

$$W = M_1\varphi_1 + M_b\varphi_b + M_c\varphi_c + \ldots,$$

$M_1, M_b, M_c, \ldots$ being undetermined constants. Then W will be a root of an irreducible equation of degree equal to the order of the group of Eq. (D.2) (§353). But on the other hand W can be considered as a function of the roots of the equation $R(x) = 0$, which function, unaltered by the substitutions in K, is evidently altered by every other substitution of G. It therefore depends on an irreducible equation whose degree is equal to v, the ratio of the orders of G and K.

Theorem IX *If K' is the largest 'normal' subgroup of G that contains K, then the group G' is simple, and conversely.*

Proof. For, if there is a 'normal' subgroup K' in G' let its order be v'. A function ψ of the roots of Eq. (D.2) unaltered by the substitutions of K' and altered by every other substitution will depend on an irreducible equation of degree $\frac{v}{v'}$ (§366), therefore the roots will be rational functions of each other (§369). But ψ can be considered as a function of the roots of $F(x) = 0$; let L be the group composed of substitutions that do not alter it. Then (1) L contains K, because the substitutions, which do not alter $\varphi_1, \varphi_b, \varphi_c \ldots$ cannot alter ψ; (2) it will be larger, because its order, being equal to that of G, divided by the degree $\frac{v}{v'}$ of the irreducible equation upon which ψ depends, is equal to $\frac{Nv'}{v}$, that of K being merely $\frac{N}{v}$; (3) finally, the roots of the equation ψ being rational functions of each other, L is a 'normal' subgroup of G.

Theorem X Let $F(x) = 0$ *be an equation whose group G is composite, and $G, H, H_1, \ldots$ a sequence of groups such that (1) each one is a 'normal' subgroup of the one before and (2) that it is as general as possible among those that satisfy this double property, and $N, \frac{N}{v}, \frac{N}{vv_1}, \ldots$ are the respective orders of the groups. The resolution of the given equation will depend on that of the successive equations and contain $v, v_1, \ldots$ substitutions respectively.*

Proof. For, let φ_1 be a function of the roots of the given equation that is unaltered by the substitutions in H. It depends on an equation of degree v (§366) the group of which is simple (§371) and of order v (§§369, 370). When this equation is solved, the group of the given equation reduces to H. Now let φ_1' be a function of the roots of the given equation that is unaltered by the substitutions in H_1. It depends on an equation of degree v_1 (§366) the group of which is simple (§371) and of order v_1 (§§369, 370), etc.

   This theorem shows that one can arrange the equations with composite groups according to the number and value of their composition factors.

§373 Now suppose that one adjoins to the given equation $F(x) = 0$ one or more of the roots of another equation $f(x) = 0$.
   The case where one adjoins several functions $\chi_1, \chi_1', \ldots$ reduces to the case where one adjoins only one: for, let G' be the group of the equation $f(z) = 0$, H_1' the group formed by those substitutions that alter none of the functions $\chi_1, \chi_1', \ldots; r_1$ a function of $z_1, \ldots, z_n$ unaltered by the substitutions of H_1 and varying under every other substitution. The adjunction of $\chi_1, \chi_1', \ldots$ reduces the group of $f(z) = 0$ to H_1', whose substitutions do not alter r_1, this function being expressible rationally. Conversely, the adjunction of r_1 reduces this group to H_1', whose substitutions do not alter $\chi_1, \chi_1', \ldots$, these functions will become rational. Therefore every rational function of r_1 can be expressed rationally in terms of $\chi_1, \chi_1', \ldots$ and conversely. Therefore it is indifferent to adjoin to an arbitrary equation either $\chi_1, \chi_1', \ldots$ or simply r_1.

§374 Therefore let us adjoin to the equation $F(x) = 0$ the unique function r_1; let $\alpha_0, \alpha_1, \ldots$ be the substitutions of H_1'; $\alpha_0\beta, \alpha_1\beta, \ldots$; $\alpha_0\gamma, \alpha_1\gamma, \ldots$ those of G'; r_1 depends on the irreducible equation

$$(X - r_1)(X - r_\beta)(X - r_\gamma)\cdots = 0. \tag{D.3}$$

Let us suppose that the adjunction of r_1 reduces the group of $F(x) = 0$ to H_1. Let, as before, $a_0, a_1, \ldots$ be the substitutions of this group, $a_0, a_1, \ldots$; $a_0b, a_1b, \ldots$; $a_0c, a_1c, \ldots$ be those of G. Finally let φ_1 be a function of the roots of $F(x) = 0$ unaltered by the substitutions of H_1 and altered by every other substitution; it will satisfy the equation

$$(Y - \varphi_1)(Y - \varphi_b)(Y - \varphi_c)\ldots = 0. \tag{D.4}$$

But by hypothesis, φ_1 is a rational function of r_1, let $\varphi_1 = \psi(r_1)$, then r_1 will be a root of the equation

$$(\psi(y) - \varphi_1)(\psi(y) - \varphi_b)(\psi(y) - \varphi_c)\ldots = 0. \tag{D.5}$$

But on the other hand r_1 satisfies the irreducible equation (D.3). One of the roots of that equation satisfies Eq. (D.5), therefore they all do (§349). Therefore the quantities $\psi(r_1), \psi(r_\beta), \psi(r_\gamma), \ldots$ all satisfy Eq. (D.4). But they satisfy equation

$$(Y - \psi(r_1))(Y - \psi(r_\beta))(Y - \psi(r_\gamma))\ldots = 0, \tag{D.6}$$

the coefficients of which, being symmetric in $r_1, r_\beta, r_\gamma, \ldots$, are rational.

The roots of Eq. (D.6) all satisfy the irreducible equation (D.4), the first member of (D.6) will be an exact power of the first member of (D.4) (§349), up to a constant factor which here reduces to unity, the two polynomials having unity for the coefficient of their first term. Let μ be the degree of this power, the sequence of terms $\psi(r_1), \psi(r_\beta), \psi_\gamma, \ldots$ has μ terms equal to φ_1, μ equal to φ_β, etc.

This done, *adjoining to the equation $F(x) = 0$ an arbitrary one of the roots of Eq. (D.4), such as r_β, the group of the equation will reduce to H_1, to the group H_b derived from the substitutions $(b^{-1}a_0b, b^{-1}a_1b, \ldots)$ to the analogous group H_c, etc according as $\psi(r_\beta)$ is equal to $\varphi_1, \varphi_b, \varphi_c, \ldots$.*[1]

In fact, for example let $\psi(r_\beta) = \varphi_b$. The adjunction of r_β making φ_b rational, the reduced group H can contain only those substitutions of G that do not alter φ_b, that is to say those of H_b. The number of these substitutions being equal to those in H_1, one sees that the order of H is at most equal to that of H_1. Conversely, starting from the rationality of the root r_β instead of that of r_1 one sees that the order of H_1 is at most equal to that of H. Therefore these orders are equal and H contains all the substitutions of H_1.

---

[1] Correcting a misprint in Jordan's text where he had $b^{-1}a_1c$.

§375 Theorem XI *In the notation of the preceding paragraph, if H_1' is a normal subgroup of G' then H_1 is a normal subgroup of G.*

Proof. In fact, H_1' being a 'normal' subgroup of G, the roots of the irreducible equation (D.3) that depend on r_1 are rational functions of each other (§369). Therefore the functions $\varphi_1, \varphi_b, \varphi_c \dots$ are respectively rational functions of r_1; therefore they are unaltered by the substitutions of H_1; but these are respectively those of $H_1, H_b, H_c, \dots$;, therefore these groups are identical and H_1 is 'normal' in G.

§376 Theorem XII *If one adjoins to the equation $F(x) = 0$, whose group is G, all the roots of the equation $f(z) = 0$ then the reduced group H_1 is a 'normal' subgroup of G.*

Proof. For, adjoining all the roots of $f(z) = 0$ simultaneously is the same as adjoining a function V_1 of these roots that alters under every substitution other than unity (§373). But then the group H_1' reduces to the single substitution 1, which is evidently normal in G', therefore H will be normal in G.

§377 Lemma If the group G is simple, it cannot be reduced by the resolution of an auxiliary equation without reducing to the single substitution 1 (the group composed of this substitution being, by definition, the only normal subgroup of G; in this case the equation $F(x) = 0$ will be completely solved.

§378 Theorem XIII *Let $F(x) = 0$ and $f(z) = 0$ be two equations whose groups are of orders N and N' respectively. If the resolution of the second equation reduces the group of the first to a group H_1 containing no more than $\frac{N}{v}$ substitutions then conversely the resolution of the first equation reduces the group of the second to a group H_1' containing no more that $\frac{N'}{v}$ substitutions. Moreover, the two equations are composed with the same auxiliary equation $F'(u) = 0$ of degree v whose group contains v substitutions.*

Proof. In fact, let $\psi(x_1, \dots, x_m)$ be a function of the roots of $F(x) = 0$, unaltered by only the substitutions of H_1. It is rationally expressible as a function of the roots $z_1, \dots, z_n$ of $f(z) = 0$. One therefore has

$$\psi(x_1, \dots, x_m) = \chi(z_1, \dots, z_n) = u.$$

This quantity $\psi(x_1, \dots, x_m)$ depends on an auxiliary irreducible equation $f'(u) = 0$ of degree μ (§366). Moreover, H_1 being 'normal' in G (§376), the roots of this equation are rational functions of each other (§369) and its group contains v substitutions (§358). The solution of this auxiliary equation reduces the group of $F(x) = 0$ to precisely the substitutions of H_1, which are $\frac{N}{v}$ in number. It likewise reduces the group of $f(z) = 0$ in such a way that it only contains $\frac{N'}{v}$ substitutions. In fact, let K' be the number of substitutions of the group G' that do not alter $\chi(z_1, \dots, z_n) = u$; u will depend on an irreducible equation of degree $\frac{N'}{K'}$. But the degree of this equation is equal to v, therefore $\frac{N'}{K'} = v$, whence $K' = \frac{M'}{v}$.

The solution of the equation $F(x) = 0$ entailing that of the auxiliary equation $f'(u) = 0$, whose roots are rational functions of $x_1, \ldots, x_m$, the group of $f(z) = 0$ will reduce in such a way that it contains at most $\frac{N'}{v}$ substitutions. On the other hand, it cannot contain a smaller number, for the reduced group H_1' contains only $\frac{N'}{\mu}$ substitutions, μ being a number greater than v, one sees on repeating all the arguments starting with the equation $f(z) = 0$ that the solution of this equation would reduce the group of $F(x) = 0$ so that it contained at most $\frac{N}{\mu}$ substitutions, a result contrary to our hypothesis. So the reduced group H_1 contains $\frac{N}{v}$ substitutions.

§379 Lemma I If the group G of the equation $F(x) = 0$ is simple it can only be solved by means of equations of which the order of the group is a multiple of the order of G.

Proof. For, let us adjoin to this equation the roots of an auxiliary equation $f(z) = 0$. If its group is reduced, it is reduced to the single substitution 1 (§377). Therefore the order of the group of $F(x) = 0$, which was N, is reduced to 1 by the adjunction of the roots of $f(z) = 0$. Therefore reciprocally the adjunction of the roots of $F(x) = 0$ to the equation $f(z) = 0$ divides the order of this group by N. Therefore its order is a multiple of N.

§380 Lemma II If the group G of the equation $F(x) = 0$ is reduced by the resolution of a simple equation $f(z) = 0$ the roots of this last equation are rational functions of $x_1, \ldots, x_n$.

Proof. For, the resolution of $F(x) = 0$ reducing the group of $f(z) = 0$ (§378) completely solves this last equation.

This proposition is an extension of a theorem of Abel's: if an equation is solvable by radicals, each of the radicals that appears in its resolution is a rational function of the roots and the roots of unity.

§381 Remark If the function $\psi(x_1, \ldots, x_m)$ is altered by every substitution operating on the roots $x_1, \ldots, x_m$ these roots are expressible rationally in terms of $\psi(x_1, \ldots, x_m)$ (§352), and consequently as functions of $z_1, \ldots, z_n$. The resolution of the equation $f(z) = 0$ therefore implies the complete resolution of the equation $F(x) = 0$. If at the same time $\chi(z_1, \ldots, z_n)$ alters under every substitution acting on the $z_1, \ldots, z_n$, the resolution of $F(x) = 0$ entails that of $f(z) = 0$. The two equations will be said to be *equivalent*.

Lemma 2 *III Equivalent equations have the same orders.*

Proof. In fact, let N be the order of the equation $F(x) = 0$, N' that of the equation $f(z) = 0$. The resolution of $f(x) = 0$ reduces the group of $F(x) = 0$ to the single substitution 1, therefore the resolution of $F(x) = 0$ reduces the group of $f(x) = 0$ to one containing no more than $\frac{N'}{N}$ substitutions (§378). But the resolution of $F(x) = 0$ entailing that of $f(z) = 0$ the group of this last equation reduces to the single substitution 1. One must therefore have $\frac{N'}{N} = 1$, whence $N = N'$.

Lemma 3 *IV Every adjunction of an auxiliary quantity which reduces the order of one of these equations by dividing its order by v reduces the order of the other by the same amount.*

Proof. In fact, the equations being equivalent before the adjunction and after, their groups cannot cease to have the same number of substitutions.

Lemma 4 *V Every equation equivalent to a composite equation is itself composed of the same auxiliary equations.*

§382 Problem Find all the irreducible equations equivalent to a given equation $F(x) = 0$.

Let $f(z) = 0$ be one of these equations. The resolution of $F(x) = 0$ implying that of $f(z) = 0$ the roots $z_1, \ldots, z_n$ of this last equation are rational functions of $x_1, \ldots, x_m$. Therefore let $z_1 = \varphi_1$: denote as before by G the group of $F(x) = 0$, by $a_0, a_1, \ldots$ the substitutions of G that do not alter the function φ_1, which substitutions from a group H_1, and by $a_0, a_1, \ldots$; $a_0 b, a_1 b, \ldots$; $a_0 c, a_1 c, \ldots$ those of G. We have seen (§366) that the irreducible equation on which φ_1 depends is

$$(Z - \varphi_1)(Z - \varphi_2)(Z - \varphi_c) \ldots = 0,$$

so the simultaneous adjunction of its roots reduces the group of the given equation to 1 (§368). But this adjunction must resolve the given equation completely; therefore K reduces to the single substitution 1; whence the result:

For an irreducible equation $f(z) = 0$ to be equivalent to $F(x) = 0$ it is necessary and sufficient that (1) one if its roots, z_1, be a rational function of the roots of $F(x) = 0$; (2) that the group H_1 formed by the substitutions of G [the group of $F(x) = 0$] that do not alter this function contain no normal subgroup of G (except that formed by the single substitution 1).

[§§383, 384, which are technical and contain a proof of theorem XIV, are presently omitted.]

§385 Theorem XIV *No irreducible equation of degree p a prime can be reduced by means of auxiliary equations of lower degree.*

Proof. For, the equation being irreducible, its group is transitive. Therefore its order is divisible by p, and one sees from theorem XIII that it continues to be divisible by p as long as one does not use an auxiliary equation whose order is divisible by p. But the order of the group of an auxiliary equation of degree $q < p$ is a divisor of $q!$; it is therefore prime to p. Therefore the order of the group of the given equation will remain divisible by p provided one only employs similar equations.

§386 Theorem XV *The general equation of degree n cannot be solved by means of equations of lower degree (except in the case when $n = 4$).*

Proof. In fact, its group is of order $n!$. By resolving an equation of degree 2 one can reduce to the alternating group whose order is $\frac{n!}{2}$. But this new group is simple (§85). Therefore the equation cannot be solved by means of an auxiliary equation unless the order of its group is at least equal to $\frac{n!}{2}$, but if q is the degree of this auxiliary equation, its order divides $q!$. Therefore q cannot be less than n.

§387 Let us consider on the contrary the fourth degree equation

$$x^4 + px^3 + qx^2 + rx + s = 0.$$

Let x_1, x_2, x_3, x_4 be its roots; H the group of order 8 obtained from the substitutions $(x_1, x_2), (x_3, x_4), (x_1, x_3)(x_2, x_4)$. A function φ_1 of the roots of the given equation and invariant under the substitutions of H depends on an equation of the third degree (§366). This solved, the group of the given equation reduces to a group K obtained from those of its substitutions which are common to the group H and its transforms by the various substitutions of G. One sees easily that these common substitutions are the four obtained from $(x_1, x_2)(x_3, x_4), (x_1, x_3)(x_2, x_4)$. But these substitutions commute with the subgroup H' of order 2, obtained from the powers of $(x_1, x_2)(x_3, x_4)$. Therefore a function of the roots that is invariant under the substitutions of H' only actually depends on an equation of the second degree. This solved, the group of the given equation reduces to H'. This equation then decomposes into two equations of the second degree, having respectively for roots x_1, x_2 and x_3, x_4 (§357). It is enough, moreover, to solve one of these equations to reduce the group of the given equation to the identity.

Among the various functions that one can take for ϕ_1, the most convenient is the function $(x_1 + x_2 - x_3 - x_4)^2$ adopted by Lagrange. One easily finds the coefficients of the equation of degree 3 on which it depends:

$$Y^3 - (3p^2 - 8q)Y^2 + (3p^4 - 16p^2q + 16q^2 + 16pr - 64s)Y - (p^3 - 4pq + 8r)^2 = 0.$$

Let v_1, v_2, v_3 be its roots, one has

$$x_1 + x_2 - x_3 - x_4 = \sqrt{v_1}; \quad x_1 - x_2 + x_3 - x_4 = \sqrt{v_2};$$

$$x_1 - x_2 - x_3 + x_4 = \sqrt{v_3}; \quad x_1 + x_2 + x_3 + x_4 = -p;$$

whence

$$x_1 = \frac{1}{4}(-p + \sqrt{v_1} + \sqrt{v_2} + \sqrt{v_3}), \ldots .$$

One also has

$$\sqrt{v_1}\sqrt{v_2}\sqrt{v_3} = (x_1 + x_2 - x_3 - x_4)(x_1 - x_2 + x_3 - x_4)(x_1 - x_2 - x_3 + x_4)$$

$$= -p^3 + 4pq - 8r,$$

whence

$$x_1 = \frac{1}{4}\left(-p + \sqrt{v_1} + \sqrt{v_2} + \frac{-p^3 + 4pq - 8r}{\sqrt{v_1}\sqrt{v_2}}\right).$$

The other roots are obtained by changing the signs of the two independent radicals $\sqrt{v_1}$, $\sqrt{v_2}$.

§388 To apply the preceding results, it is necessary to know how to determine the group of a given equation.

The route to follow to deal with this question is this: 1) construct the various groups of possible substitutions $G, G', \dots$ of the roots $x_1, \dots, x_n$ of the equation; 2) let G be one of these groups, chosen arbitrarily: one will be assured that it does or does not contain the group of the equation by forming a function φ of the roots that is invariant under the substitutions of G and varies under any other substitution, calculating by the method of symmetric functions the equation that has for its roots the various values of φ, and examining whether this equation does or does not have a rational root. Among the groups in the sequence $G, G', \dots$ which also contains the group of the equation, the smallest will be the group itself.

This method, which is theoretically satisfactory, will be impracticable if one wants to apply it to a numerical equation given at random. But the equations that one meets in analysis always have special properties that allow us to determine their group more easily, often from the coefficients themselves. We shall give several examples in the following chapters.

D.3   Jordan: The Quintic Is Not Solvable by Radicals

Jordan's argument that the quintic is not solvable by radicals went as follows:

§517 An equation is said to *solvable by radicals* if its roots can be made rational by the adjunction of the roots of a sequence of binomial equations.

One can suppose that the auxiliary binomial equations are all of prime degree; for, the resolution of a binomial equation of degree pq evidently reduces to the successive resolution of two binomial equations, of degrees p and q.

We call *solvable groups* those that characterise the equations that are solvable by radicals.

§518 Theorem I *Every abelian equation of prime degree is solvable by radicals.*
In fact, let X be such an equation;

$$x_0, x_1 = \varphi(x_0), \dots, x_{p-1} = \varphi(x_{p-2}) = \varphi^{p-1}(x_0)$$

its roots, θ a pth root of unity, r an arbitrary integer: the function

$$\psi = [x_0 + \theta^r x_1 + \cdots + \theta^{(p-1)r} x_{p-1}]^p$$

is evidently unaltered by the substitution $(x_0, x_1, \ldots, x_{p-1})$, whose powers form the group of X. It is therefore expressed rationally as a function of θ and the coefficients of X. One easily obtains its values on replacing $x_0, \ldots, x_{p-1}$ by their values as a function of x_0, which reduces it to the form $\psi(x_0, \theta)$. One therefore has the equality

$$\psi = \psi(x_0, \theta),$$

which is not affected if one performs on its two members one of the powers of the substitution $(x_0, x_1, \ldots x_{p-1})$: one therefore has

$$\varphi = \psi(x_0, \theta) = \psi(x_1\theta) = \cdots = \frac{\psi(x_0, \theta) + \psi(x_1\theta) + \cdots}{p} = u_r,$$

u_r being a symmetric function of $(x_0, x_1, \ldots x_{p-1})$.

This function being calculated, it becomes

$$x_0 + \theta^r x_1 + \cdots + \theta^{(p-1)r} x_{p-1} = \sqrt[p]{u_r}.$$

Let us set successively $r = 1, 2, \ldots, p - 1$: one will have $p - 1$ distinct linear equations, which, together with this one:

$$x_0 + x_1 + \cdots + x_{p-1} = P$$

(where P is the known function of the coefficients of X that represents the sum of the roots) allows us to determine all the roots.

To obtain x_ρ one adds these equation, multiplied respectively by $\theta^\rho, \theta^{-r\rho}, \ldots, 1$ then dividing by p it becomes

$$x_\rho = \frac{P + \theta^{-\rho}\sqrt[p]{u_1} + \cdots + \theta^{-r\rho}\sqrt[p]{u_r}}{p}.$$

This expression contains $p - 1$ radicals of degree p; but these radicals are all rational functions of a single one of them in such a way that the value of this one being taken arbitrarily that of the others is completely determined. In fact, the function

$$\sqrt[p]{u_r}\left(\sqrt[p]{u_1}\right)^{p-r} = (x_0 + \theta^r x_1 + \cdots)(x_0 + \theta x_1 + \cdots)^{p-r},$$

being unaltered by the substitution $(x_0, x_1, \ldots, x_{p-1})$, is rational, and can be calculated at once to be the function φ. Let a_r be its value, one will have

$$\sqrt[p]{u_r} = \frac{a_r}{u_1} \left(\sqrt[p]{u_1} \right)^r .$$

§519 Theorem II *For an equation to be solvable by radicals, it is necessary and sufficient that its resolution reduces to that of a sequence of abelian equations of prime degree.*

This condition is necessary. For each of the binomial equations of prime degree that arises in the resolution of the proposed equation can be resolved by means of two successive abelian equations (§417) of which one is of prime degree. As for the other, its resolution leads to that of a sequence of abelian equations of prime degree.

This condition is sufficient: for, the abelian equations on which the resolution of the proposed equation depends being solvable by radicals, the proposed equation will be.

Here are two other statements of the same theorem.

§520 Theorem III *For an equation to be solvable by radicals, it is necessary and sufficient that its composition factors all be prime.*

For, let X be an equation solvable by radicals; G its group; N its order. Let us successively adjoin the roots of the abelian equations that lead to its resolution; its group reduces successively until it contains only the substitution 1. Let $H, K, \ldots$ be the successive reduced groups; $\frac{N}{\lambda}, \frac{N}{\lambda\mu}, \ldots$ their orders; Z the abelian equation of degree p such that the adjunction of its roots reduces the given group to H. Reciprocally, the adjunction of the roots of X to the equation Z will reduce the order of this group to this latter on dividing by λ (§378). But this order is equal to p (§358); and for it to be divisible by λ it is necessary that $\lambda = p$. One sees likewise that $\mu, \ldots$ are all prime numbers. Moreover, the substitutions of G permute with those of H, those of H with those of K, etc.[2] Therefore $\lambda, \mu, \ldots$ are composition factors of G.

Reciprocally, if the equation X has composition factors $\lambda, \mu, \ldots$ all prime let $G, H, K, \ldots$ be successive groups of orders $N, \frac{N}{\lambda}, \frac{N}{\lambda\mu}, \ldots$ such that each one is contained in the preceding one and commutes with its substitutions. A function of the roots of X, invariant under the substitutions of H, depends on an equation Z of degree λ and order λ (§366 and §370). But by hypothesis λ is prime; the group of Z therefore reduces to powers of a cyclic substitution of order λ. This equation will therefore be abelian and of prime degree. Finally its resolution will reduce the given group to H.

This done, a function of the roots of the given equation that is invariant under the substitutions of K will depend on an abelian equation of degree μ, whose resolution will reduce the group of the given equation to K, etc.

---

[2]We would say that H was a normal subgroup of G.

§521 Lemma I Every group Γ contained in a solvable group G is itself solvable.

For, its composition factors, dividing those of G (§392) which are prime are themselves prime.

§522 Lemma II The general equation of degree n is not solvable by radicals if $n > 4$.

For, its composition factors, 2 and $\frac{n!}{2}$ are not all prime.

§523 Lemma III The general equation of degree 3 is solvable by radicals.

For, its composition factors, 2 and 3, are prime.

[Jordan then outlined Lagrange's resolution of this equation; I have omitted this material.]

D.4   Netto's Review

Eugen Netto, a student of Kronecker's in Berlin, reviewed Jordan's book in the abstracting journal of the day, the *Jahrbuch über die Fortschritte der Mathematik*, in these terms (you may want to consult the notes that follow at some points):

> In this work Mr. C. Jordan presents the investigations on substitutions and algebraic equations that have been previously described by him in this connection. The fact that some works by German researchers in this field are not used, he explains, partly by their aphoristic form, partly by the variety of methods they used, and partly by the hope that one day it will be possible that each complete theory can be brought fully to light. Several sections, extracts, fragments and results of the work have previously appeared in various periodicals; that so many are still being published, shows, of course, that the whole theory is not yet as finished and complete, as the author would like: there are still many gaps left to fill, some arguments to shorten, some proofs to correct.
>
> The great size of the book makes it impossible to enter into the minutest details here. We must be content to follow the progress overall, and to highlight the main results.
>
> In the first book the basics of the theory of congruences is explained, and at the end the Galois imaginaries that are expected to provide the solution to congruences of higher degrees, are briefly introduced.
>
> The first chapter of the second book deals with the elements of the theory of substitutions, in particular substitution groups. In this connection, the words: "transitive", "non-primitive", and "composite" are subjected to a closer examination.
>
> [Netto then described Jordan's definitions of linear, orthogonal, abelian, and hypo-abelian groups.]
>
> All these are analyzed and their order and composition factors determined. Furthermore, in this chapter a canonical form for linear substitutions is given, as are explicit methods for constructing special linear groups, for finding elements of each of the groups treated that commute with each other, and a theory of linear fractional substitutions is described.
>
> In the first chapter of the third book the concept of group of an equation is established, which brings together the theory of equations and that of substitutions, and which, for example, reduces the solubility of algebraic equations to the composition of their groups. By adjunction of certain irrationalities one can bring about a reduction of the equations, and the detailed investigation of these relationships leads to the theorems on solvability of equations, etc. The second chapter gives the theory of Abelian and Galois equations. The first are those in which all the roots can be expressed rationally in terms of any one of

them, the latter those in which all can be expressed in terms of any two. The group of these equations contains only substitutions that commute with each other; the binomial equations form a particular case.

[(*))] The geometric applications in the third chapter refer to the inflection points of the curves of the third degree; those third-order curves, whose intersection with a given curve of fourth order coincide in fours; Kummer's surface; the lines on a cubic surface; and contact problems. The application to the theory of transcendents forms the content of the fourth chapter. In it the circular, elliptic and hyperelliptic functions, the division of the periods, the modular equations, etc. are discussed. In the part on "the solution of equations by transcendents" we emphasize the following theorem, which is proved after an account of the solution methods for the equations of the third, fourth, and fifth degree. The resolution of the general equation

$$X = x^q + ax^{q-1} + \cdots = 0$$

will, after the adjunction numerical irrationalities that depend only on q, be reduced to the equation which describes the twofold division of the periods of those hyperelliptic functions in terms of $\sqrt{(X)}$.

Galois showed that an irreducible equation of a prime degree is solvable if and only if its group is linear. The construction of the group arises here that can be described without great difficulty. This is different for equations of arbitrary degree.

Galois introduced a most important property of the groups of solvable equations (*Traité* IV 1: Theorem 4), but did not give the construction for them and said, by restricting himself to primitive equations, that their degree can only be a prime power, a view Mr Jordan refuted earlier when he showed that all solvable primitive equations of a degree are of only one type.

In the first chapter of the fourth book the author now gives other criteria for solvable groups, in addition to the above-mentioned one due to Galois, and one of them is based on its construction. This is carried out in the second chapter, and the problem for the general case is reduced to a consideration of special cases. A general formula is not given but rather a method, by means of which the problem for equations of higher degree is reduced to those of much lower degree. The expressions for the substitutions of the desired groups contain imaginary magnitudes, namely the roots of irreducible congruences of higher degrees. However this is only apparent, although the transformation to a real form requires further work. The given method provides all the desired groups, the proof given in the last chapter that all the groups just found are generally distinct from each other takes up a lot of space and seemingly involves many difficulties.

This is not altogether intelligible. The reference to the Galois groups of binomial equations ($X^n - 1 = 0$) being cyclic and therefore a special case of abelian groups marks a step on the road by which the term 'abelian group' was introduced into mathematics.

The so-called Galois imaginaries were introduced in Galois's (1830). In modern terms they are elements of a finite degree extension of a finite field. They were briefly discussed in Sect. 23.2.

The paragraph marked with a star picks up two remarkable things: the range of geometrical examples known to Jordan, and the fact that he gave a method, involving what are called hyper-elliptic functions, for solving any polynomial equation. This method is, of course, not algebraic. There then follows a brisk indication of what Jordan did in creating a theory of Galois groups.

Appendix E
Klein's *Erlanger Programm*, Groups and Geometry

E.1 Introduction

Another source of ideas for modern algebra, specifically group theory, was the rapidly changing field of geometry. By 1870 projective geometry was being seen as the most elementary and therefore most fundamental form of geometry, and, thanks to the publication of Beltrami's (1868), the new subject of non-Euclidean geometry was starting to gain acceptance among mathematicians—but not philosophers. It was in this context that, in 1871, the next major mathematical contribution was made, and the subjects of projective and non-Euclidean geometry started to come together. The so-called Kleinian view of geometry, as presented by Klein in 1872, revolves around the idea of a mother geometry and sub-geometries: projective geometry is the fundamental geometry and the other geometries (for example, Euclidean and non-Euclidean) are special cases.[1]

Klein saw a role for groups in several ways in geometry, and this chapter also briefly indicates how he used group theory to study the symmetries and other mathematical properties of the regular icosahedron.

E.2 Felix Klein

Young German mathematicians in the nineteenth century were encouraged to travel to other universities to enlarge their education, and in autumn 1869 Klein travelled to Berlin to take part in the lively mathematical seminars of Kummer and Weierstrass, in which the members lectured on topics they had chosen themselves. The trip was

---

[1]For more detail on the history of geometry in the nineteenth century, and Klein's *Erlanger Programm* in particular, see Gray (2011).

© Springer Nature Switzerland AG 2018
J. Gray, *A History of Abstract Algebra*, Springer Undergraduate Mathematics Series,
https://doi.org/10.1007/978-3-319-94773-0

not a success, but while there he heard of the existence of non-Euclidean geometry, and, he wrote later, immediately conjectured that it was closely connected to the so-called Cayley metric (to be described below) and "I made sure of this idea despite the most vehement opposition".

In Berlin he met the somewhat older Norwegian mathematician Sophus Lie, and travelled with him to Paris to learn group theory from Camille Jordan. His trip was cut short by the outbreak of the Franco-Prussian War, but barely a year after his return, on Clebsch's recommendation, Klein was appointed as a full Professor of Mathematics in Erlangen in autumn 1872, at the astonishing age of only 23.

For his Inaugural Address he lectured on the importance of pure and applied mathematics and how they should be taught. Of more lasting importance was a short paper that he circulated at the time, in which he gave a uniform account of the existing directions of geometrical research and sorted them into a system: the famous 'Erlanger Programm'.[2] In the 1890s this became a retrospective guideline for his research.[3]

In 1871 and 1873 he published in two papers entitled 'On the so-called non-Euclidean geometry' that probably did more to convey the message of the Programm than did his obscurely published pamphlet.[4] Cayley (1859) had noticed a way in which one can start with projective geometry and smuggle in the idea of Euclidean distance, but only Klein saw the generality of the idea. It rests on the use of cross-ratio.

Consider a chord PQ of a circle passing through two points A and B. The cross-ratio of the four points P, A, B, Q is defined to be $CR(P, A, Q, B) = \frac{PA}{PB} \cdot \frac{QB}{QA}$. It follows that if C is another point on the same line then

$$CR(P, A, Q, B) \cdot CR(P, B, Q, C) =$$

$$\frac{PA}{PB} \cdot \frac{QB}{QA} \cdot \frac{PB}{PC} \cdot \frac{QC}{QB} = \frac{PA}{PC} \cdot \frac{QC}{QA} = CR(P, A, Q, C).$$

So Klein defined $d(AB) = -\frac{1}{2} \log CR(P, A, Q, B)$, which gave him

$$d(AB) + d(BC) = d(AC),$$

whenever A, B, and C lie in that order on a line. Therefore one can consider $d(AB)$ as the length from A to B (the minus sign is to make lengths positive).

---

[2] See Klein (1872), English translation (1893).

[3] See also Lê (2015) for more about the context.

[4] Note the '*so-called*'—one presumes that Klein did not want ill-informed criticism from non-mathematicians.

Exercise.

1. Check Klein's argument with coordinates. Let $P = 1, Q = -1, A = 0, B = b$
   and $C = c$. Then $d(AB) = -\frac{1}{2}\log\left(\frac{1-b}{1+b}\right)$, and the Möbius transformation $z \mapsto$
   $\frac{z-b}{1-bz}$ which preserves cross-ratio (proof by computation, but to be taken on trust
   here!) sends b to 0 and c to $c' = \frac{c-b}{1-bc}$, and indeed $d(AB) + d(BC) = d(AC)$.
   (As a check on the arithmetic, note that $\frac{1-c'}{1+c'} = \frac{(1+b)(1-c)}{(1+c)(1-b)}$.)

2. To explain the 'fudge factor' of $-\frac{1}{2}$, note that $a' = -\frac{1}{2}\log\left(\frac{1-a}{1+a}\right) = \frac{1}{2}\log\left(\frac{1+a}{1-a}\right)$
   implies that $a = \frac{e^{a'}-e^{-a'}}{e^{a'}+e^{-a'}} = \tanh a'$. So a point a Euclidean distance of a from
   the origin is a non-Euclidean distance of $\tanh^{-1} a$ from the origin. Note that
   $0 < a < 1$ is consistent with $0 < \tanh a' < 1$.

Why does this work? For any definition of length we require that the length of a
segment does not alter as it is moved around. In particular, the segment cannot return
to only a part of its original position. Invariance is guaranteed in the present case,
because cross-ratio is invariant under projective transformations. That is to say, if a
projective transformation sends A, B, P, Q to A', B', P', Q' respectively, then

$$CR(P, A, Q, B) = CR\left(P', A', Q', B'\right).$$

Therefore a projective transformation mapping the circle to itself and the line AB
meeting the circle at points P and Q to the line $A'B'$ meeting the circle at points P'
and Q', satisfies

$$d(AB) = d\left(A'B'\right)$$

so the above definition is in fact satisfactory.

Klein showed in this way in his (1871) how to derive the fundamental formulae
of two-dimensional non-Euclidean geometry by using projective ideas. In his (1873)
he extended these arguments to n dimensions. Ever the scholar, and ever eager to
impress, he also cited a formidable range of predecessors: not only Lobachevskii,
as Beltrami had, but Bolyai (in the original and in an Italian translation), Beltrami,
and Cayley.

Klein was pleased to bring non-Euclidean geometry and projective geometry
together, for in his view geometry had become too fragmented. Even so, he missed
affine geometry, which is a geometry of the plane in which transformations send
lines to lines and preserve ratios of lengths. He did, however, mention that inversion
in circles is geometrically interesting, and this gave him another geometry.

He brought about this sense of unity through his innovative use of the group
concept. His presentation, as he himself noted when the work was reprinted in the
1890s, was less than perfect—the only property of a group that he insisted upon was
closure—but the point Klein was making was important. He shifted attention from
figures to transformations, and argued that geometry was about groups as well as

the invariant properties of shapes. He defined a geometric property to be one that was invariant under all the operations of the group associated to that geometry. So cross-ratio is a property of projective geometry because it is not altered by projective transformations, and it does not need to be built up out of lengths, which are not, indeed, projective invariants.

The Erlanger Programm in the 1890s

Klein proceeded to work on other topics, until in the 1890s he found himself back at Göttingen, now as a senior professor with an empire to build. He arranged for visitors to translate the Erlanger Programm into English, Italian, French, Russian, Polish and Hungarian, and presented it as a research Programm uniting geometry and group theory. Indeed, those two subjects had not only progressed considerably in the intervening 20 years and had indeed grown closer together, but it would be hard to argue that Klein himself had worked intensively on drawing out its implications. It was the situation in the 1890s that gave the Erlanger Programm a prominence that it continues to enjoy. The use of group theory to classify geometries did indeed bring a unity to the subject that had begun to disappear.

The question of the influence of Klein's Erlanger Programm has been much debated. Many universities that teach geometry teach the Kleinian view of geometry, which is exactly the view that various different geometries, all the ones where we represent the transformations with matrices, form into a hierarchy. But is this because of the Programm? Earlier historians of mathematics, and many mathematicians, Garrett Birkhoff among them, believe that it was of great influence.[5] More recently, a number of historians have argued convincingly against this view.[6] They argue that group theory was established between 1870 and 1890 by Camille Jordan, Sophus Lie, and Henri Poincaré without the Erlanger Programm having any impact.[7] The Programm is a very convenient peg on which to hang a number of ideas, but it was probably not their source or inspiration.

E.3   Geometric Groups: The Icosahedral Group

Klein promoted visual, geometric thinking at every opportunity, as a way of understanding problems and as a way of generating new research. Here we look briefly at his geometrical account of how the group A_5 enters the study of the quintic

---

[5]See Birkhoff and Bennett (1988).
[6]See Hawkins (1984) and Hawkins (2000).
[7]For Poincaré, see Gray (2012).

equation, which Klein described very fully in his famous book *On the icosahedron* (1884).

The regular icosahedron *Icos* has $V = 12$ vertices, $E = 30$ edges, and $F = 20$ faces. Note that $V - E + F = 2$, the Euler characteristic of the sphere. It is easier to work with its dual, the regular dodecahedron, which has its vertices at the midpoints of the faces of the icosahedron and vice versa. So it has 20 vertices, 30 edges, and 12 faces. Every face has a diametrically opposite face, so let us put the dodecahedron with its top face horizontal, label that face a, and the faces around it b, c, d, e, f. We label the diametrically opposite faces a', b', c', d', e', f'. The point O equidistant from all the vertices is called the centre of the dodecahedron.

We now consider the symmetry group of the icosahedron, G_I. To specify a rotation it is enough to say where one face goes, say a—this puts it in one of 12 positions—and where an adjacent face goes, say b—for which there are 5 choices— so G_I has at least 60 elements. Each one of these can be composed with a reflection in a plane through the centre and an edge of the dodecahedron. Any symmetry of the dodecahedron is either a rotation or a rotation composed with a reflection, so G_I has 120 elements, and the group of direct symmetries or rotations, G_I^+, has 60 elements.

If we regard each element of G_I^+ as a permutation of the 12 faces we get a map from G_I to S_{12}, the group of permutations of 12 objects. This map is far from being onto, and it is more convenient to look at the action of G_I^+ on the six body diagonals of the dodecahedron—the lines joining the midpoint of a face to the midpoint of the opposite face. These are permuted by G_I^+ and so we obtain a map from G_I^+ to S_6.

Next, a look at where we are heading. The first important group of order 60 that we meet is surely A_5, the alternating group on five letters. It is natural to wonder if G_I^+ is isomorphic to this group. But currently we have it represented in S_6, and there is another interesting group around, to which we now turn.

To obtain this group we start with the field of five elements, $\mathbb{F}_5$, denoted $\{0, 1, 2, 3, 4\}$ and form the plane (the two-dimensional vector space) with these coordinates, $\mathbb{F}_5^2$, and we look at the linear maps from this plane to itself that fix the origin and have determinant 1. We choose the obvious basis: $(1, 0)$ and $(0, 1)$ and we observe that there are 6 directions for the image of $(1, 0)$: $(1, 0), (0, 1), (1, 1), (1, 2)$, $(1, 3), (1, 4)$. In each direction there are four choices for the image, which we denote (a, b). There are now 5 choices for the direction of the image of $(0, 1)$—all ensure that the linear transformation we are constructing is invertible—we denote the image of $(0, 1)$ by (ct, dt). The condition that the determinant be 1 forces $(ad - bc)t = 1$ and because the map is known to be invertible and so $(ad - bc) \neq 0$ we have $t = (ad - bc)^{-1}$. So the maps we seek are specified by the image of $(1, 0)$ and the direction for $(0, 1)$. This gives us a group of $6 \cdot 4 \cdot 5 = 120$ elements, known as $2I$, the binary icosahedral group, or $SL(2, 5)$. If we factor out by the centre—the maps that commute with every element of this group, and which consists of two elements $\begin{pmatrix} 1 & 0 \\ 0 & 1 \end{pmatrix}$ and $\begin{pmatrix} 4 & 0 \\ 0 & 4 \end{pmatrix}$—we obtain a quotient group of order 60, known as $PSL(2, 5)$, the projective special linear group (of the 2-dimensional vector space over the field of 5 elements). Klein called this group I or the icosahedral group.

There is another way of looking at this group that helps explain the modern name. Every time we have a vector space of dimension say $n + 1$ over a field $\mathbb{F}$ we can form the corresponding projective space of dimension n whose points are the 1-dimensional vector subspaces (lines through the origin). When we do that with $\mathbb{F}_5^2$ we obtain the 1-dimensional projective space $\mathbb{F}_5 P^1$ with 6 points, represented in homogeneous coordinates by $[0, 1], [1, 0], [1, 1], [1, 2], [1, 3], [1, 4]$, which is convenient to denote $\infty, 0, 1, 2, 3, 4$. The group action we have is given by

$$ z \mapsto \frac{az + b}{cz + d}, \quad a, b, c, d \in \mathbb{F}_5, \ ad - bc \neq 0. $$

We shall often represent this map by the matrix $\begin{pmatrix} a & b \\ c & d \end{pmatrix}$ but when we do notice that the maps $z \mapsto \frac{az+b}{cz+d}$ and $z \mapsto \frac{taz+tb}{tcz+td}$ have the same effect, so we factor out the matrices of the form $\begin{pmatrix} t & 0 \\ 0 & t \end{pmatrix}$, and once again obtain the group $PSL(2, 5)$.

Now, it is apparent from its construction that $PSL(2, 5)$ is represented as a subgroup of S_6 in the same way that we obtained G_I, so it is likely that they are isomorphic. The questions then become: how do we show this, and what have they to do with A_5?

It is easy to see that $PSL(2, 5)$ acts doubly transitively on the set of six directions. It is enough to show that the directions ∞ and 0 can be sent to the directions a and $b \neq a$. The cases $a = \infty, b = 0$ and $a = 0, b = \infty$ are trivial, so we may assume a and b are chosen from $1, 2, 3$, and 4. Consider the element $\begin{pmatrix} a & b \\ 1 & 1 \end{pmatrix}$. It sends the directions ∞ and 0 to the directions a and $b \neq a$, and it has determinant $a - b$, so the required element is $\begin{pmatrix} a & a - 1 \\ 1 & 1 \end{pmatrix}$.

Likewise, G_I^+ acts doubly transitively on the set of six axes of the dodecahedron that join diametrically opposite mid-face point. However, it is not clear that A_5 acts on a set of six things, nor that $PSL(2, 5)$ or G_I^+ act on five. We can easily compute that each of the three groups has 15 elements of order 2, 20 elements of order 3, and 24 elements of order 5, which makes it certain that the groups are isomorphic, but leaves the isomorphisms between them obscure. A better way to proceed is to look at the subgroups of these groups and let the each group act on its subgroups by conjugation.

One way to understand the group A_5 is that it has 5 subgroups of order 12, each of which is isomorphic to A_4. These are the groups that fix exactly one of the five objects that are permuted by A_5—for example, the even permutations that fix the object 5. These five subgroups are permuted by conjugation in A_5.

Within each copy of A_4 is a unique copy of the Klein group, for example the group consisting of the identity element and the elements

$$ (12)(34), \ (13)(24), \text{ and } (14)(23). $$

This suggests that we should look for elements of Klein groups in the other two groups. With the group $PSL(2, 5)$ we can make a start by looking for elements that switch the directions ∞ and 0. This is done by a matrix of the form $\begin{pmatrix} 0 & b \\ a & 0 \end{pmatrix}$. The determinant must be 1, so we have $ab = -1$, and this gives us the elements $\begin{pmatrix} 0 & 2 \\ 2 & 0 \end{pmatrix}$ and $\begin{pmatrix} 0 & 1 \\ 4 & 0 \end{pmatrix}$.

The first of these corresponds to this permutation of the directions: $(\infty 0)(23)$. The second corresponds to this permutation of the directions: $(\infty 0)(14)$.

Each of these transformations fixes two directions, and this suggests that we look at the transformations that fix ∞ and 0. The non-identity transformation that does this turns out to be $\begin{pmatrix} 2 & 0 \\ 0 & 3 \end{pmatrix}$, which is the product of the previous two transformations, and corresponds to the permutation $(23)(14)$.

It is clear from their representation as permutations that these transformations and the identity together define a Klein group. How many such groups are there in $PSL(2, 5)$? We may pair ∞ with any other element and argue as before, but we find that when we do we can make no further choices. So there are five of these Klein groups, and $PSL(2, 5)$ acts on them by conjugation.

To find Galois's five groups of order 12 in $PSL(2, 5)$ we observe that this group will necessarily have element of orders 2, 3, and 5. So we find an element τ of order 5, for example $z \rightarrow z + 1$, and an element σ of order 2, say $z \rightarrow -1/z$. We then check that their composite $\tau\sigma$, which is $z \rightarrow (z - 1)/z$, has order 3 and that these elements generate a group of order 12, as required. Conjugation produces four other groups of this kind, and we are done.

Fig. E.1 A square in a dodecahedron

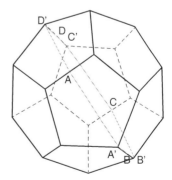

Now we turn to identifying the five Klein groups in G_I^+, the group of the dodecahedron. In Fig. E.1 we look at lines like AA' joining a mid-edge point A to its opposite vertices, marked as A' and D'. The pentagonal faces are congruent, so these line segments are equal (say to α). Likewise, all the lengths from a vertex along an edge to the midpoint of that edge (such as $A'B$) are equal, say to β. Therefore

$$AB = BC = CD = DA,$$

each being $\alpha + \beta$. The symmetry of the dodecahedron also forces the angles $AA'B$, $BB'C$, $CC'D$ and $DD'A$ to be equal. Therefore $ABCD$ is in fact a square, and if we denote the centre of the dodecahedron by O the four angles AOB, BOC, COD, DOA are right angles, and the points A and C are diametrically opposite, as are B and D.

Fig. E.2 Relabelling the faces of a dodecahedron

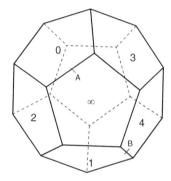

We are now able to follow the effect of a rotation through π about the axes AC and BD on the faces of the dodecahedron. Label the front face in Fig. E.2 ∞ and the others as shown, and label the faces round the back (with some dashed edges) with the label of their diametrically opposite face. Then rotation through π about the axis AC is the permutation $(\infty0)(23)$, and rotation through π about the axis BD is the permutation $(\infty0)(14)$. The composition of these rotations is $(23)(14)$, as required.

The dodecahedron has 20 vertices, occurring in 10 diametrically opposite pairs. The above construction pairs the pairs (AC with BD, for example), so there are five squares of the kind described and so five Klein groups in G_I^+. Again, G acts on them by conjugation.

We deduce that the three groups $PSL(2, 5)$, G, and A_5 are isomorphic because they are even permutations on five objects.

A prettier way to understand how G_I^+ permutes 5 things is to see that there are the five cubes that share vertices with the icosahedron. Each cube has twelve edges, and the icosahedron has twelve faces, so one edge (of a cube) goes in each face, and it goes in as a diagonal of the pentagon (note that there are indeed five of these, so each cube has one edge in each face of the dodecahedron). I leave it as an exercise to find these cubes.

We need one final representation of G_I^+, which is obtained by inscribing it in a sphere and then looking at a suitably symmetric image of it on the plane after stereographic projection. The edges of the icosahedron are arcs of great circles on the sphere, so they project to circular arcs in the plane, and so elements of G_I^+ will be represented as Möbius transformations

$$z \mapsto \frac{az+b}{cz+d}, \quad a,b,c,d \in \mathbb{C}, \quad ad - bc \neq 0.$$

Of these, one is a rotation through an angle of $2\pi/5$. If, with Klein, we place one vertex at $z = 0$ and another at $z = \infty$, and the remaining vertices at the points $\rho^k(\rho + \rho^4)$, and $\rho^k(\rho^2 + \rho^3)$, where $\rho = e^{2\pi i/5}$ is a fifth root of unity, this rotation is $z \mapsto \rho z$. Another element is a rotation through an angle of $2\pi/5$ about the point ρ, and these rotations generate the group.[8]

E.4   The Icosahedral Equation

A typical point on a face of the icosahedron is moved to a total of 60 positions under the action of the rotation group G_I^+. But there are points with smaller orbits: the 12 centres of faces, the 20 vertices, and the 30 mid-edge points. Klein considered each of these in turn. He looked at a fixed image of the icosahedron on the plane. The 12 centres of faces are 12 points of the plane, and so they satisfy a polynomial of degree 12. By looking at their coordinates and knowing how they are found, Klein was able to write down this polynomial; it is

$$f(z_1, z_2) = z_1 z_2 (z_1^{10} + 11 z_1 z_2 - z_2^{10}),$$

where Klein used homogeneous coordinates to handle the point at infinity. We can regard $z_1/z_2 = z$. This polynomial is invariant under the action of the group.

From this polynomial Klein extracted another by means of a piece of the theory of forms that we have not discussed. It is the so-called *Hessian* of f, denoted $H = H(f)$, and which is of degree 20 in this case:

$$H(z_1, z_2) = -(z_1^{20} + z_2^{20}) + 228(z_1^{15} z_2^5 - z_1^5 z_2^{15}) - 494 z_1^{10} z_2^{10},$$

and its zeros correspond to the 20 vertices of the icosahedron.[9] This polynomial is also invariant under the action of the group.

---

[8]Klein went on in 1878 to investigate the next group in the sequence after $PSL(2, 5)$. This is $PSL(2, 7)$, a famous group of order 168. There is now an extensive literature on the group, which plays an important role in the study of automorphism groups of Riemann surfaces among other topics, and one may begin by consulting the book *The Eightfold Way, The Beauty of Klein's Quartic Curve* (1999). For a historical introduction, see Gray (1999).

[9]The Hessian of a homogeneous function of two variables is the determinant of its second partial derivatives.

A further use of the same pre-existing theory gave Klein a homogeneous polynomial of degree 30 that is also invariant under the action of the group.[10] The zeros of this form are the mid-edge points of the icosahedron:

$$T(z_1, z_2) = (z_1^{30} + z_2^{30}) + 522(z_1^{25} z_2^5 - z_1^5 z_2^{25}) - 10005(z_1^{20} z_2^{10} + z_1^{10} z_2^{20}).$$

Moreover, these three homogeneous polynomials satisfy an equation (wonderfully known as a syzygy!):

$$T^2 + H^3 - 1728 f^5 = 0.$$

Klein set

$$Z : Z - 1 : 1 = H^3 : -T^2 : 1728 f^5,$$

and wrote

$$q(z) = \frac{H(z_1, z_2)^3}{1728 f(z_1, z_2)^5} = \frac{H(z, 1)^3}{f(z, 1)^5},$$

and called the equation

$$q(z) = u$$

the *icosahedral equation*. It is an equation of degree 60 that depends on a single parameter u.

He now set himself a number of tasks, of which these interest us:

- to explain how any irreducible quintic equation can be written in this form;
- to explain how the icosahedral group appears geometrically (and not just abstractly) as the Galois group of the general quintic equation;
- to explain why certain equations of degree 6 reduce to the quintic equation (an old question raised by Galois and solved in the late 1850s by Hermite, Kronecker, and Brioschi).

---

[10]This is the so-called functional determinant of the form f and its Hessian, which is the determinant of the first partial derivatives of the two forms.

Appendix F
From Dedekind's 11th Supplement (1894)

Dedekind explained at the end of the preceding section of his Supplement XI *Über die Theorie der algebraischen Zahlen*, (1894) that the aim was to use his theory of modules over a quadratic field Ω to present a general theory that included the composition of binary forms.

> The division of binary quadratic forms into genera (Supplement IV) can be carried over easily to the theory of ideals, and both this question and proof of the reciprocity theorem that rests on the enumeration of two-sided classes (§§152 to 154) acquires a far simpler form in the new presentation, whose production we must, however, leave to the reader. On the other hand, we would like to introduce the general theory of modules for quadratic fields, because this includes the composition theory for binary quadratic forms and is of great significance for many other questions, e.g. for the theory of complex multiplication of elliptic functions.[1]

He had already defined a module as a subset of Ω that is closed under subtraction; it is exactly what we would call a $\mathbb{Z}$ module: a subset closed under addition, subtraction, and multiplication by rational integers (but not, as an ideal is, closed under multiplication by all integers in Ω).

I briefly summarise the opening paragraphs of this section, which is the last part of a very long paper. Dedekind began by observing that he had already shown that every finite (i.e. finite-dimensional) module **m** over a quadratic field Ω has a basis of the form $[\alpha, \beta]$, where α and β are independent. Straight-forward calculations now showed that in fact there is a rational number (i.e. a fraction of two ordinary integers) m and an irrational number ω such that the module is $m[1, \omega]$. Furthermore, ω satisfies the quadratic equation

$$a\omega^2 - b\omega + c = 0,$$

---

[1](Dedekind Supplement XI, §186).

© Springer Nature Switzerland AG 2018
J. Gray, *A History of Abstract Algebra*, Springer Undergraduate Mathematics Series,
https://doi.org/10.1007/978-3-319-94773-0

where a, b, c are rational integers without common divisor and $a > 0$. Dedekind denoted the fundamental number of Ω by D, and set

$$\theta = \frac{D + \sqrt{D}}{2}, \quad \mathbf{o} = [1, \theta],$$

so $a\omega$ is an algebraic integer of the form

$$a\omega = h + k\theta = \frac{b + \sqrt{d}}{2},$$

where h, k are rational integers and

$$d = b^2 - 4ac = \Delta(1, k\omega) = Dk^2.$$

All references to sections of the paper before this one are to material covered in the notes. It will help to remember the mantra 'to divide is to contain'. As for notation, the expression $\mathbf{a} > \mathbf{b}$ means that $\mathbf{b}$ *divides* $\mathbf{a}$—be careful! Given two modules $\mathbf{m}$ and $\mathbf{m}'$ Dedekind defined the quotient of $\mathbf{m}'$ by $\mathbf{m}$ to be the module $\{\nu : \nu\mathbf{m} > \mathbf{m}'\}$, and denoted it $\frac{\mathbf{m}'}{\mathbf{m}}$. Given modules or ideals $\mathbf{a}, \mathbf{b}$ and $\alpha \in \mathbf{a}$ one can defined the conjugacy class of all elements in $\mathbf{a}$ conjugate mod $\mathbf{b}$ to α, that is, the set $\{\alpha' \in \mathbf{a} \mid \alpha - \alpha' \in \mathbf{b}\}$. Then $(\mathbf{a}, \mathbf{b})$ denotes the number of such conjugacy classes. In the setting where $\mathbf{a}$ and $\mathbf{b}$ are two modules with least common multiple $\mathbf{m} = \mathbf{a} - \mathbf{b} = \mathbf{a} \cap \mathbf{b}$, Dedekind denoted the number of congruence classes of elements in $\mathbf{a}$ modulo $\mathbf{m}$ by $(\mathbf{a}, \mathbf{m})$. So $(\mathbf{a}, \mathbf{b}) = 1$ if and only if $\mathbf{a}$ is divisible by $\mathbf{b}$.

At one point in what follows we also need Dedekind's definition of a proper module. You will see below that given a module $\mathbf{m}$ he defined its order, $\mathbf{m}^0$, to be $\{\nu : \mathbf{m} \supset \nu\mathbf{m}\}$. He defined (§170) $\mathbf{m}^{-1}$ to be the quotient $\frac{\mathbf{m}^0}{\mathbf{m}}$, and $\mathbf{m}^{-n}$ to be the quotient $\frac{\mathbf{m}^0}{\mathbf{m}^n}$. He said a module was proper if $\mathbf{mm}^{-1} = \mathbf{m}^0$. He then proved that for proper modules $\mathbf{a}$ and $\mathbf{b}$

$$(\mathbf{a}^{-1})^0 = \mathbf{a}^0, \ (\mathbf{a}^{-1})^{-1} = \mathbf{a}, \ (\mathbf{ab})^0 = \mathbf{a}^0\mathbf{b}^0, \ (\mathbf{ab})^{-1} = \mathbf{a}^{-1}\mathbf{b}^{-1}.$$

I have tacitly omitted all but one of the footnotes Dedekind supplied because they would take us too far afield to explain.

The Extract, §187, pp. 207–216

By the *order* $\mathbf{m}^0$ of a module $\mathbf{m}$, which we shall more briefly denote by $\mathbf{n}$, we shall understand, as before (§170) the totality of all numbers ν for which $\mathbf{m}\nu$ is divisible by $\mathbf{m}$. It obviously follows from this definition that if η denotes an arbitrary nonzero number then $\mathbf{n}$ is at the same time the order of $\eta\mathbf{m}$; if we therefore maintain the previous designations then the sought-for numbers ν are all those for which $[\nu, \nu\omega]$ is divisible by $[1, \omega]$, and for this it is necessary and sufficient that both numbers ν and $\nu\omega$ are contained in $[1, \omega]$. Therefore we must have $\nu = x + y\omega$, where x and

y denote rational integers; then $v\omega = x\omega + y\omega^2$, and since $x\omega$ is in $[1, \omega]$ this must also be true for $y\omega^2$; consequently (3) becomes

$$y\omega^2 = \frac{y(b\omega - c)}{a},$$

consequently both products by and cy must be divisible by a; but the numbers a, b, c have no common factor, so it follows that y must be divisible by a, so it must be the case that $y = az$ and $v = x + za\omega$ where z is likewise a rational integer; and since conversely every such number $x + za\omega$ has the requisite property we obtain the result that

$$(7) \qquad \mathbf{n} = [1, a\omega] = [1, k\theta] = \mathbf{o}k + [1].$$

Every order $\mathbf{n}$ is therefore a module which contains only integers and among these the number 1 and therefore every rational integer (cf. §173, III); conversely it is clear that every such module $\mathbf{n}$ (in our case of the quadratic fields) is certainly also an order, namely the order of $\mathbf{n}$ itself. For the discriminant, index, and conductor of an order $\mathbf{n}$ we readily deduce from (4), (6), and (7) that

$$(8) \qquad \Delta(\mathbf{n}) = d, \ (\mathbf{o}, \mathbf{n}) = k, \ \frac{\mathbf{n}}{\mathbf{o}} = \mathbf{o}k,$$

and it is clear that every order $\mathbf{n}$ is completely determined by its index k.[2]

It is clear that every module $\mathbf{m}$ is an ideal then and only then, when it is divisible by $\mathbf{o}$, and $\mathbf{n} = \mathbf{o}$, so $k = 1$, and m is an integer divisible by a. This allows us, therefore to extend the idea of the norm to an arbitrary module $\mathbf{m}$, and indeed we will here understand by the quotient

$$(9) \qquad N(\mathbf{m}) = \frac{(\mathbf{n}, \mathbf{m})}{(\mathbf{m}, \mathbf{n})},$$

which in fact, when $\mathbf{m}$ is an ideal reduces to the earlier definition of the corresponding value $(\mathbf{o}, \mathbf{m})$ (§180). Since the basis numbers for $\mathbf{m}$ are related to those of $\mathbf{n}$ by the linear equations

$$m = m \cdot 1 + 0 \cdot a\omega, \ m\omega = 0 \cdot 1 + \frac{m}{a} \cdot a\omega$$

the result follows [from §175, (10)] that

$$(10) \qquad N(\mathbf{m}) = \begin{vmatrix} m & 0 \\ 0 & \frac{m}{a} \end{vmatrix} = \frac{m^2}{a}.$$

---

[2]The material in this sentence was presented differently in (1879).

[For material that was given here in (1871) but not in (1879) or (1894), see Note 1 at the end of this Appendix.]

If α is an arbitrary number in the field Ω and one generally denotes by α' the conjugate number that α goes to under the non-identity permutation of the field, then

$$(11) \qquad a(\omega + \omega') = b, \ a\omega\omega' = c;$$

as μ runs through all the numbers in the module $\mathbf{m}$ the numbers μ' form the conjugate module $m[1, \omega']$ to $\mathbf{m}$, which we will denote by $\mathbf{m}'$. If we retain the above choice of basis, then we have to set

$$(12) \qquad \mathbf{m}' = m[1, -\omega']$$

and since

$$a(-\omega')^2 - (-b)(-\omega') + c = 0$$

the transition from $\mathbf{m}$ to $\mathbf{m}'$ is easily carried out by replacing b with $-b$ while m, a, c, k, d remain unaltered. Thus $\mathbf{m}$ is naturally conjugate to $\mathbf{m}'$ and both modules have the same order $\mathbf{n} = \mathbf{n}'$ and the same norm; so they are then only identical if b is divisible by a, so $b \equiv 0$ or $\equiv a \pmod{2a}$, and in this case $\mathbf{m}$ can be considered as a two-sided module (cf. §58).

Every number μ contained in the module $\mathbf{m}$ is of the form

$$(13) \qquad \mu = m(x + y\omega),$$

where x, y denote rational integers; from this it follows that

$$N(\mu) = \mu\mu' = m^2(x + y\omega)(x + y\omega')a$$

and on carrying out the multiplication this gives

$$(14) \qquad N(\mu) = \mu\mu' = N(\mathbf{m})(ax^2 + bxy + cy^2);$$

therefore to every module $\mathbf{m}$ there corresponds, if we maintain the above rules for the choice of basis, an original binary quadratic form $(a, \frac{1}{2}b, c)$ or rather to a definite family of infinitely many such parallel forms [i.e. equivalent forms, JJG] in which b runs through all the individuals in a definite number class with respect to the modulus $2a$ and whose discriminant $b^2 - 4ac$ is simultaneously the discriminant d of the order $\mathbf{n}$; to the conjugate module $\mathbf{m}'$ corresponds the corresponding family $(a, -\frac{1}{2}b, c)$. [At this point in (1879) Dedekind explained the concept of multiplication of modules.] Evidently the same family $(a, \frac{1}{2}b, c)$ corresponds to all and only those modules of the form **m** n, where n is any non-zero rational number. Since further the numbers $1, a\omega$ form a basis for the order **n**, and

$$a\omega\mu = m(-cy + (ax+by)\omega)$$

$$\begin{vmatrix} x & y \\ -cy & ax+by \end{vmatrix} = ax^2 + bxy + cy^2$$

so the form $(a, \frac{1}{2}b, c)$ coincides exactly with the one that we earlier (on p. 156 of this paper) said corresponded with the module **m**.

[Here I insert some results from §170 that Dedekind proceeds to ask his readers to recall:

- (Theorem V): a module **a** is a proper module if and only if it is a factor of its order $\mathbf{a}^0$, i.e. when there is a module **n** that satisfies the condition $\mathbf{an} = \mathbf{a}^0$, and consequently $\mathbf{a}^{-1} = \mathbf{na}^0$.
- (Theorem VII) If **a** is a proper, and **b** an arbitrary module, then

$$(36) \qquad \frac{\mathbf{ab}}{\mathbf{a}} = \mathbf{ba}^{-1}.$$

- (Theorem VIII) If **a** and **b** are proper modules then so is their product **ab** and

$$(37) \qquad (\mathbf{ab})^0 = \mathbf{a}^0\mathbf{b}^0, \ (\mathbf{ab})^{-1} = \mathbf{a}^{-1}\mathbf{b}^{-1}.]$$

As we now turn to the multiplication of modules, we first recall two general results proved in §170 (p. 72)

$$(15) \qquad \mathbf{mn} = \mathbf{m}, \mathbf{n}^2 = \mathbf{n},$$

which can be derived easily by actual multiplication from (2) and (7).[3] Of particular importance is the construction of the product **m m′** of two conjugate modules. From (2) and (12) we at once obtain by multiplication

$$\mathbf{mm}' = m^2[1, \omega, \omega', \omega\omega'].$$

If one adds the second basis number to the third, one obtains

$$\mathbf{mm}' = \frac{m^2}{a}[a, a\omega, b, c]$$

and since $[a, b, c] = [1]$ we obtain the result

$$(16) \qquad \mathbf{mm}' = \frac{m^2}{a}[1, a\omega] = \mathbf{n}N(\mathbf{m});$$

---

[3]These are the equations (2) $\mathbf{m} = [m, m\omega] = m[1, \omega]$, and (7) $\mathbf{n} = [1, a\omega] = [1, k\theta] = \mathfrak{o}k + [1]$.

in which $\mathbf{m}$ (after §170 V) is a proper module, and so it immediately follows that

$$(17) \qquad \mathbf{m}' = \mathbf{m}^{-1} N(\mathbf{m}).$$

We now consider the product of two arbitrary modules $\mathbf{m}$ and $\mathbf{m}_1$, and set

$$(18) \qquad \mathbf{m}\mathbf{m}_1 = \mathbf{m}_2;$$

since $\mathbf{m}_2$ consists of all numbers μ_2 of the form $\sum \mu\mu_1$ the conjugate module $\mathbf{m}_2'$ consists of all numbers μ_2 of the form $\sum \mu'\mu_1'$, and consequently

$$\mathbf{m}'\mathbf{m}_1' = \mathbf{m}_2' = (\mathbf{m}\mathbf{m}_1)'.$$

On multiplying these two equations, one obtains as a result

$$\mathbf{n}\mathbf{n}_1 N(\mathbf{m})N(\mathbf{m}_1) = \mathbf{n}_2 N(\mathbf{m}_2),$$

where $\mathbf{n}_1$, $\mathbf{n}_2$ denote the orders of $\mathbf{m}_1$, $\mathbf{m}_2$; now since the product $\mathbf{n}\mathbf{n}_1$ contains only integers and evidently also the number 1 it is, by what was said above, an order; the previous equation therefore shows that if one cancels the rational numbers entering each side one obtains the theorem

$$(19) \qquad N(\mathbf{m})N(\mathbf{m}_1) = N(\mathbf{m}_2) = N(\mathbf{m}\mathbf{m}_1)$$

whence the following:

$$(20) \qquad \mathbf{n}\mathbf{n}_1 = \mathbf{n}_2;$$

the norm of a product is therefore equal to the product of the norms of the factors, and likewise the order of a product is equal to the product of the orders of the factors (cf. §170, VIII).

Since the number 1 is contained in each order, the product $\mathbf{n}\mathbf{n}_1$ is a common divisor of $\mathbf{n}$ and $\mathbf{n}_1$, and indeed, as will now be shown, its greatest common divisor. If k, k_1, k_2 denote the indices of the orders $\mathbf{n}, \mathbf{n}_1, \mathbf{n}_2$, then $\mathbf{n} = [1, k\theta]$, $\mathbf{n}_1 = [1, k_1\theta]$ and so

$$\mathbf{n}\mathbf{n}_1 = [1, k\theta, k_1\theta, kk_1\theta^2];$$

but since $\theta^2 = D\theta - D_1$, where D_1 is a rational integer, the last basis number $kk_1\theta^2$ can be dropped, because it is a sum of multiples of the first two, and one obtains

$$(21) \qquad \mathbf{n}\mathbf{n}_1 = [1, k\theta, k_1\theta] = \mathbf{n} + \mathbf{n}_1,$$

as was required. Since, from (20), the same product is also $= [1, k_2\theta]$, it follows that k_2, the index of the product, is the greatest common divisor of k, k_1, the indices of the factors. If furthermore d, d_1, d_2 denote the discriminants of $\mathbf{n}, \mathbf{n}_1, \mathbf{n}_2$, then $d = Dk^2, d_1 = Dk_1^2, d_2 = Dk_2^2$, and consequently the discriminant of the product is also the greatest common divisor of the discriminants of the factors.

The last theorem also comes about in the following way, where we let the letters $m_1, \omega_1, a_1, b_1, c_1$ and $m_2, \omega_2, a_2, b_2, c_2$ have the same meaning for the modules $\mathbf{m}_1$ and $\mathbf{m}_2$ that m, ω, a, b, c had for $\mathbf{m}$. Then it follows from (20) that

$$[1, a_2\omega_2] = [1, a\omega][1, a_1\omega_1] = [1, a\omega, a_1\omega_1, aa_1\omega\omega_1],$$

and therefore (from §172) there are four equations of the form (22)

$$1 = 1 \cdot 1 + 0 \cdot a_2\omega_2$$

$$a\omega = f \cdot 1 + e \cdot a_2\omega_2$$

$$a_1\omega_1 = f_1 \cdot 1 + e_1 \cdot a_2\omega_2$$

$$aa_1\omega\omega_1 = f_2 \cdot 1 + e_2 \cdot a_2\omega_2,$$

where the eight coefficients on the right are rational integers such that the six determinants formed from them,

$$e, e_1, e_2, fe_1 - ef_1, fe_2 - ef_2, f_1e_2 - e_1f_2$$

have no common factor; but since every common factor of the first three also goes into the following ones, it follows that e, e_1, e_2 have no common factor. So it further follows from (22) that

$$(f + ea_2\omega_2)(f + e_1a_2\omega_2) = (f + e_2a_2\omega_2),$$

so

$$ee_1(a_2\omega_2)^2 - (e_2 - ef_1 - e_1 f)(a_2\omega_2) + ff_1 - f_2 = 0;$$

if one compares this with the equation

$$(a_2\omega_2)^2 - b_2(a_2\omega_2) + a_2c_2 = 0,$$

then it follows that

$$(23) \qquad e_2 = ef_1 + e_1f + ee_1b_2, \quad f_2 = ff_1 - ee_1a_2c_2;$$

from the first of these equations it follows that every common divisor of e, e_1 also goes into e_2; since it was shown above that these three numbers have no common divisor so e, e_1 are relatively prime numbers. If now (5) is used to replace the quantities $a\omega, a_1\omega_1, a_2\omega_2$ with

$$\frac{b + k\sqrt{D}}{2}, \frac{b_1 + k_1\sqrt{D}}{2}, \frac{b_2 + k_2\sqrt{D}}{2},$$

it follows that

$$(24) \qquad k = ek_2, \; k_1 = e_1 k_2, \; (\mathbf{n_1}, \mathbf{n}) = e, \; (\mathbf{n}, \mathbf{n_1}) = e_1,$$

so also

$$(25) \qquad d = d_2 e^2, \; d_1 = d_2 e_1^2,$$

and moreover

$$(26) \qquad f = \frac{b - b_2 e}{2}, \; f_1 = \frac{b_1 - b_2 e}{2};$$

similarly one obtains from the last of equations (22) or by substituting the above expressions in (23),

$$(27) \qquad e_2 = \frac{be_1 + b_1 e}{2}, \; f_2 = \frac{bb_1 + d_2 ee_1 - 2b_2 e_2}{4}.$$

From (24) and (25) it follows once again that k_2 is the greatest common divisor of k, k_1, as similarly d_2 is for d, d_1.

So if two modules $\mathbf{m}, \mathbf{m_1}$ are given one finds the numbers e, e_1, k_2, d_2 from (24) and (25) under the condition that e, e_1 must be relatively prime, and in this way, because of (27) e_2 is also found. We shall now see how to determine the module $\mathbf{m_2}$ completely by obtaining the numbers m_2, a_2, b_2, c_2 from the given data. Since the product $\mathbf{m_1}$ is contained in $\mathbf{m_2}$ and consequently also in $[m_2]$, one can immediately set

$$(28) \qquad mm_1 = pm_2, \; m_2 = \frac{mm_1}{p}$$

where p denotes a natural number; one now replaces the norms entering theorem (19) by their expressions in (10), to obtain

$$(29) \qquad aa_1 = p^2 a_2, \; a_2 = \frac{aa_1}{p^2},$$

so the determination of m_2 and a_2 is reduced to that of p. One next replaces the modules $\mathbf{m}, \mathbf{m_1}, \mathbf{m_2}$ by the expressions (2), so that the equation $\mathbf{m_2} = \mathbf{mm_1}$ takes the form

$$(30) \qquad [1, \omega_2] = p[1, \omega][1, \omega_1] = p[1, \omega_1, \omega, \omega\omega_1];$$

so (after §172) one can set (31)

$$p = p \cdot 1 + 0 \cdot \omega_2$$

$$p\omega_1 = p' \cdot 1 + q' \cdot \omega_2$$

$$p\omega = p'' \cdot 1 + q'' \cdot \omega_2$$

$$p = \omega\omega_1 = p''' \cdot 1 + q''' \cdot \omega_2$$

where the eight coefficients on the right-hand side are rational integers such that the six determinants formed from them

$$pq', \ pq'', \ pq''', \ p'q'' - q'p'', \ p'q''' - q'p''', \ p''q''' - q''p''',$$

like the three numbers q', q'', q''' have no common divisor. If one now substitutes for $\omega, \omega_1, \omega\omega_1$ the expressions from (22) one obtains the equations

$$p(f_1 + e_1 a_2 \omega_2) = a_1(p' + q'\omega_2)$$

$$p(f + e a_2 \omega_2) = a_1(p'' + q''\omega_2)$$

$$p(f + e_2 a_2 \omega_2) = a a_1(p'' + q''\omega_2),$$

which, since ω_2 is irrational, break into the following

$$(32) \qquad pe_1 a_2 = a_1 q', \ pea_2 = aq'', \ pe_1 2 a_2 = a a_1 q''',$$

$$(33) \qquad pf_1 = a_1 p', \ pf = ap''', \ pf_2 = a a_1 p'''.$$

If one substitutes the corresponding expression for a_2 from (29) one obtains

$$(34) \qquad ae_1 = pq', \ a_1 e = pq'', \ e_2 = pq''',$$

and since q', q'', q''' as remarked above, have no common divisor, p is evidently determined as the greatest (positive) common divisor of the three numbers ae_1, ae, e_2, and the same therefore holds for three numbers q', q'', q''' as well as for the two numbers m_2, a_2 which are given by Eqs. (28) and (29). If one further multiplies Eq. (33) by $2a, 2a_1, 1$ and replaces aa_1 by $p^2 a_2$, one obtains from (34), on replacing f_1, f, f_2 by the expressions given for them in (26) and (27), the equations

$$\frac{ab_1}{p} - q'b_2 = 2a_2 p', \quad \frac{a_1 b}{p} - q''b_2 = 2a_2 p'',$$

$$\frac{bb_1 + d_2 e e_1}{2p} - q'''b_2 = 2a_2 p''',$$

and the congruences (35) (mod $2a_2$):

$$q'b_2 \equiv \frac{ab_1}{p}, \quad q''b_2 \equiv \frac{a_1b}{p}, \quad q'''b_2 \equiv \frac{bb_1 + d_2ee_1}{2p};$$

in which the number b_2 is completely determined with respect to the modulus $2a_2$, because q', q'', q''' have no common divisor (cf. §145); and from this it finally follows that c_2 is given by the equation

$$(36) \qquad c_2 = \frac{b_2^2 - d_2}{4a_2}.$$

This completes the determination of the product $\mathbf{m}_1$ from the two factors $\mathbf{m}, \mathbf{m}_1$, and we have now only to make the following remark. Since the existence of the module $\mathbf{m}_2 = \mathbf{mm}_1$ is known from what was said earlier, we must conclude that the numbers entering (26), (27), (29), (35), and (36) in the form of fractions are really integers, that furthermore the three congruences (35) are really the same, and that the resulting numbers a_2, b_2, c_2 have no common factors; this can easily be shown directly, and we leave it to the reader.[4]

We now denote by x, y and x_1, y_1 two systems of independent variables and construct the bilinear functions (37)

$$x_2 = pxx_1 + p'xy_1 + p''yx_1 + p'''yy_1$$

$$y_2 = \quad q'xy_1 + q''yx_1 + q'''yy_1$$

further one sets

$$\mu = m(x + y\omega), \quad \mu_1 = m(x_1 + y_1\omega_1), \quad \mu_2 = m_2(x_2 + y_2\omega_2),$$

then it follows from (28) and (31) that $\mu_2 = \mu\mu_1$, and so also for rational values of the variables $N(\mu_2) = N(\mu)N(\mu_1)$; if one looks at these norms by their expressions in (14) and looks back at (19) it follows that

$$(38) \qquad a_2x_2^2 + b_2x_2y_2 + c_2y_2^2 = (ax^2 + bxy + cy^2)(a_1x_1^2 + b_1x_1y_1 + c_1y_1^2);$$

one therefore says that the form $(a_2, \frac{1}{2}b_2, c_2)$ passes by means of the bilinear substitution (37) into the product of the two forms $(a, \frac{1}{2}b, c)$ and $(a_1, \frac{1}{2}b_1, c_1)$, and calls the first form the compound of the latter two; evidently (38), in virtue of (37), is an identity that holds for all values of the independent variables.

---

[4]See Arndt (1859).

Note 1

From the correctness of this, one also convinces oneself easily of the theorem that the greatest common divisor of both modules is

$$\mathbf{d} = [m, m\omega, 1, a\omega].$$

To reduce this to a basis of two terms, we set

$$m = \frac{t}{u}, \quad \frac{m}{a} = \frac{v}{w},$$

where t, u and vw are two pairs of relatively prime numbers; then evidently

$$[m, 1] = \left[\frac{t}{u}, \frac{u}{u}\right] = \left[\frac{1}{u}\right]$$

$$[m\omega, a\omega] = a\omega\left[\frac{v}{w}, \frac{w}{w}\right] = \left[\frac{a\omega}{w}\right]$$

so with

$$\mathbf{d} = \left[\frac{1}{u}, \frac{a\omega}{w}\right]$$

one now expresses the basis numbers of $\mathbf{m}$ and $\mathbf{n}$ linearly in terms of those of $\mathbf{d}$, and so (after §165, (2) and (22)) it is the case that

$$[\mathbf{n}, \mathbf{m}] = [\mathbf{d}, \mathbf{m}] = tv, \quad [\mathbf{m}, \mathbf{n}] = [\mathbf{d}, \mathbf{n}] = uw,$$

so

$$N(\mathbf{m}) = \frac{tv}{uw},$$

which agrees with (10).

Appendix G
Subgroups of S_4 and S_5

G.1 The Subgroups of S_4

The subgroups of S_4 can all be found by looking at the possible types of permutations that can occur. By Lagrange's theorem, the possible orders of subgroups of
S_4, which has order 24, are 1, 2, 3, 4, 6, 8, and 12, and it is easy to see that groups
of each order can occur (but not every non-isomorphic group of each order). The
cycle types give us cyclic subgroups of orders 1, 2, 3, and 4, corresponding to the
possible orders of the elements of the group S_4. The two non-isomorphic groups
of order 4 both occur, the cyclic one and the non-cyclic one where all the non-
identity elements have order 2 (the Klein group, K). The three conjugate copies
of the cyclic one are generated by four-cycles: $\{e, (1234), (13)(24), (1432)\}$ is an
example. There are four non-cyclic ones: $K = \{e, (12)(34), (13)(24), (14)(23)\}$,
the symmetry group of a rectangle with vertices labelled $\{1, 2, 3, 4\}$, and three of
which $\{e, (12), (34), (12)(34)\}$ is an example. These three are mutually conjugate,
but K is a normal subgroup of S_4.

Of the two non-isomorphic groups of order 6, any copy of S_3 as a subgroup of
S_4 will do for the non-commutative one, but the cyclic group of order 6 cannot be a
subgroup of S_4 because there is no element of order 6 in S_3.

There are five non-isomorphic groups of order 8, but it turns out that the
symmetry group of the square with vertices labelled $\{1, 2, 3, 4\}$ is the only subgroup
of S_4 of order 8.

The group A_4 of even permutations is the only subgroup of S_4 of order 12.

A subgroup H of S_4 is normal if and only if it contains all the elements of a
given cycle type, because conjugation preserves cycle type, and any element of a
given cycle type is conjugate to any other element of that type (think of conjugating
as relabelling the elements). So the only normal subgroups of S_4 are A_4 and K.
Certainly K is normal in A_4, and it is the only subgroup of A_4 of order four. All
three elements of A_4 of order 2 are contained in K, and each trivially generates a

© Springer Nature Switzerland AG 2018
J. Gray, *A History of Abstract Algebra*, Springer Undergraduate Mathematics Series,
https://doi.org/10.1007/978-3-319-94773-0

normal subgroup C_2 of K, but none of these subgroups of A_4 is normal in A_4, so we have exhausted the possibilities for the composition series, which must be

$$S_4 \triangleright A_4 \triangleright K \triangleright C_2 \triangleright \{e\}$$

The striking theorem found by Jordan is that if we look at the orders of successive quotients then the numbers we get are independent of the choice of groups in the sequence. In the present case we had no choice. Note that the numbers in the above sequence are 2, 3, 2, 2, whose product is 24, as it should be.

G.2   The Subgroups of S_5

In this case we shall not attempt to find all the different subgroups, but merely examine each possible order. By Lagrange's theorem, the orders of the subgroups of S_5 are restricted to: 1, 2, 3, 4, 5, 6, 8, 10, 12, 15, 20, 24, 30, 40, and 60. How many of these actually exist? Evidently, the cases 1, 2, 3, 4, 6, 8, 12, and 24 do, because we may stay within one of the S_4s that are the subgroups of S_5 that fix one object.

Notice, however, that elements with cycle type $(\ldots)(..)$ have order 6, so the cyclic subgroup of order 6 is a subgroup of S_5.

We are left with the orders 5, 10, 15, 20, 30, and 40. Of these, the element (12345) in S_5 clearly generates a subgroup of order 5. The symmetries of a regular pentagon with vertices labelled $\{1, 2, 3, 4, 5\}$ provide a subgroup of S_5 of order 10.

To see that there is a subgroup of order 20 and that it is isomorphic to the group we denoted G_{20} at the end of Sect. 10.2, we can argue by filling in the gaps in a statement of Galois's. At one point he looked at the field (as we would call it) of five elements $\{0, 1, 2, 3, 4 \bmod 5\}$, which we shall call k_5 for the moment, and at maps from k_5 to itself of the form $j \mapsto aj + b$, $a, b \in k_5, a \neq 0$. Note that $aj + b = a'j + b'$ if and only if $a = a'$ and $b = b'$.

We check, which Galois did not, that this is a group as follows

- the composite of $j \mapsto aj + b$ followed by $j \mapsto a'j + b'$ is $a'(aj + b) + b' = a'aj + a'b + b'$, which is of the given form;
- the identity element is $j \mapsto j$;
- the inverse transformation to $j \mapsto aj + b$ is $j \mapsto a^{-1}j - a^{-1}b$.

This group has 20 elements: there are 4 choices for a and 5 for b, and they are all distinct maps. To see that it is not commutative it is enough to find a, a', b and b' such that $a'b + b' \neq ab' + b$, as is the case with $a = b = b' = 1, a' = 2$.

This group is generated by $t(j) = j + 1$ and $d(j) = 2j$, as we see by finding $t^b d^a(j)$ explicitly. We need to find a and b such that $t^b d^a(j) = a'j + b'$ for given a' and b'. We have

$$t^b(j) = j + b, d^a(j) = 2^a j$$

so

$$t^b(d^a(j)) = t^b(2^a j) = 2^a j + b = a' j + b',$$

where $a' = 2^a$ and $b' = b$.

Finally, we see that this group is a subgroup of S_5 by writing the effect of t and d as permutations on $\{1, 2, 3, 4, 5\}$. I leave this as an exercise (but see some earlier remarks).

The remaining orders (15, 30, and 40) are more difficult. In a long paper of 1815 Cayley showed that the number of values a rational expression in n variables can take is either 2 or it is greater than or equal to the largest odd prime that divides n. In only slightly more modern language, Cayley showed that the size of an orbit of a subgroup of the group S_n is either 2 or it is greater than or equal to p, the largest odd prime dividing n. The size of the orbit is the index of the stabiliser of a point of that orbit, and so the size of the stabiliser is either $n!/2$ or it is less than or equal to $n!/p$. This rules out subgroups of index 3 or 4 in S_5, so it has no subgroups of orders 40 or 30. To see that there is no subgroup of order 15 it is simplest to argue that, up to isomorphism, there is only one group of order 15 – the cyclic group of that order, which cannot be a subgroup of S_5.

It is a famous fact that the only normal subgroup of S_5 is A_5 and A_5 is simple.[1] As a consequence, the composition series for S_5 is 60, 2. To prove it, we can start by listing the cycle types in A_5 as follows:

- the identity (.),
- (...), and there are $5 \cdot 4 \cdot 3/3 = 20$ of these,
- (..)(..), and there are $5 \cdot 4 \cdot 3 \cdot 2/2 \cdot 2 \cdot 2 = 15$ of these,
- (.....), and there are $5 \cdot 4 \cdot 3 \cdot 2/5 = 24$ of these,

The question now is whether these cycle types coincide with conjugacy classes. To see that any two elements with cycle type (...) are conjugate in A_5 we observe that the elements have either 1, 2, or 3 symbols in common. In the first case, we may write the elements as (123) and (145), and a suitable conjugating element is (24)(35).

In the second case, we may write the elements as (123) and (124), or as (123) and (142). The element (123) is conjugated into (124) with (345), and the element (123) is conjugated into (142) with (243).

In the third case, we may write the elements as (123) and (132) and use (23)(45).

I leave it as exercises to show that any two elements with cycle type (..)(..) are conjugate in A_5. Moreover, we have

$$(123)(124) = (13)(24),$$

---

[1] For a survey of several proofs that apply to all A_n, $n \geq 4$, see Keith Conrad's article at http://www.math.uconn.edu/~kconrad/blurbs/grouptheory/Ansimple.pdf.

and

$$(13)(12) = (123),$$

so any subgroup of A_5 that is a union of conjugacy classes must contain all elements of types $(\ldots)$ and $(..)(..)$, a total of 35, which is greater than 30, and so the subgroup must be all of A_5, and so A_5 is a simple group, as claimed.[2]

However, some of the subgroups of S_5 have interesting composition series. For example, G_{20} has the normal subgroup $S(pentagon)$ of order 10 consisting of elements of the form $x \mapsto \pm x + c \pmod 5$, which in turn has a normal subgroup C_5 consisting of elements of the form $x \mapsto x + c \pmod 5$. The composition series here goes $2, 2, 5$.

---

[2]For definiteness, the cycle type $(\ldots..)$ splits into two conjugacy classes, exemplified by $(abcde)$ and $(abced)$. Using this information, we can deduce that no subgroup of A_5 can be written as a union of conjugacy classes, and so A_5 is simple.

Appendix H
Curves and Projective Space

H.1 Intersections and Multiplicities

We say that a plane curve is of degree k if it can be defined as the zero set of a polynomial in two variables of degree k. Mathematicians in the eighteenth century were aware that curves of degrees k and m should meet in km points. But they were also aware that their intuition could not be turned into a theorem without being more precise about the meaning of the terms involved, because otherwise the claim was obviously false. Here we look at the problems and some of the work involved in solving them.

Consider a line (a curve of degree 1) and a curve of degree k. They should meet in k points. However, the parabola $y = x^2$ and the line $y - 9 = 6(x - 3)$ meet in only one point, the point $(3, 9)$ where the line is tangent to the curve. So points of tangency have to be treated with care.

The line $x = 3$ also only meets the parabola in one point, the point $(3, 9)$. Now the only way out is to postulate a further intersection point 'at infinity'.

The line $y = -1$ does not meet the parabola at all. The remedy now must be to introduce the idea of points with complex coordinates.

How can these problems be dealt with properly? The simplest solution to the problem of tangents is to count intersection points with multiplicity, so a point of tangency is counted twice. But this raises a problem when the curve has points of self-intersection.

Consider now the *folium of Descartes* depicted in Fig. H.1, the curve with equation

$$x^3 + y^3 = 3xy.$$

The curve crosses itself at the origin; we say it has a double point there. A line through the origin with equation $y = mx$, $m > 0$, meets the curve again where

© Springer Nature Switzerland AG 2018
J. Gray, *A History of Abstract Algebra*, Springer Undergraduate Mathematics Series,
https://doi.org/10.1007/978-3-319-94773-0

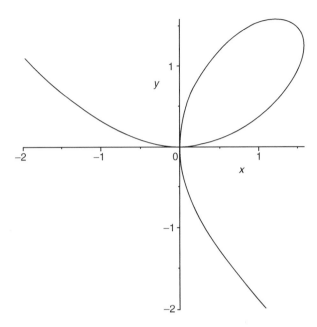

Fig. H.1 The folium of Descartes, $x^3 + y^3 = 3xy$

$x^3(1 + m^3) = 3mx^2$, i.e. at $\left(\frac{m}{1+m^3}, \frac{m^2}{1+m^3}\right)$. The other roots of this cubic are both $x = 0$, which means that the line through the origin meets the folium twice there. So the algebraic test for tangents with which we began also picks up double points on curves. The horizontal and vertical lines through the origin meet the curve triply there, and are the only ones that look like tangents.

The recourse to counting points with multiplicity therefore requires attention be paid to multiple points on curves.

In the course of the nineteenth century several attempts were made to say what can be meant by points 'at infinity', and the algebraic solution that finally emerged was to move the problem into projective space.

Real projective (two-dimensional) space, as a set, consists of all the Euclidean lines through the origin in $\mathbb{R}^3$; each Euclidean line is a *point* of projective space. Each line through the origin contains the Euclidean points (at, bt, ct), $t \in \mathbb{R}$, so the projective point corresponding to this Euclidean line is represented by the coordinates $[a, b, c]$. Notice that we cannot have $a = 0 = b = c$, and that this is the only restriction on a, b, c. Notice also that for all non-zero k the projective point/Euclidean line $[a, b, c]$ and $[ak, bk, ck]$ are the same. For this reason projective coordinates are often called homogeneous coordinates.

The projective point/Euclidean line $[a, b, c]$ meets the Euclidean plane $z = 1$ where $ct = 1$, i.e. at the Euclidean point $\left(\frac{a}{c}, \frac{b}{c}\right)$, unless $c = 0$, in which case the Euclidean line and plane are parallel. This gives us a way of 'seeing' almost every projective point/Euclidean line as a point, and if we also use the planes $x = 1$ and

$y = 1$ we can obtain pictures of every projective point/Euclidean line as a point if we wish.

Suppose now we embed the Euclidean plane in Euclidean three-space as the set of points $(x, y, 1)$, thus

$$(x, y) \mapsto (x, y, 1).$$

We consider the Euclidean point $(x, y, 1)$ as defining the projective point $[x, y, 1]$, and if we write $x = \frac{X}{Z}, y = \frac{Y}{Z}$, we can think of this projective point as the point $[X, Y, Z]$. Projective points with coordinates $[X, Y, 0]$ are not the images of Euclidean points by the above map, and capture the intuition of points 'at infinity', as we shall now see.

The equation of the parabola $y = x^2$ in homogeneous coordinates is $YZ = X^2$. The projective points on it with coordinates $[X, Y, Z]$, $Z \neq 0$, have coordinates $[\frac{X}{Z}, \frac{Y}{Z}, 1]$, and these correspond to Euclidean points in the plane $z = 1$ with coordinates (x, y) that satisfy the equation $y = x^2$. But there are also projective points on the projectivised parabola with coordinates $[X, Y, 0]$ for which $YZ = X^2$. In this case, this is the point $[0, 1, 0]$.

The equation of the line $x = 3$ in its projectivised form is $X = 3Z$, and the projective point $[0, 1, 0]$ lies on this line. So in the projective plane the projective line with equation $X = 3Z$ meets the parabola in two points, of which $[X, Y, Z] = [3, 9, 1]$ is one and $[0, 1, 0]$ is the other.

In this way, the intuition of points 'at infinity' was eventually made rigorous.

Complex projective space is defined in the same way, but with the real numbers $\mathbb{R}$ replaced by the complex numbers $\mathbb{C}$.

Intersections

A topic of increasing importance in nineteenth century mathematics was understanding curves through the common points of two given curves, say $F(x, y) = 0$ and $G(x, y) = 0$. The hope—often expressed as a theorem—was that any such curve has an equation of the form

$$A(x, y)F(x, y) + B(x, y)G(x, y) = 0,$$

for some polynomials $A(x, y)$ and $B(x, y)$ determined by the curve.

For example, when $f(x, y) = y - x^2$ and $g(x, y) = y - 9$ their common points are $(3, 9)$ and $(-3, 9)$. The curve $y - x^3 + 9x - 9 = 0$ passes through these two points, and indeed

$$y - x^3 + 9x - 9 = -x(x^2 - y) + (y - 9)(1 - x) = -xf(x, y) + (1 - x)g(x, y).$$

More generally, if the given curves are Φ with equation $F = 0$ and Γ with equation $G = 0$, and a curve Θ passes through their common points. It was, and is, tempting to suppose that the equation for Θ is a linear combination of the equations for Φ and Ψ, where the coefficients are some polynomials in x and y. More precisely, this condition is evidently sufficient, and one might guess that it is necessary.

However, we get into trouble immediately with the parabola Φ with equation $F(x, y) = y - x^2 = 0$ and the line Γ with equation $G(x, y) = y = 0$. They meet at the origin, but consider the line with equation $x - y = 0$—or any line with equation $y - ax = 0$, $a \neq 0$. This line passes through the origin. But equations of the form

$$a(x, y)F + b(x, y)G = a(x, y)(y - x^2) + b(x, y)y$$

all have no term in x alone, unlike $x - y$. So the claim seems to fail.

Of course, what has happened is that the parabola Φ and the line Γ are tangent at the origin. If we ask for curves that pass through the origin and have a common tangent with X and Y there—in this case the line Y, of course—we recover the claim. Let us check that.

The condition that a curve with equation

$$H(x, y) = 0 = c_{00} + c_{10}x + c_{01}y + \cdots + c_{jk}x^j y^k + \cdots$$

passes through the origin is that $c_{00} = 0$. What is the condition that it has the line $y = 0$ as a tangent there? From the equation $H = 0$ we deduce that $H_x dx + H_y dy = 0$, where H_x and H_y denote partial derivatives of H with respect to x and y respectively. This means that

$$\frac{dy}{dx} = -\frac{H_x}{H_y},$$

so the condition is that $H_x(0, 0) = 0$. But

$$H_x(x, y) = 0 = c_{10} + \cdots \sum_{j > 0, k} j c_{jk} x^{j-1} y^k,$$

so

$$H_x(0, 0) = c_{10},$$

and we deduce that the condition is precisely that H also has no term in x alone.

This leads to the conjecture that the curves that pass with the 'right multiplicity' through the common points of two plane curves X and Y with equations $F = 0$ and $G = 0$ respectively are the curves with equations of the form $AF + BG = 0$, where A and B are polynomials in x and y.

Similar problems occur when Φ and Γ are allowed to be curves that have multiple points.

Appendix I
Resultants

I.1 Netto's Theorem

In the notation of Chap. 24, Netto could write

$$R_1(x) = (x - x_1)^{\mu_1}(x - x_2)^{\mu_2} \ldots (x - x_k)^{\mu_k},$$

where $x_1, x_2, \ldots, x_k$ are the x-coordinates of the singular points and the corresponding μ values are the degrees of the intersections. There was also a similar equation for the y coordinates, with the same values for the exponents μ. His theorem was that the highest value of the exponents $\mu_1, \mu_2, \ldots \mu_k$ is the necessary value of r.

For example, let $f(x, y) = xy$ and $f_1(x, y) = y - 2x + x^2$, and consider the curve with equation $y = 0$ that passes through their common intersections. It turns out that while y cannot be written in the required form, y^2 can be. Indeed, $g(x) = 2 - x$ and $g_1(x, y) = y$ satisfy

$$(2 - x)xy + y(y - 2x + x^2) = y^2.$$

To find this solution, we have to solve the equations

$$\left(\sum_{j,k} a_{jk} x^j y^k \right)(x^2 - y^2) + \left(\sum_{j,k} b_{jk} x^j y^k \right)(x - y + x^2) = y,$$

for unknown constants a_{jk} and b_{jk}. In these equations, the coefficient of $x^j y^k$ on the left-hand side is

$$a_{j-2,k} - a_{j,k-2} + b_{j-1,k} - b_{j,k-1} + b_{j-2,k},$$

© Springer Nature Switzerland AG 2018

J. Gray, *A History of Abstract Algebra*, Springer Undergraduate Mathematics Series,
https://doi.org/10.1007/978-3-319-94773-0

with the convention that a_{pq} or b_{pq} is zero if either $p < 0$ or $q < 0$. In the case at hand the coefficient of $x^j y^k$ is 1 if $j = 0, k = 1$ and zero otherwise. So whatever else, we have a system of linear equations for the unknown coefficients. Therefore, we have to have enough equations to have a chance of solving them, and—the tricky part—enough of the equations must be linearly independent. In the present case, it is enough to let the polynomials $g(x, y)$ and $g_1(x, y)$ be linear.

Now let's try a harder case: let $f(x, y) = x^2 - y^2$ and $f_1(x, y) = x - y + x^2$, and consider the curve with equation $2y - x^2 = 0$ that passes through their common intersections, which are at $(0, 0)$ and $(-2, 2)$. Now, if we try to solve

$$\left(\sum_{j,k} a_{jk} x^j y^k \right) (x^2 - y^2) + \left(\sum_{j,k} b_{jk} x^j y^k \right) (x - y + x^2) = y - x^2,$$

we find that the equations for x and y say, respectively

$$b_{00} = 0, \quad b_{00} = 1.$$

So the equation cannot be satisfied.

Netto's theorem asks us to solve

$$\left(\sum_{j,k} a_{jk} x^j y^k \right) (x^2 - y^2) + \left(\sum_{j,k} b_{jk} x^j y^k \right) (x - y + x^2) = (y - x^2)^r$$

for some integer r, so we could now try $r = 2$, but it too would lead to an incompatible system of equations, and we would only succeed with $r = 3$. But by now the algebra is becoming long and difficult, and we are guided by our understanding of the intersections of plane curves. The corresponding theory for polynomials in more variables could not appeal to any such understanding of the behaviour of surfaces or higher-dimensional varieties, so Hilbert's work was that much more impressive.

I.2   Resultants

Netto's technique for finding the common zeros of two polynomials involved what is called their resultant. Let $f(x) = a_0 x^k + a_1 x^{k-1} + \cdots + k$ and $f_1(x) = b_0 x^m + b_1 x^{m-1} + \cdots + b_m$ be two polynomials of degrees k and m respectively. The key result (which I shall not prove) is that f and f_1 have a common factor if and only if there are non-zero polynomials $A(x)$ of degree at most $m - 1$ and $B(x)$ of degree at most $k - 1$ such that $Af + Bf_1 = 0$.

The thing to note is that the equation $Af + Bf_1 = 0$ amounts to a system of linear equations for the unknown coefficients of A and B, and so they exist if and only if the following determinant vanishes

$$\begin{vmatrix} a_0 & a_1 & \ldots & a_k & 0 & 0 & \ldots & 0 \\ 0 & a_1 & a_2 & \ldots & a_k & 0 & \ldots & 0 \\ & & & \ldots & & & & \\ 0 & 0 & \ldots & 0 & a_1 & a_2 & \ldots & a_k \\ b_0 & b_1 & \ldots & b_m & 0 & 0 & \ldots & 0 \\ 0 & b_1 & b_2 & \ldots & b_m & 0 & \ldots & 0 \\ & & & \ldots & & & & 0 \\ 0 & 0 & \ldots & 0 & b_1 & b_2 & \ldots & b_m \end{vmatrix}.$$

Think of this as m rows of as followed by k rows of bs. It is called the resultant of f and f_1, written $Res(f, f_1, x)$.

In fact, polynomials A and B always exist so that

$$Af + Bf_1 = Res(f, f_1, x).$$

To give an example, let us take $f(x) = x^2 + 5x + 6$ and $f_1(x) = x^2 + x - 2$. The determinant we must consider is

$$\begin{vmatrix} 1 & 5 & 6 & 0 \\ 0 & 1 & 5 & 6 \\ 1 & 1 & -2 & 0 \\ 0 & 1 & 1 & -2 \end{vmatrix}.$$

This vanishes, and so f and f_1 must have a common factor—as indeed they do: $x + 2$.

Netto knew that these results can be extended to polynomials in two variables, by treating either variable as a parameter. In the first example above, we have $f(x, y) = xy$ and $f_1(x, y) = y - 2x + x^2$. We get a surprise: the resultants of these work out to be

$$R_1(x) = xy, \quad R_2(y) = xy^2.$$

But if we make the change of variable replacing x by x and y by $y + 2x$, so $f(x, y)$ becomes $x(y + 2x)$ and $f_1(x, y)$ becomes $y + x^2$, their resultants are

$$R_1(x) = x^2(2 - x), \quad R_2(y) = y^2(4 + y).$$

These correctly locate the coordinates of the common points of the curves: $(0, 0)$ (twice) and $(2, -4)$. What went wrong was that in the first expressions f was not of maximal degree in x.

In our second example there is no need to transform the variables, and we find the resultants of $f(x, y) = x^2 - y^2$ and $f_1(x, y) = x - y + x^2$ are

$$R_1(x) = -x^3(x + 2), \quad R_2(y) = y^3(y - 2).$$

This confirms that there is a triple zero at $(0, 0)$ and a further zero at $(2, -2)$.

Appendix J
Further Reading

J.1 Other Accounts of the History of Galois Theory

Galois theory is a difficult subject for beginners, and traditionally appears as an option for final year students. It has therefore attracted a number of treatments, several of which have much to offer their readers, and it is surely likely that some of the charm of these books is the effort that the authors have made to explain this or that part of history and the mathematics properly to themselves.

All recent histories of the subject have taken their cue from the long and well-researched paper (Kiernan 1971). This paper more or less mapped out a journey from Lagrange to Artin, and so from the first substantial study of the quintic equation to the creation of the modern theory as a branch of structural, modern algebra. Among the important critical comments have been those of van der Waerden (1972), who argued for a different interpretation of Artin's work, Dean (2009), who argued that Dedekind had already presented the 'Galois correspondence' in the 1870s, and Brechenmacher (2011), who argued for a closer reading of Galois's work, as discussed in Chap. 14 above.

The constraints of the course forced me to suppress most eighteenth century developments. Among books largely or wholly devoted to the history of Galois theory, the account in Tignol (2001) is particularly good on the eighteenth century part of the story. It is, for example, richly informative about Lagrange, and also Vandermonde, who is omitted in my account, and valuable for its proof of some mathematical facts about the symmetric functions that I have taken for granted. Galuzzi's historically sensitive *Lectures* (2014) are also very helpful for the mathematics involved in the work of Lagrange and Galois.

Abel's contribution is always difficult to describe precisely, and a gentle account that nicely brings out its importance is Pesic (2003). Another good account aimed more at mathematicians is Cooke (2008).

© Springer Nature Switzerland AG 2018

J. Gray, *A History of Abstract Algebra*, Springer Undergraduate Mathematics Series,
https://doi.org/10.1007/978-3-319-94773-0

The mathematical side, including much of the eighteenth century material but also the reduction of the quintic and the computation of Galois groups is well presented in Bewersdorff (2006).

J.2   Other Books on the History of Algebraic Number Theory

The definitive work here, as I have already remarked, is Goldstein, Schappacher, and Schwermer's *The Shaping of Arithmetic after C.F. Gauss's Disquisitiones Arithmeticae* (2007).

Weil's *Number theory: An approach through history from Hammurapi to Legendre* (1984) stops with Legendre, although he also wrote some interesting essays on cyclotomy, and is richly informative about the period before Gauss.

Scharlau and Opolka (1984) remains the best introduction to Dirichlet's work on number theory that we have, but it is only part of a small book with a large agenda and we still lack a full-length account of this most important mathematician. There is the mathematical biography by Elstrodt (2007), and the collection of sources produced by Biermann (1959) that are in German and out of print.

Goldman (1998) is very helpful on the topics it covers, and it goes much further than I have done.

In the last few years there has been some very good work on different aspects of Dedekind's work: Avigad (2006), Ferreirós (2007), Dean (2009), Reck (2012) and the references cited there, and Hafner (forthcoming).

Kronecker has long had a friend in Harold Edwards, whose works should always be consulted.

Although there is an abundance of literature on Emmy Noether, it is likely that Colin McLarty's forthcoming book, the first to consider all of her work, will greatly change our picture of her and add to our sense of her achievements.

Lemmermeyer (2000) takes a major strand of number theory from the mid-18th to the mid-nineteenth century in more mathematical detail than I have been able to, and has many interesting things to say on the way. His and Schappacher's introduction to the English edition of Hilbert's *Zahlbericht* is also valuable.

Robert Langlands, one of the leading figures in number theory in the second half of the twentieth century, has published a series of lectures on number theory from a rich historical perspective at www.math.duke.edu/langlands.

References

Abel, N.H.: Mémoire sur les équations algébriques, ou l'on démontre l'impossibilité de la résolution de l'équation générale du cinquième degré, vol. 1, pp. 28–34. Grundahl, Christiania (1824)

Abel, N.H.: Beweis der Unmöglichkeit der algebraischen Auflösung der allgemeinen Gleichunge von höheren Graden als dem vierten aufzulösen, [etc.], J. Math. **1**, 65–84 (1826); tr. as Démonstration de l'impossibilité de la résolution algébrique des équations générales qui passent le quatrième degré in Oeuvres complètes, 1, 66–87, republished with an appendix in Bulletin de sciences, mathematiques, astronomiques, [etc.] (Férussac's Bulletin) 6, 1826, 347–354

Abel, N.H.: Mémoire sur une classe particulière d'équations résolubles algébriquement, vol. 1, pp. 478–507 (1829)

Arndt, F.: Auflösung einer Aufgabe in der Komposition quadratischen Formen. J. Math. **56**, 64–71 (1859)

Artin, E.: Galois Theory. University of Notre Dame Press, Notre Dame (1942)

Avigad, J.: Methodology and metaphysics in the development of Dedekind's theory of ideals. In: Ferreirós, J., Gray, J.J. (eds.) The Architecture of Modern Mathematics, pp. 159–186. Oxford University Press, New York (2006)

Ayoub, R.G.: Paolo Ruffini's contributions to the quintic. Arch. Hist. Exact Sci. **23**, 253–277 (1980)

Barrow-Green, J., Gray, J.J., Wilson, R.: The History of Mathematics: A Source-Based Approach, vol. 1. American Mathematical Society, Providence (2018)

Bertrand, J.: Mémoire sur le nombre des valeurs que peut prendre une fonction, etc. J. Écol. Polytech. **18**, 123–140 (1845)

Betti, E.: Sopra la risolubilità per radicali delle equazioni algebriche di grado primo. Ann. sci. mat. fis. **2**, 5–19 (1851); in Opere matematiche 1, 1903, 17–27

Betti, E.: Sulla risoluzione delle equazioni algebriche. Ann. sci. mat. fis. **3**, 49–115 (1852); in Opere matematiche 1, 1903, 31–80

Betti, E.: Sopra l'abbassamanto delle equazioni modulari delle funzioni ellitiche. Ann. sci. mat. fis. **3**, 81–100 (1853); in Opere matematiche 1, 1903, 81–95

Bewersdorff, J.: Galois Theory for Beginners. A Historical Perspective (transl. D. Kramer). Student Mathematical Library, vol. 35. American Mathematical Society, Providence (2006)

Biermann, K.-R.: Johan Peter Gustav Lejeune Dirichlet: Dokumente für sein Leben und Wirken. Akademie-Verlag, Berlin (1959)

Biermann, K.R.: Die Mathematik und ihre Dozenten an der Berliner Universität, 1810-1920. Stationen auf dem Wege eines mathematischen Zentrums von Weltgeltung. Akademie-Verlag, Berlin (1973/1988)

© Springer Nature Switzerland AG 2018
J. Gray, *A History of Abstract Algebra*, Springer Undergraduate Mathematics Series,
https://doi.org/10.1007/978-3-319-94773-0

Birkhoff, G., Bennett, M.K.: Felix Klein and his "Erlanger Programm". History and Philosophy of Modern Mathematics. Minnesota Studies in the Philosophy of Science, XI, pp. 145–176. University of Minnesota Press, Minneapolis (1988)

Bliss, G.A.: The reduction of singularities of plane curves by birational transformations. Bull. Am. Math. Soc. **29**, 161–183 (1923)

Blumenthal, O.: Lebensgeschichte. In: Hilbert, D. (ed.) Gesammelte Abhandlungen, vol. 3, pp. 388–429. Springer, Berlin (1935)

Bolza, O.: On the theory of substitution-groups and its applications to algebraic equations. Am. J. Math. **13**, 59–96, 97–144 (1890–1891)

Bolza, O.: Review of Netto, E. The theory of substitutions and its applications to algebra (transl. F.N. Cole, 1892). Bull. Am. Math. Soc. **2**, 83–106 (1893)

Borel, É., Drach, J.: Introduction à l'étude de la théorie des nombres et à l'algèbre supérieure. Librarie Nony, Paris (1895)

Bottazzini, U., Gray J.J.: Hidden Harmony – Geometric Fantasies: The Rise of Complex Function Theory. Springer, New York (2013)

Boucard, J.: Louis Poinsot et la théorie de l'ordre: un chaînon manquant entre Gauss et Galois? Rev. Hist. Math. **17**, 41–138 (2011)

Brechenmacher, F.: Self-portraits with Évariste Galois (and the shadow of Camille Jordan). Rev. Hist. Math. **17**, 273–372 (2011)

Bring, E.S.: Meletemata quaedam mathematematica circa transformationem aequationum algebraicarum, Lund, repr. Q. J. Math. **6**, 1864 (1786)

Brioschi, F.: Sulle equazioni del moltiplicatore per le trasformazione delle funzioni ellittiche. Ann. Mat. pure appl. **1**, 175–177 (1858a); in Opere matematiche 1, XLIX, 321–324, Hoepli, Milan, 1901

Brioschi, F.: Sulla risoluzione delle equazioni del quinto grado. Ann. Mat. pure appl. **1**, 256–259 (1858b); in Opere matematiche 1, LII, 335–341, Hoepli, Milan, 1901

Brioschi, F.: Sul metodo di Kronecker per la risoluzione delle equazioni di quinto grado. Atti dell' Istituto Lombardo di scienze, lettere ed arti **1**, 275–282 (1858c); in Opere matematiche 3, CXI, 177–188, Hoepli, Milan, 1901

Brioschi, F.: Ueber die Auflösung der Gleichungen vom fünften Grade. Math. Ann. **13**, 109–160 (1878); The paper was omitted from the *Opere matematiche* because there was no Italian translation, and readers were referred to Appendice terza: La risoluzione delle equazioni delle funzioni ellittiche, in Opere matematiche 4, 260–322

Brill, A., Noether, M.: Die Entwickelung der Theorie der algebraischen Functionen in älterer und neuerer Zeit. Jahresber. Deutsch. Math. Verein. **3**, 107–566 (1894)

Brown, E.: The first proof of the quadratic reciprocity law, revisited. Am. Math. Mon. **88**, 257–264 (1981)

Burkhardt, H.: Die Anfänge der Gruppentheorie und Paolo Ruffini. Z. Math. **37**(suppl.), 119–159 (1892)

Burnside, W.: Theory of Groups of Finite Order, 2nd edn. Cambridge University Press, Cambridge (1911)

Cantor, G.: Beitrage zur Begründung der transfiniten Mengenlehre. Math. Ann. **46**, 481–512 (1895)

Cauchy, A.L.: Sur le nombre des valeurs qu'une fonction peur acquérir, etc. J. Écol. Polytech. **10**, 1–27 (1815); in *Oeuvres complètes* (2) 1, 64–90

Cauchy, A.L.: Mémoire sur les arrangements que l'on peut former avec des letters donnés, [etc.]. Exercises d'analyse et de physique mathématique **3**, 151–252 (1844); in *Oeuvres complètes*, (2) 13, 171–282

Cayley, A.: A sixth memoir on quantics. Philos. Trans. R. Soc. Lond. **149**, 61–90 (1859); in Mathematical Papers 2, 561–592

Châtelet, A.: Leçons sur la théorie des nombres. Gauthier-Villars, Paris (1913)

Cooke, R.: Classical Algebra: Its Nature, Origins, and Uses. Wiley, Hoboken (2008)

Corry, L.: Modern algebra and the rise of mathematical structures. Science Networks. Historical Studies, vol. 17, 2nd edn. 2004. Birkhäuser, Basel (1996)

Corry, L.: From *Algebra* (1895) to *Moderne algebra* (1930): changing conceptions of a discipline –
a guided tour using the *Jahrbuch über die Fortschritte der Mathematik*. In: Gray, J.J., Parshall,
K.H. (eds.): Commutative Algebra and Its History: Nineteenth and Twentieth Century, pp. 221–
243. American Mathematical Society/London Mathematical Society, Providence (2007)

Corry, L.: Number crunching vs. number theory: computers and FLT, from Kummer to SWAC
(1850–1960), and beyond. Arch. Hist. Exact Sci. **62**, 393–455 (2008)

Corry, L.: A Brief History of Numbers. Oxford University Press, Oxford (2015)

Cox, D.A.: Primes of the Form $x^2 + ny^2$. Wiley, New York (1989)

Cox, D.A.: Galois Theory, 2nd edn. 2012. Wiley Interscience, New York (2004)

de Séguier, J.A.M.J.: Théorie des Groupes Finis: Eléments de la Théorie des Groupes Abstraits.
Gauthier-Villars, Paris (1904)

Dean, E.T.: Dedekind's Treatment of Galois theory in the *Vorlesungen*. Carnegie Mellon University, Department of Philosophy. Paper 106 (2009). http://repository.cmu.edu/philosophy/106

Dedekind, R.: Beweis für die Irreduktibilität der Kreisteilungs-Gleichungen. J. Math. **54**, 27–30
(1857); in Gesammelte Mathematische Werke 1, 68–71

Dedekind, R.: Vorlesungen über Zahlentheorie von P. Lejeune-Dirichlet (1863); 1st. ed. Braunschweig. 2nd. ed. 1871, 3rd. ed. 1879, 4th. ed 1894, rep. in Gesammelte Mathematische Werke
3, 1–222

Dedekind, R.: Anzeige, rep. in Gesammelte Mathematische Werke 3, 408–420 (1873)

Dedekind, R.: Sur la théorie des nombres entiers algébriques. Bull. sci. math. **1**, 17–41 (1877);
69–92; 114–164; 207–248, and separately published, Gauthier-Villars, Paris, transl. J. Stillwell
as Theory of Algebraic Integers, Cambridge U.P. 1996

Dedekind, R.: Über die Theorie der ganzen algebraischen Zahlen. Nachdruck des elften Supplements (Lejeune-Dirichlet 1893) (1964); preface by B. L. van der Waerden, Vieweg,
Braunschweig

Dedekind, R., Weber, H.: Theorie der algebraischen Functionen einer Veränderlichen. J. Math. **92**,
181–291 (1882); in *Gesammelte Mathematische Werke* 1, 238–350

Del Centina, A.: Unpublished manuscripts of Sophie Germain and a revaluation of her work on
Fermat's last theorem. Arch. Hist. Exact Sci. **62**, 349–392 (2008)

Del Centina, A., Fiocca, A.: The correspondence between Sophie Germain and Carl Friedrich
Gauss. Arch. Hist. Exact Sci. **66**, 585–700 (2012)

Dickson, L.E.: Linear Groups, with an Exposition of the Galois Field Theory. Teubner/Dover,
Leipzig (1901); repr. 1958

Dirichlet, P.G.L.: Sur la convergence des séries trigonométriques. J. Math. **4**, 157–169 (1829); in
Werke, I, 117–132

Dirichlet, P.G.L.: Recherches sur diverses applications de l'analyse infinitésimale à la théorie des
nombres. J. Math. 19, 324–369; 20, 1–12; 134– 155 (1839/1840); in *Gesammelte Werke* 1,
411–496

Dirichlet, P.G.L.: De formarum binariarum secundi gradi compositione, Berlin. Gesammelte Werke
2, 105–114 (1851)

Dirichlet, P.G.L.: Gesammelte Werke, 2 vols, L. Fuchs and L. Kronecker (eds.), Berlin (1889/1897)

Dirichlet, P.G.L.: Vorlesungen über Zahlentheorie, transl. J. Stillwell as Lectures on Number
Theory, HMath 16, London and American Mathematical Societies, 1999 (1863)

Dugac, P.: Richard Dedekind et les Fondements des Mathématiques. Vrin, Paris (1976)

Dyck, W.: Gruppentheoretische Studien. Mathematische Annalen, vol. 20, pp. 1–45. Leipzig,
Teubner (1882)

Dyck, W.: Gruppentheoretische Studien II. Ueber die Zusammensetzung einer Gruppe directer
Operationen, über ihre Primitivität und Transitivität. Math. Ann. **22**, 70–108 (1883)

Edwards, H.M.: Fermat's Last Theorem: A Genetic Introduction to Algebraic Number Theory.
Springer, New York (1977)

Edwards, H.M.: Dedekind's invention of ideals. Bull. Lond. Math. Soc. **15**, 8–17 (1983)

Edwards, H.M.: Galois Theory. Springer, New York (1984)

Edwards, H.M.: Divisor Theory. Birkhäuser, Boston (1990)

Edwards, H.M.: The construction of solvable polynomials. Bull. Am. Math. Soc. **46**, 397–411 (2009); corrigenda 703–704

Ehrhardt, C.: Évariste Galois and the social time of mathematics. Rev. Hist. Math. **17**, 175–210 (2011)

Eisenbud, D.E.: Commutative Algebra. With a View Towards Algebraic Geometry. Springer, New York (1995)

Elstrodt, J.: The life and work of Gustav Lejeune Dirichlet (1805–1859), Analytic Number Theory. Clay Mathematics Proceedings ,vol. 7, pp. 1–37. American Mathematical Society, , Providence (2007)

Euler, L.: Recherches sur les racines imaginaires des équations. Mem. Acad sci. Berlin **5**, 222–288 (1751); in Opera Omnia (1), 6, 78–150 (E 170)

Euler, L.: De numeris, qui sunt aggregata duorum quadratorum. Novi Commentarii academiae scientiarum Petropolitanae **4**, 3–40 (1758); in Opera Omnia (1) 2, 295–327 (E 228)

Euler, L.: Demonstratio theorematis Fermatiani omnem numerum primum formae $4n + 1$ esse summam duorum quadratorum. Novi Commentarii academiae scientiarum Petropolitanae **5**, 3–13 (1760); in Opera Omnia (1) 2, 328–337 (E 241)

Euler, L.: *Vollständige Einleitung zur Algebra*. In: Opera Omnia (1), vol. 1 (1770); transl. John Hewlett as Elements of Algebra, rep. Springer, New York, 1984 (E 387)

Euler, L.: Leonhard Euler. Correspondence. In: Lemmermeyer, F., Mattmüller, M. (eds.) Opera Omnia, (4) A: Commercium Epistolicum, Vol. IV, parts I and II. Birkhäuser, Boston (2015)

Ferreirós, J.: Labyrinth of Thought. A history of Set Theory and its Role in modern Mathematics, 2nd edn. Birkhäuser, Boston (2007)

Ferreirós, J., Gray, J.J. (eds.): The Architecture of Modern Mathematics. Oxford University Press, Oxford (2006)

Fraenkel, A.: Über die Teiler der Null und die Zerlegung von Ringen. J. Math. **145**, 139–176 (1914)

Frei, G.: On the development of the genus of quadratic forms. Ann. Sci. Math. Qué. **3**, 5–62 (1979)

Frei, G. (ed.): Der Briefwechsel David Hilbert – Felix Klein (1886–1918). Vandenhoeck & Ruprecht, Göttingen (1985)

Frei, G.: Gauss's Unpublished Section Eight of the Disquisitiones Arithmeticæ. Vandenhoeck & Ruprecht, Göttingen (2006)

Galuzzi, M.: Lectures on the history of algebra (2014). http://www.mat.unimi.-it/users/galuzzi/Lezioni_2014.pdf

Galois, É.: Sur la théorie des nombres. Bull. Sci. Math. Phys. Chim. **13**, 428–435 (1830); rep. in (Galois 1846, 398–405) and in (Neumann 2011, 61–75)

Galois, É.: Oeuvres Mathématiques d'Évariste Galois, notes by J. Liouville. J. Math. **11**, 381–444 (1846). (See (Neumann 2011) below.)

Gauss, C.F.: Disquisitiones Arithmeticae. G. Fleischer, Leipzig (1801); in Werke I. English transl. W.C. Waterhouse, A.A. Clarke, Springer, 1986

Gauss, C.F.: (3rd Proof) Theorematis arithmetici demonstratio nova. Comment. Soc. regiae sci. Gottingen. In: Werke, vol. 2, pp. 1–8 (1808)

Gauss, C.F.: (4th Proof) Summatio serierum quarundam singularium. Comment. Soc. regiae sci. Gottingen. In: Werke, vol. 2, pp. 9–45 (1811)

Gauss, C.F.: Demonstratio nova altera theorematis omnem functionem algebraicam rationalem integram unius variabilis in factores reales primi vel secundi gradus resolvi posse. Comm. Recentiores (Gottingae) **3**, 107–142 (1816). In Werke, vol. 3, pp. 31–56

Gauss, C.F.: (5th Proof) Theorematis fundamentalis in doctrina de residuis quadraticis demonstrationes et amplicationes novae. In: Werke, vol. 2, pp. 47–64 (1818a)

Gauss, C.F.: (6th Proof) Theorematis fundamentalis in doctrina de residuis quadraticis demonstrationes et amplicationes novae. In: Werke, vol. 2, pp. 47–64 (1818b)

Gilain, Chr.: Sur l'histoire du théorème fondamental de l'algèbre: théorie des équations et calcul intégral. Arch. Hist. Exact Sci. **42**, 91–136 (1991)

Goldman, J.R.: The Queen of Mathematics. A Historically Motivated Guide to Number Theory. A.K. Peters, Ltd, Wellesley (1998)

Goldstein, C.: Un théorème de Fermat et ses lecteurs. Presses Universitaires de Vincennes, Saint-Denis (1995)

Goldstein, C.: The Hermitian form of the reading of the *Disquisitiones*. In: Goldstein, C., Schappacher, N., Schwermer, J. (eds.) The Shaping of Arithmetic After C.F. Gauss's Disquisitiones Arithmeticae, pp. 377–410. Springer, Berlin (2007)

Goldstein, C.: Charles Hermite's stroll through the Galois fields. Rev. Hist. Math. **17**, 211–272 (2011)

Goldstein, C., Schappacher, N.: Several disciplines and a book (1860–1901). In: Goldstein, C., Schappacher, N., Schwermer, J. (eds.) The Shaping of Arithmetic After C.F. Gauss's Disquisitiones Arithmeticae, pp. 67–104. Springer, Berlin (2007)

Goldstein, C., Schappacher, N., Schwermer, J.: The Shaping of Arithmetic After C.F. Gauss's Disquisitiones Arithmeticae. Springer, Berlin (2007)

Gray, J.J.: From the history of a simple group. In: Levy, S. (ed.): The Eightfold Way, The Beauty of Klein's Quartic Curve. MSRI Publications, vol. 35, pp. 115–131. Cambridge University Press, Cambridge (1999)

Gray, J.J.: Worlds out of Nothing; a course on the history of geometry in the 19th century, 2nd rev. edn. Springer, Berlin (2011)

Gray, J.J.: Poincaré and the idea of a group. Nieuw Archief voor Wiskunde **13**, 178–186 (2012)

Gray, J.J.: Depth – A Gaussian tradition in mathematics. Philos. Math. **23**(2), 177–195 (2015)

Gray, J.J., Parshall, K.H. (eds.): Commutative Algebra and Its History: Nineteenth and Twentieth Century. American Mathematical Society/London Mathematical Society, Providence (2006)

Hadamard, J., Kürschàk, J.: Propriétés générales des corps et des variétés algébriques'. Encylopédie des sciences mathématiques pures et appliquées **1**(2), 235–385 (1910/1911)

Hawkins, T.: The Erlanger Programm of Felix Klein: reflections on its place in the history of mathematics. Hist. Math. **11**, 442–470 (1984)

Hawkins, T.: Emergence of the Theory of Lie Groups. An Essay in the History of Mathematics 1869–1926. Sources and Studies in the History of Mathematics and Physical Sciences. Springer, Berlin (2000)

Hawkins, T.: The Mathematics of Frobenius in Context. Springer, Berlin (2013)

Hermite, Ch.: Sur la résolution de l'équation du cinquième degré. Comptes rendus **46**, 508–515 (1858); in Oeuvres 2, 5–12

Hermite, Ch.: Sur l'équation du cinquième degré. Comptes rendus vol. 61, 877, 965, 1073, and vol. 62, 65, 157, 245, 715, 919, 959, 1054, 1161, 1213 (1865/66); in Oeuvres 2, 347–424

Hilbert, D.: Ueber die Theorie der algebraischen Formen. Math. Ann. **36**, 473–534 (1890); in Gesammelte Abhandlungen 2, 199–257

Hilbert, D.: Ueber die vollen Invariantensysteme. Math. Ann. **42**, 313–373 (1893); in Gesammelte Abhandlungen 2, 287–344

Hilbert, D.: Die Theorie der algebraischen Zahlkörper (Zahlbericht). Jahresbericht der Deutschen Mathematiker-Vereinigung **4**, 175–546 (1897); in Gesammelte Abhandlungen 1, 63–363, transl. I. Adamson, The theory of algebraic number fields, Berlin, Springer, 1998

Hölder, O.: Zurückführung einer beliebigen algebraischen Gleichung auf eine Kette von Gleichungen. Math. Ann. **34**, 26–56 (1889)

Hölder, O.: Galois'sche Theorie mit Anwendungen. Encyklopädie der mathematischen Wissenschaften **1**, 480–520 (1899)

Hollings, C.D.: 'Nobody could possibly misunderstand what a group is'; a study in early twentieth-century group axiomatics. Arch. Hist. Exact Sci. **71**, 409–481 (2017)

Hulpke, A.: Galois groups through invariant relations. Groups St. Andrews 1997 Bath, II. London Math. Soc. Lecture Note Series, vol. 261, pp. 379–393. Cambridge University Press, Cambridge (1999)

Jordan, C.: Commentaire sur le Mémoire de Galois. Comptes rendus **60**, 770–774 (1865); in Oeuvres 1, 87–90

Jordan, C.: Commentaire sur Galois. Math. Ann. **1**, 142–160 (1869); in Oeuvres 1, 211–230

Jordan, C.: Traité des Substitutions et des Équations Algébriques. Gauthier-Villars, Paris (1870); rep. Gabay 1989

Katz, V. (ed.): The Mathematics of Egypt, Mesopotamia, China, India, and Islam: A Sourcebook. Princeton University Press, Princeton (2007)

Khinchin, A.Y.: Continued Fractions. University of Chicago Press, Chicago (1964)

Kiernan, B.M.: The development of Galois theory from Lagrange to Artin. Arch. Hist. Exact Sci. **8**, 40–154 (1971)

Klein, C.F.: Ueber die sogenannte Nicht-Euklidische Geometrie. Math. Ann. **4**, 573–625 (1871); in Gesammelte Mathematische Abhandlungen I, (no. XVI) 254–305

Klein, C.F.: Vergleichende Betrachtungen über neuere geometrische Forschungen, Programm zum Eintritt in die philosophische Facultät und den Senat der Universität zu Erlangen, Deichert, Erlangen (1872); in Gesammelte Mathematische Abhandlungen I (no. XXVII) 460–497

Klein, C.F.: Ueber die sogenannte Nicht-Euklidische Geometrie. (Zweiter Aufsatz). Math. Ann. **6**, 112–145 (1873); in Gesammelte Mathematische Abhandlungen I (no. XVIII) 311–343

Klein, C.F.: Ueber die Transformation siebenter Ordnung der elliptischen Functionen. Math. Ann. **14**, 428–471 (1879a); in Gesammelte Mathematische Abhandlungen III (no. LXXXIV) 90–134

Klein, C.F.: Über die Auflösung gewisser Gleichungen vom siebenten und achten Grade. Math. Ann. **15**, 251–282 (1879b); in Mathematische Annalen 15, 251–282, in Gesammelte Mathematische Abhandlungen 2, 390–425

Klein, C.F.: Vorlesungen über das Ikosaeder und die Auflösung der Gleichungen vom fünften Grade. Teubner, Leipzig (1884); English transl. *Lectures on the Icosahedron* G.G. Morrice 1888, Dover reprint 1956. New edition, with introduction and commentary P. Slodowy, Birkhäuser 1993

Klein, C.F.: A comparative review of recent researches in geometry. Bull. N. Y. Math. Soc. **2**, 215–249 (1893); English translation by M.W. Haskell of (Klein 1872)

Klein, C.F.: Über die Auflösung der allgemeinen Gleichungen fünften und sechsten Grades. J. Math. **129**, 151–174 (1905); in Gesammelte Mathematische Abhandlungen 2, 481–506

Klein, C.F.: Vorlesungen über die Entwicklung der Mathematik im 19. Jahrhundert, Springer, Berlin (1926–1927); Chelsea rep. New York

König, G.: Einleitung in die allgemeine Theorie der algebraischen Gröszen. Teubner, Leipzig (1904)

Koreuber, M.: Emmy Noether, die Noether-Schule und die moderne Algebra. Zur Geschichte einer kulturellen Bewegung. Springer, Berlin (2015)

Kosman-Schwarzbach, Y.: The Noether Theorems: Invariance and Conservation Laws in the Twentieth Century: Invariance and Conservation Laws in the 20th Century, B. Schwarzbach (transl.). Springer, Berlin (2011)

Kronecker, L.: Ueber die algebraisch auflösbaren Gleichungen, pp. 365–374. Akademie der Wissenschaften, Berlin, Monatsberichte der Königlich Preussischen Akademie der Wissenschaften zu Berlin (1853); in Werke, IV, 1–11

Kronecker, L.: Ueber die algebraisch auflösbaren Gleichungen, II, pp. 203–215. Akademie der Wissenschaften, Berlin, Monatsberichte der Königlich Preussischen Akademie der Wissenschaften zu Berlin (1856); in Werke, IV, 25–38

Kronecker, L.: Sur la résolution de l'équation du cinquième degré. Comptes rendus **46**, 1150–1152 (1858); in Werke vol. 4, VI, 43–48

Kronecker, L.: Mittheilung über algebraische Arbeiten (Über Gleichungen fünften Grades), pp. 609–617. Monatsber K. Preuss Akad Wiss Berlin (1861); in Werke 4, VIII, 53–62

Kronecker, L.: Auseinandersetzung einiger Eigenschaften der Klassenzahl idealer complexen Zahlen. Akademie der Wissenschaften, Berlin (1870); in Werke, I, 271–282

Kronecker, L.: Grundzüge einer arithmetischen Theorie der algebraischen Grössen. Festschrift Reimer, Berlin (1882); and Journal für Mathematik 92, 1–123, in Werke 2, 237–388

Kronecker, L.: In: K. Hensel (ed.) Vorlesungen über Zahlentheorie. Springer, Berlin (1901)

Kummer, E.E.: Zur Theorie der complexen Zahlen. J. Math. **35**, 319–326 (1847); in Collected Papers 1, 203–210

Kummer, E.E.: Über die den Gaussischen Perioden der Kreistheilung entsprechenden Congruenzwurzeln. J. Math. **53**, 142–148 (1857); in Collected Papers 1, 945–954

Krull, W.: Zur Theorie der allgemeinen Zahlringe. Math. Ann. **99**, 51–70 (1927)

Lacroix, S.: Compéments des Element d'Algèbre, 6th edn. Bachelier, Paris (1835)

Lagrange, J.-L.: Solution d'un Problème d'Arithmétique *Miscellanea Taurinensia* 4 (1769); in *Oeuvres de Lagrange* 1, 671–731, J.-A. Serret (ed.) Paris

Lagrange, J.-L.: Réflexions sur la résolution algébrique des équations. Nouv. Mém de l'Académie des Sciences, Berlin, pp. 222–259 (1770/71); in Oeuvres de Lagrange 3, 205–404, J.-A. Serret (ed.) Paris

Lagrange, J.-L.: Sur la forme des racines imaginaires des équations. Nouveaux mémoires de l'Académie de Berlin, pp. 222–258 (1772); in Oeuvres de Lagrange 3, 479–516, J.-A. Serret (ed.) Paris

Lagrange, J.-L.: Recherches d'arithmétique. Nouv. Mém de l'Académie des Sciences Berlin, pp. 265–365 (1773/1775); in *Oeuvres de Lagrange* 3, 695–795, J.-A. Serret (ed.) Paris

Lagrange, J.-L.: Traité de la résolution des équations numériques de tous les degrés, Paris (1st ed. 1798, 3rd ed. 1826) (1808); in *Oeuvres de Lagrange* 8, J.-A. Serret (ed.) Paris

Landsberg, G.: Algebraische Gebilde. Arithmetische Theorie algebraischer Grössen. Encyclopädie der mathematischen Wissenschaften **1**, 283–319 (1899)

Lasker, E.: Zur Theorie der Moduln und Ideale. Math. Ann. **60**, 20–115 (1905)

Laubenbacher, R., Pengelley, D.: Voici ce que j'ai trouvé: Sophie Germain's grand plan to prove Fermat's Last Theorem. Hist. Math. **37**, 641–692 (2010)

Lê, F.: "Geometrical Equations": Forgotten Premises of Felix Klein's Erlanger Programm. Hist. Math. **42**, 315–342 (2015)

Lê, F.: Alfred Clebsch's "Geometrical Clothing" of the theory of the quintic equation. Arch. Hist. Exact Sci. **71**, 39–70 (2017)

le Rond D'Alembert, J.: Recherches sur le calcul intégral. Mém. Acad. Berlin **1746**, 182–224 (1748)

Lemmermeyer, F.: Reciprocity Laws. From Euler to Eisenstein. Springer Monographs in Mathematics, Springer, Berlin (2000)

Lemmermeyer, F.: The development of the principal genus theorem. In: Goldstein, C., Schappacher, N., Schwermer, J. (eds.) The Shaping of Arithmetic After C.F. Gauss's Disquisitiones Arithmeticae, pp. 529–562, Springer, Berlin (2007)

Levy, S. (ed.): The Eightfold Way, The Beauty of Klein's Quartic Curve. MSRI Publications, vol. 35, Cambridge University Press, Cambridge (1999)

Lützen, J.: Joseph Liouville, 1809–1882. Master of Pure and Applied Mathematics. Springer, Berlin (1990)

Lützen, J.: Why was Wantzel overlooked for a century? The changing importance of an impossibility result. Hist. Math. 374–394 (2009)

Macaulay, F.S.: The Algebraic Theory of Modular Systems. Cambridge University Press, Cambridge (1916)

Magnus, W.: Preface in (Dickson 1901) (2007); rep. Dover 2007

McLarty, C.: Theology and its discontents: David Hilbert's foundation myth for modern mathematics. In: Doxiades, A., Mazur, B. (eds.) Circles Disturbed: The Interplay of Mathematics and Narrative, pp. 105–129. Princeton University Press, Princeton (2012)

Merz, J.T.: History of European Thought in the Nineteenth Century, 4 vols. Blackwood, Edinburgh (1896–1914)

Minkowski, H.: Geometrie der Zahlen. Teubner, Leipzig (1896)

Molk, J.: Sur une notion qui comprend celle de la divisibilité et la théorie générale de l'élimination. Acta Math. **6**, 1–166 (1885)

Montucla, J.-É.: Histoire des recherches sur la quadrature du cercle. Jombert, Paris (1754)

Moore, E.H.: Concerning Jordan's linear groups. Bull. Am. Math. Soc. **2**, 33–43 (1893)

Moore, E.H.: A doubly-infinite system of simple groups. Chicago Congress, Mathematical Papers, pp. 208–242 (1896)

Moore, G.H.: Zermelo's Axiom of Choice. Its Origins, Development, and Influence. Studies in the History of Mathematics and Physical Sciences, vol. 8. Springer, Berlin (1982)

Morley, H.: Life of Cardan, 2 vols. Chapman and Hall, London (1854)

Nagell, T.: Introduction to Number Theory. Wiley, New York (1951)

Netto, E.: Substitutionentheorie und ihre Anwendungen auf die Algebra. Teubner, Leipzig (1882); English transl. The Theory of Substitutions and its Applications to Algebra. P.N. Cole (transl.) The Register Publishing Company, 1892

Netto, E.: Zur Theorie der Elimination. Acta Math. **7**, 101–104 (1885)

Netto, E.: Ueber die arithmetisch-algebraischen Tendenzen Leopold Kronecker's. Chicago Congress Papers, pp. 243–257 (1896)

Netto, E.: Vorlesungen über Algebra. Teubner, Leipzig (1896/1900)

Neumann, P.M.: The concept of primitivity in group theory and the second memoir of Galois. Arch. Hist. Exact Sci. **60**, 379–429 (2006)

Neumann, P.M.: The Mathematical Writings of Évariste Galois. Heritage of European Mathematics. European Mathematical Society, Zürich (2011)

Noether, E.: Idealtheorie in Ringbereichen. Math. Ann. **83**, 24–66 (1921); in Gesammelte Mathematische Abhandlungen, N. Jacobsen (ed.) Springer, 1983, 354–396. English transl. Daniel Berlyne, arXiv:1401.2577 [math.RA]

Noether, E.: Abstrakter Aufbau der Idealtheorie in algebraischen Zahl- und Funktionenkörpern. Math. Ann. **96**, 26–61 (1926); in *Gesammelte Mathematische Abhandlungen*, N. Jacobsen (ed.) Springer, 1983, 493–528

Noether, M.: Ueber die algebraischen Functionen und ihre Anwendung in der Geometrie. Math. Ann. **7**, 269–316 (1874)

Noether, M.: Rationale Ausführung der Operationen in der Theorie der algebraischen Functionen. Math. Ann. **23**, 311–358 (1884)

Parshall, K., Rowe, D.E.: The Emergence of the American Mathematical Research Community, 1876–1900: J. J. Sylvester, Felix Klein, and E. H. Moore. American Mathematical Society/London Mathematical Society, Providence (1994)

Pesic, P.: Abel's Proof. MIT Press, Cambridge (2003)

Petri, B., Schappacher, N.: From Abel to Kronecker: Episodes from 19th Century Algebra. In: Laudal, O.A., Piene, R. (eds.) The legacy of Niels Henrik Abel: The Abel bicentennial 2002, pp. 261–262. Springer, Berlin (2002)

Plofker, K.: Mathematics in India. Princeton University Press, Princeton (2009)

Purkert, W.: Ein Manuskript Dedekinds über Galois-Theorie. NTM Schr. Geschichte Naturwiss. Tech. Medizin **13**, 1–16 (1976)

Reck, E.: In: Zalta, E.N. (ed.) Dedekind's Contributions to the Foundations of Mathematics, The Stanford Encyclopedia of Philosophy (Winter 2012 Edition) (2012) http://plato.stanford.edu/archives/win2012/entries/dedekind-foundations/

Reed, D.: Figures of Thought: Mathematics and Mathematical Texts. Routledge, London (1994)

Reid, L.W.: The Elements of the Theory of Algebraic Numbers, With an Introduction by David Hilbert. Macmillan, New York (1910)

Remmert, R.: The Fundamental Theorem of Algebra. In: Ebbinghaus, H.-D. et al. (eds.) Numbers, pp. 97–122. Springer, Berlin (1990)

Rosen, M.I.: Niels Hendrik Abel and Equations of the Fifth Degree. Am. Math. Mon. **102**, 495–505 (1995)

Rothman, T.: Science à la Mode. Princeton University Press, Princeton (1989)

Ruffini, P.: Teoria generale delle equazioni, 2 vols. Bologna (1799)

Sartorius von Waltershausen, W.: Gauss zum Gedächtnis, 2nd edn. Martin Sändig (1856)

Scharlau, W., Opolka, H.: From Fermat to Minkowski. Springer, Berlin (1984)

Schlimm, D.: On abstraction and the importance of asking the right research questions: could Jordan have proved the Jordan-Hölder theorem? Erkenntnis **68**(3), 409–420 (2008)

Serret, J.-A.: Cours d'algèbre supèrieure, 1st edn. Gauthier-Villars, Paris (1849); 3rd ed. 1866

Smith, D.E.: Source Book in Mathematics. Dover, New York (1959)

Smith, H.J.S.: Report on the theory of numbers, British Association for the Advancement of Science. In: Collected Mathematical Papers 1894, vol. 1, 38–364 (1859); Chelsea rep., New York, 1965

Speiser, A.: Theorie der Gruppen von endlicher Ordnung. Springer, Berlin (1923)

Sommer, J.: Vorlesungen über Zahlentheorie: Einführung in die Theorie der algebraischen Zahlkörper. Teubner, Leipzig (1907)

Stedall, J.A.: Catching Proteus: The Collaborations of Wallis and Brouncker. I. Squaring the Circle. Notes and Records of the Royal Society of London, vol. 54, pp. 293–316 (2000a, b); and II. Number Problems, ibid 317–331

Steinitz, E.: Algebraische Theorie der Körper. J. Math. **137**, 167–309 (1910); rep. 1930 with additional material by R. Baer and H. Hasse, de Gruyter, Leipzig

Szénássy, B.: History of Mathematics in Hungary Until the 20th Century. Springer, Berlin (1992)

Taton, R.: Les relations d'Evariste Galois avec les mathématiciens de son temps. Revue d'histoire des sciences et de leurs applications **1**, 114–130 (1947)

Taylor, R., Wiles, A.: Ring-theoretic properties of certain Hecke algebras. Ann. Math. **141**, 553–572 (1995)

Tignol, J.-P.: Galois' Theory of Algebraic Equations. World Scientific, Singapore (2001)

Tunnell, J.: A classical Diophantine problem and modular forms of weight 3/2. Invent. Math. **72**(2), 323–334 (1983)

Vogt, H.: Leçons sur la résolution algébrique des équations, avec une préface de J. Tannery. Nony et Cie, Paris (1895)

van der Waerden, B.L.: Moderne Algebra. Springer, Berlin (1931); transl. J.R. Schulenberger as *Algebra*, Springer, 1970

van der Waerden, B.L.: Die Galois-Theorie von Heinrich Weber bis Emil Artin. Arch. Hist. Exact Sci. **9**, 240–248 (1972)

van der Waerden, B.L.: On the sources of my book *Moderne Algebra*. Hist. Math. **2**, 31–40 (1975)

Walker, R.J.: Algebraic Curves. Princeton University Press, Princeton (1950); Dover Reprint, New York, 1962

Wantzel, P.L.: Recherches sur les moyens de reconnaitre si un problème de géométrie peut se resoudre avec la règle et le circle. J. math. pures appl. **2**, 366–372 (1837)

Wantzel, P.L.: Classification des nombres incommensurables d'origine algbrique. Nouvelles Annales de Mathématiques pures et appliquées (1) **2**, 117–127 (1843)

Wantzel, P.L.: De l'impossibilité de résoudre les équations algébriques avec les radicaux. Nouvelles Annales de Mathématiques pures et appliquées (1) **4**, 57–65 (1845)

Weber, H., Beweis des Satzes, dass jede eigentlich primitive quadratische Form unendlich viele Primzahlen darzustellen fähig ist. Math. Ann. **20**, 301–329 (1882)

Weber, H.: Die allgemeinen Grundlagen der Galois'schen Gleichungstheorie. Math. Ann. **43**, 521–549 (1893)

Weber, H.: Lehrbuch der Algebra, 3 vols. (1895–1896); 2nd. ed. 1904–1908

Weil, A.: Number Theory for Beginners. Springer, Berlin (1979)

Weil, A.: Number Theory from Hammurapi to Legendre. Birkhäuser, Boston (1984)

Weyl, H.: Algebraic Theory of Numbers. Annals of Mathematics Studies, vol. 1. Princeton University Press, Princeton (1940)

Wiles, A.: Modular elliptic curves and Fermat's Last Theorem. Ann. Math. **141**, 443–551 (1995)

Wiman, A.: Endliche Gruppen linearen Substitutionen. Encyklopädie der mathematischen Wissenschaften **1**, 522–544 (1900)

Wussing, H.: The Genesis of the Abstract Group Concept, transl. A. Shenitzer. MIT Press, Cambridge (1984)

Index

© Springer Nature Switzerland AG 2018

J. Gray, *A History of Abstract Algebra*, Springer Undergraduate Mathematics Series, https://doi.org/10.1007/978-3-319-94773-0

Printed in the United States
By Bookmasters